Problem Solving
and Structured
Programming with
ForTran 77

Problem Solving and Structured Programming with ForTran 77

MARTIN O. HOLOIEN
Moorhead State University

ALI BEHFOROOZ
Moorhead State University

Brooks/Cole Publishing Company
Monterey, California

To my parents   *M. H.*
To Farideh and Amir   *A. B.*

Brooks/Cole Publishing Company
A Division of Wadsworth, Inc.

Printed in the United States of America

10 9 8 7 6 5 4 3 2

Library of Congress Cataloging in Publication Data

Holoien, Martin O., date
     Problem solving and structured programming with
ForTran 77.

     Includes index.
     1. FORTAN (Computer program language)   2. Struc-
tured programming.   3. Problem solving—Data processing.
I. Behforooz, Ali, date.   II. Title.
QA76.73.F25H63   1983   001.64'24   82-24436
ISBN 0-534-01275-2

Photo credits:

Page 2, Figure 1.1, Courtesy of IBM.
Page 3, Figure 1.2, Courtesy of IBM.
Page 6, Figure 1.4, Courtesy of IBM.
Page 7, Figure 1.5, Courtesy of IBM.
Page 8, Figure 1.6, Courtesy of Sperry Corporation.
Page 9, Figure 1.7, Courtesy of Sperry Corporation.
Page 22, Figure 1.23, Courtesy of Sperry Corporation.
Page 25, Figure 1.25, Courtesy of IBM.

Subject Editor: *Michael V. Needham*
Production Editor: *Suzanne Ewing*
Manuscript Editor: *Kirk Sargent*
Interior and Cover Design: *Vicki Van Deventer*
Illustrations: *Robert Inlow*
Typesetting: *TriStar Graphics, Minneapolis, Minnesota*

Cover Photo: *Stan Rice*

Preface

This book was written to meet specific needs in a ForTran course offered at the university where both authors teach. We found that we needed a textbook for a ForTran course that included the following:

- A large number of easy-to-follow programming examples
- Techniques for students to learn how to solve a given problem, and then develop a ForTran program
- Information about using both batch and interactive computer systems

We have incorporated these features into this book.

Our approach taken in teaching problem solving and the process of developing a ForTran program is of particular interest. We have found that most students first learning programming have trouble deciding how to begin writing the program. This difficulty is related to knowing exactly what the problem is, what facts are given, and how to begin analyzing this information to be able to arrive at its solution.

In this book we stress the process of developing procedures for solving problems. We discuss a thorough process of analysis during which one develops a set of abbreviated English statements (called pseudocode) identifying the steps that must be completed to obtain the desired results. Successively more detailed iterations of the steps are developed until the level of detail in the English statements is such that each one of them suggests one or two ForTran statements. This process makes the transition from English statements to ForTran statements easy even for the first-time programmer. Of course, the English statements developed are derived knowing that a program in ForTran is the ultimate goal. That is, the English statements are determined by the programming language being taught.

The book is intended for a first computer science course in ForTran. Beginning college algebra should have been completed. The topical organization allows the teacher to omit those aspects of the ForTran language associated with advanced ForTran programing without destroying continuity. Appropriately identified exercises are provided that apply ForTran to other fields such as economics, sociology, mathematics, and engineering. Therefore the book is suited for a first course in ForTran regardless of the student's major. However, by teaching the entire book, and by assigning those exercises specifically developed to teach computer science concepts, the book can be used in a ForTran course for computer science majors.

As a guide for using this textbook, other than proceeding directly from beginning to end, we suggest the following alternatives:

1. If the goal is to begin programming within the first days of the course, Chapter 1 may be omitted.
2. If problem-solving skills are not to be emphasized, Chapter 4 may be treated superficially without causing harm to the later chapters.
3. Chapters 5 and 6 may be interchanged if it is desired to introduce formatted input and output before studying DO loops and arrays; understanding any of the concepts would not be hindered.
4. If the book is used in a ForTran course for non-computer science majors, Chapter 8 may be omitted.
5. If the course is for computer science majors or minors, all eight chapters plus Appendix A should be studied.

The features of ForTran 77 are especially emphasized. But a few aspects of the ForTran language that have been used so long in several dialects have been included (and carefully noted) even though ForTran 77 does not use them. This book may be used effectively by those not having access to a ForTran 77 compiler. The differences among ForTran 77 and some commonly used ForTran compilers are summarized in Appendix C.

Each chapter summary provides a quick reference to the concepts in that chapter. The extensive index makes it easy to locate all places in the book where a given topic is treated.

Acknowledgments

We wish to thank the following people who reviewed the manuscript in its various stages: Joyce Blair, Eastern Kentucky University; Dwight Caughfield, Abilene Christian University; Charles P. Downey, University of Nebraska; George Friedman, Jr., University of Illinois; Joe Grimes, California Polytechnic State University; H. T. Lau, Vanderbilt University; T. K. Lim, Drexel University; Ken Nygard, North Dakota State University; Charles Pfleeger, University of Tennessee; Lonny Winrichs, University of Wisconsin.

Martin O. Holoien
Ali Behforooz

Contents

Problem Solving and Structured Programming with
ForTran 77

CHAPTER 1
Some Introductory Information

INTRODUCTION

With the thought that some background information will help you understand computers better, we shall begin our book with a review of some historical facts related to the development of the computer. The computer was not the result of a single invention; it was preceded by other inventions that quite naturally led to its development. Because inventions are products of people's minds, the historical review of computation will tell something about people and their struggles to achieve their creative goals.

We shall start with some concepts that must have been known thousands of years ago. Human beings have been interested in methods of organizing information, especially numeric information, for as long as any human records have existed. Computing aids such as fingers, sticks, tallies, pebbles, counting boards, abacuses, and the like have been around for centuries. But perhaps the earliest device that can honestly be called a computing *machine* was Pascal's calculating machine.

Blaise Pascal

Blaise Pascal, the son of a French government official, was born on June 19, 1623. There is no record that Blaise was ever inside a school building or that he even had any contact with children his own age. All that he learned came through his interactions with his father and his father's scientist friends. Consequently Blaise, unusually intelligent as he was, became something of a prodigy in the fields of science and mathematics.

When Blaise was in his late teens, his father was appointed the Royal Commissioner for the French Tax Service, in Normandy, some hundred miles northwest of Paris. In this job with the Tax Service the elder Pascal was responsible for computing the taxes for all residents of the region. Not only did this require an almost endless *amount* of calculations but also very *accurate* results. Blaise, who had in many ways demonstrated his considerable computational ability, was able to help with the calculation of tax bills.

To one who has encountered the frontiers of theoretical mathematics, the tedium of summing, multiplying, subtracting, and dividing is almost more than can be tolerated. Such was the case with Blaise. As he worked on the endless arithmetic assigned to him by his father, Blaise found himself thinking more and more about the possibility of designing a machine to carry out this detestable work. By the time he was 19, he had completed the design of his calculating machine. (See Figure 1.1.)

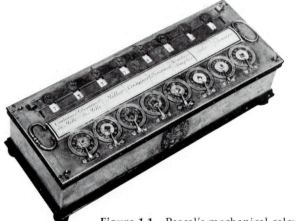

Figure 1.1   Pascal's mechanical calculator

And a remarkable machine it was, capable of performing all four arithmetic operations at speeds never before imagined. By 1645, Blaise had built a model of his machine that he felt was good enough for public presentation. But for all of his dreams, Pascal never realized wealth through this invention nor any other. Potential customers were scared off by the belief that the machine could be operated and repaired only by Pascal or one of his men. Furthermore, the cost was great enough to deter many potential customers.

Charles Babbage

Charles Babbage (Figure 1.2) was born on December 26, 1792, in Totnes, on the southeastern coast of England. When he reached school

Figure 1.2 Charles Babbage

age, he was sent to a church school to be given more personal attention from the minister/teacher.

When he was 19, Charles was sent to Cambridge to begin his university study. His love for mathematics developed early and it did not take him long to find out that he knew more about the subject than his Cambridge professors. This was especially true in the area of calculus. Although this discovery and the persistent attitude of superiority displayed by his professors soured him on Cambridge, Babbage continued studying there and earned his master's degree in 1817.

Although his formal education stopped with the completion of this degree, Babbage continued his study of mathematics, particularly the writings of Leibnitz on the representation of functions as infinite series. Since the evaluation of even a small number of the terms in the series representation of a function requires a great deal of arithmetic, it was probably this interest that led Babbage to begin work on his calculating machines. The first of these he called "the Difference Engine."

While perfecting his calculating machine, Babbage discovered that by adding a few more gears he could return to the machine some of the information computed by the machine itself. This led him to conclude that the machine might be given instructions on how to proceed once certain computed information was available, a concept so revolutionary that he realized he was on the way toward developing a whole new machine. He called this later brainchild "the Analytical Engine" (Figure 1.3).

It seems incredible, but more than 100 years before the first computer was built, Babbage had thought of many of its capabilities. Here is what he wrote in his book *Life of a Philosopher*, published in 1864:

Figure 1.3   Printing mechanism for the Analytical Engine

The Analytical Engine consists of two parts:

1st. The store in which all the variables to be operated upon, as well as all those quantities which have arisen from the result of other operations, are placed.

2nd. The mill into which the quantities about to be operated upon are always brought.

Every formula which the Analytical Engine can be required to compute consists of certain algebraical operations to be performed upon given letters, and of certain other modifications depending on the numerical value assigned to those letters.

There are therefore two sets of cards, the first to direct the nature of the operations to be performed—these are called operation cards: the other to direct the particular variables on which those cards are required to operate—these latter are called variable cards. Now the symbol of each variable or constant is placed at the top of a column capable of containing any required number of digits.

Under this arrangement, when any formula is required to be computed, a set of operation cards must be strung together, which contain the series of operations in the order in which they occur. Another set of cards must then be strung together, to call in the variables into the mill, the order in which they are required to be acted upon.

The Analytical Engine is therefore a machine of the most general nature. Whatever formula it is required to develop, the law of its development must be communicated to it by two sets of cards. When these have been placed, the engine is special for that particular formula. The numerical value of its constants must then be put on the columns of wheels below them, and on setting the Engine in motion it will calculate and print the numerical results of that formula.

This excerpt clearly expresses two of the fundamental ideas of today's digital computers: (1) the use of a previously prepared set of instructions for the machine (today called the "program"), and (2) a place for storing intermediate as well as final results (today called the computer memory). Remember, Babbage formulated these concepts around 1835, one hundred years before any kind of computer had ever been constructed.

Almost none of Babbage's contemporaries understood the significance of his inventions, but there was one person who did. Ada Augusta Lovelace, wife of the Earl of Lovelace and daughter of the English poet Lord Byron, was a friend of Babbage who had become very much interested in Babbage's machines. When a scientific paper written in French by an Italian named Manea discussing Babbage's Analytical Engine appeared, Lady Lovelace produced an English translation with copious notes. This translation was the first clear description of Babbage's work written in his own language. The explanatory notes so lucidly described the functions of the machine that it was clear to all who read them that Ada Lovelace was probably the second most knowledgeable person about the Analytical Engine. In fact, in her notes she suggested how the machine might be given instructions to perform a sequence of arithmetic operations. Thus she has recently been credited as being the world's first programmer.

A programming language called *Ada* has been designated as the standard language for computers used by the United States Department of Defense. This action by an influential governmental agency virtually guarantees the widespread use of the language named in honor of Babbage's contemporary, Ada Augusta Lovelace.

Charles Babbage died on October 18, 1871, at the age of 78. Little notice of his death was taken in the newspapers of London and only one friend besides his family and the undertakers were present at the funeral. What a different attitude toward Babbage today, now that his Analytical Engine is used everywhere in the form of the electronic digital computer. In fact, it is widely agreed that Charles Babbage deserves recognition as "the father of the computer."

Herman Hollerith

Herman Hollerith (Figure 1.4) was born on February 29, 1860 in Buffalo, New York. As a child, Hollerith was a brighter student than most, evidenced by his being awarded the bachelor's degree from Columbia University's School of Mines in 1879 at age 19.

Figure 1.4   Herman Hollerith

One of Hollerith's professors at Columbia was a consultant for the United States Census Bureau while it was preparing for the 1880 census. Hollerith was asked to work with him on a census-related project. So, in October 1879, Herman Hollerith went to work for the Census Bureau. Dr. John Billings, Director of Vital Statistics at the Bureau, discussed with Hollerith the concept of processing census information by machine. The idea was apparently of interest to Hollerith because less than five years later the first patent for a card processing machine was issued to Herman Hollerith on March 31, 1884 (see Figure 1.5). At this time he resigned from his position with the Census Bureau. James Powers was hired to succeed Hollerith. In time for the 1910 census, Powers had developed a card punch, a card sorter, and a tabulating machine that could perform simple counting operations.

Powers formed his own company, the Powers Accounting Machine Company, and left the Census Bureau in 1911. In 1927 the Powers Company became The Tabulating Machines Division of the Remington-Rand

Figure 1.5   Hollerith's card processing machine

Corporation. Still later, in 1955, The Remington-Rand Corporation became the Sperry-Rand Corporation and the Tabulating Machines Division became the Univac Division.

Meanwhile, Hollerith continued active in his company, improving his machines to make them increasingly efficient. In 1911 his company merged with three others to make the Computing, Tabulation, Recording Company (CTR). That firm was renamed the International Business Machines Corporation (IBM) in 1924. Hollerith died on November 17, 1929 at the age of 69.

Aiken and the MARK I

Howard H. Aiken was born on March 9, 1900 at Hoboken, New Jersey. He earned the bachelor's degree from the University of Wisconsin in 1923. In 1931 he began his graduate study in physics at Harvard University. He was awarded a Ph.D. from Harvard in 1939. During the eight years at Harvard, Aiken designed and built several calculating machines, each for solving a specific physics problem associated with his doctoral research. Between 1939 and 1943 he worked with four IBM engineers, during which time they invented the first electromechanical computer, later dubbed the MARK I. This machine was about 8 feet high, 51 feet long, and 3 feet deep. It consisted of relays, switches, wheels, vacuum tubes, and other electrical components numbering 760,000 in all.

The MARK I handled numbers in decimal form and did its arithmetic to twenty-three digits of accuracy. The speed of the MARK I was three additions per second. Later models of this computer were built with the MARK II being completed in 1947 and MARK III and IV shortly thereafter.

Howard Aiken accepted a faculty position as Professor of Information Technology at the University of Miami, Florida in 1947. He died in March 1973.

The first electronic computer

About the same time that Aiken was developing his computer, three other men were working on calculating machines destined to be even more powerful than the MARK I. These three were John V. Atanasoff, John W. Mauchly, and J. Presper Eckert. (See Figure 1.6; photograph of Atanasoff not available.)

Figure 1.6   John Mauchly (left) and J. Presper Eckert

Atanasoff, working as a professor at Iowa State University at Ames, Iowa, designed and developed a computer that contained vacuum tubes as the main switching mechanism rather than electromechanical relays as in the MARK I. The result was a machine hundreds of times faster than

Aiken's computers. Atanasoff did his work with a graduate student whose name was Berry; hence, the name they gave their machine, completed in 1942, was the ABC (for Atanasoff-Berry Computer).

Meanwhile, Mauchly and Eckert, engineering professors at the University of Pennsylvania, constructed a machine similarly based on the action of vacuum tubes, which they completed in February 1946. At its dedication this machine was named ENIAC, for Electronic Numerical Integrator and Calculator. It was a huge thing consisting of forty-seven sections of electronic parts and wiring. Each section was 2 feet wide, 9 feet high, and 1 foot deep. The entire machine weighed thirty tons and occupied 15,000 square feet of floor space. It contained 19,000 vacuum tubes; 70,000 resistors; 10,000 capacitors; and was said to have 500,000 soldered joints. It was a remarkable machine for its time, capable of performing 5000 additions per second. It continued functioning reliably for nine years. (See Figure 1.7.)

Figure 1.7   The ENIAC

The first commercial computer

All the machines discussed to this point were research machines or were built for special governmental applications. With the purchase of the Eckert-Mauchly Computer Company, the Remington-Rand corporation planned to produce computers for profit. The first commercial computer was delivered by Remington-Rand in 1951. It was called UNIVAC for Universal Automatic Computer. UNIVAC I was capable of processing both numeric and alphabetic information. Its speed was about 4000 additions of five-digit numbers in one second.

The IBM Corporation was not long in coming with its own version of a commercial electronic computer and in the spring of 1953 delivered its first 701 computer, calling it the "Defense Computer." A year later IBM delivered its first 650 computer, the one destined to be industry's workhorse for many years.

Interestingly enough, from one commercial computer in 1951, by September of 1962 there were an estimated 16,200 computers installed in the United States alone. By 1978, this figure was over 100,000, and if microcomputers are included, by 1982 there were more than two million of these machines in use.

Computer generations

The first electronic computers used vacuum tubes as their basic counting device. The ABC, ENIAC, and UNIVAC I were the first of these computers. This group of computers became known as "the first generation" of computers. Their computation speed was in the range of 3000 to 5000 additions per second.

In 1948 an electronic device called a transistor was invented at Bell Laboratories in New Jersey. This device could perform most of the functions that a vacuum tube could perform but much faster, with only a fraction of the energy consumption, and with the generation of only a small amount of heat. It seemed to be the answer for computer designers of the late 1950's and early 1960's who had been fighting the problem of keeping computer rooms cool enough that these electronic marvels would operate properly. Thus, during the period from 1958 to 1964 a large number of such computers were built and marketed. Computers utilizing transistors came to be known as second-generation computers. Second-generation computers had calculating speeds ranging from 2000 to 500,000 additions per second with an average speed of around 100,000.

In 1957 another event occurred that was to have a great influence on computer technology. In that year the first integrated circuit was produced. An integrated circuit (IC) is a device that incorporates all of the capabilities of many transistors and other circuit components into one tiny "chip" of solid material about 1/16 square inch. This marvelous device is manufactured by a microphotographic process much of which is mechanized; hence the cost of production is very low. Computers utilizing IC's as the basic counting device are referred to as third-generation

Figure 1.8   Vacuum tube, transistor, integrated circuit, and a dime

computers and have been marketed from about 1964 to the present time. Third-generation computers perform as many as ten million additions per second. (See Figure 1.8.)

STRUCTURE OF A COMPUTER

Now that we have some concept of how computers came into existence, let's find out something about their organization—that is, the physical equipment that goes together to make up a computer.

Computer functions

In a broad sense, a modern computer consists of inter-connected devices capable of performing the following functions:

1. Input.
2. Output.
3. Memory.
4. Arithmetic/logic.
5. Control.

A block diagram will help explain the relationships among these functions (see Figure 1.9).

The input device converts information from a form understood by human beings to one usable by the machine. The output device reverses this process; that is, information in a form processable by the machine is converted to a form usable by human beings. Some examples of forms

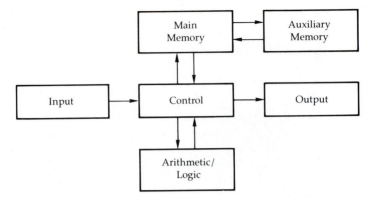

Figure 1.9   Block diagram of computer organization

usable by human beings are punched cards, keys on a typewriter-like device, printed paper, images on a TV-like screen, drawings on paper, and readings on a meter. The computer-usable form consists of electric pulses or magnetic fields. Information in such a form that it can be placed directly into an input device is said to be "machine readable."

The control function has the task of directing machine-usable information to the appropriate device of the computer for processing. The arrows connecting the rectangles in Figure 1.9 represent the paths that information may take within a computer.

The memory function must store information as directed by the control function. Once stored, such information is available for further processing by other functions of the computer. The distinction between main memory and auxiliary memory (see Figure 1.9) is related to the availability of the stored information. Main memory is needed to make information available very quickly, whereas auxiliary memory stores information that is not needed immediately. Information from auxiliary memory must always be brought into main memory before it can be processed by the other modules of the computer. Thus main memory is the working memory.

The arithmetic/logic function performs arithmetic operations on appropriate information and conducts comparisons of stored information so as to be able to make logical decisions based on such comparisons.

Control, and arithmetic/logic modules are usually housed in the same cabinet and are referred to as the central processing unit (*cpu*) of the computer. Figure 1.10 shows a complete, medium-scale computer system.

Input and output devices

Just as the earliest data entry devices were punched-card machines, the earliest input device included as a part of a computer system was a

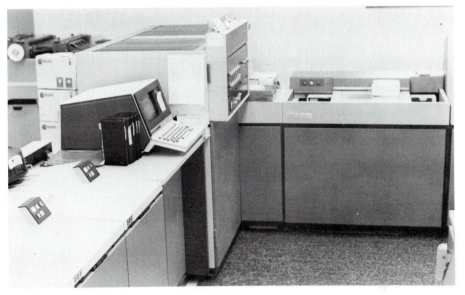

Figure 1.10 A complete digital electronic computer system

card reader, a machine that could sense the holes punched into cards by keypunch machines.

Card readers.   Card readers are generally of 2 types: (1) mechanical sensing of holes; or (2) photoelectric sensing. Mechanical sensing occurs whenever cards properly positioned in the card reader pass between a metal roller and metal fingerlike devices called brushes (see Figure 1.11). Since the paper of which cards are made does *not* conduct electricity, the current in circuits of which each of the brushes is a part stops momentarily whenever the card separates brush from roller. However, whenever a hole in the card permits a given brush to make contact with the roller, a surge of electricity results. The number and placement of punched holes causes a specific pattern of electrical pulses that is interpreted by the central processor as a specific character (digit, letter, or special character).

Photoelectric sensing is accomplished by passing the punched card between a high-intensity light and eighty photoelectric cells, each cell corresponding to a column on the card. A hole in any column allows light to pass through to the corresponding photoelectric cell, causing the transmission of an electric pulse. The pattern of holes determines the pattern of activation of photoelectric cells, which in turn determines the pattern of electrical pulses that are interpreted by the central processor as a specific character. (See Figure 1.12.)

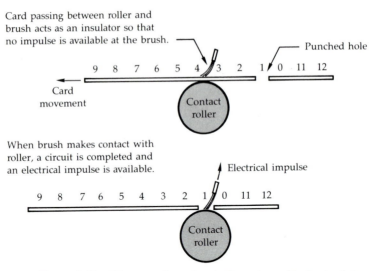

Figure 1.11 Diagram of mechanical sensing of holes in data processing cards

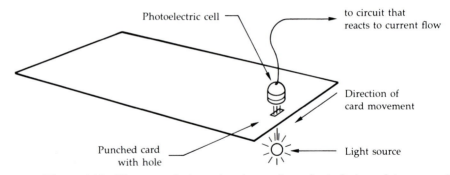

Figure 1.12 Diagram of photoelectric sensing of a hole in a data processing card

Magnetic tape units.   Magnetic tape units have been parts of computer systems from the very early days of computers. These units are capable of sensing magnetized spots on plastic tape coated with an iron oxide compound. The oxide has the property of being magnetizable. When appropriate magnetic fields of adequate strength are applied to such tape, the particles in the oxide coating align themselves in accordance with the magnetic fields. (See Figure 1.28.) The pattern of alignment remains fixed until the tape is again subjected to magnetic fields of similar intensity. Such patterns represent information. Figure 1.13 shows a typical magnetic tape unit.

The process of magnetizing the oxide coating is usually referred to as "writing" on the tape while the process of sensing magnetized spots

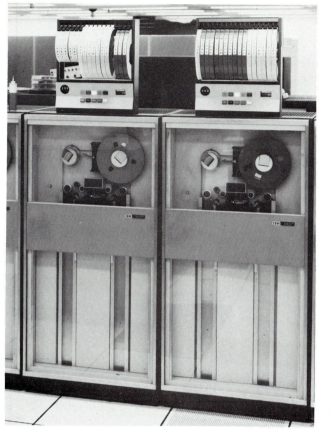

Figure 1.13 Magnetic tape units

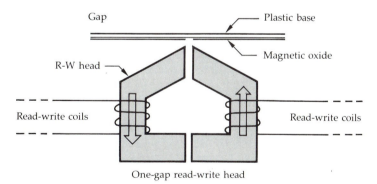

Figure 1.14 Read-write head of magnetic tape unit

already on a magnetic tape is called "reading" the tape. The specific part of a magnetic tape unit that performs these two functions is called the "read-write head." Figure 1.14 shows the read-write head of a tape drive and a detailed diagram of it.

On many of today's microcomputers magnetic tape units are tape cassette units, no larger than the portable tape cassette units with which we are all familiar. The method of storing and retrieving information is the same as for the much larger reel tape units, but the cost of the tape unit as well as the magnetic tape is considerably less.

Magnetic disk units.   Another input device often found on computer systems is a magnetic disk unit. (See Figure 1.15.)

Figure 1.15   Magnetic disk unit

The principle of storing and retrieving information using a magnetic disk is exactly the same as for magnetic tapes. The difference is that instead of using a flexible ribbon of plastic tape coated with oxide as the storage medium, the medium now is a collection of rigid, steel platters attached to the same central post, the steel platters being coated with oxide. Figure 1.16 shows a removable disk pack and a diagram of how information is organized when stored on such a magnetic disk.

Printers.   A device found as a part of most computer systems is a printer. A printer is capable of accepting information from a computer's memory and producing corresponding printed characters on paper. The variety of printers available on computer systems is great, as are the methods by which they produce printed material. Figure 1.17 shows a typical line printer as found on most medium and large-scale computers. Also shown is a typical printer often used as a part of minicomputer systems.

The line printer is so called because its printing mechanism prints an entire line of (typically) 132 characters at one time. Such printers

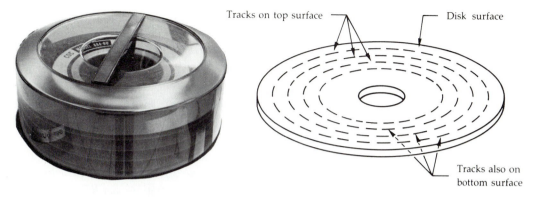

Tracks on top surface — Disk surface

Tracks also on bottom surface

Figure 1.16 Magnetic disk pack and diagram of the pattern of storing information on it

Figure 1.17 Line printer (left) and matrix printer (right)

ABCDEFGHIJKLM
NOPQRSTUVWXYZ
0123456789—.:
&/+$*!%@=(+)

Figure 1.18 Print characters formed by dot matrix of seven rows and five columns

produce from 400 to 2000 or more lines per minute and are almost always a part of medium and large-scale computer systems.

Matrix printers produce an array (matrix) of data so as to form a single character at a time. (See Figure 1.18.) Printing speeds range from ten characters per second to sixty characters per second, obviously a much slower speed than line printers. Of course, the cost is also comparatively less.

Batch terminal.   If the program instructions and data are submitted on cards so that the computer accepts *all* such input in a given batch of cards before it does *any* processing, the input device is called a *batch* device. Thus a card reader is a batch input device. There are machines made that incorporate the card reading function and the printing function in the same metal cabinet. When such a device is connected to a cpu but physically located somewhere other than in the same room as the cpu, it is called a *remote-job-entry terminal,* or an *RJE terminal.* It may also be called a *batch terminal,* since punched cards are read in batches; then the input is processed to produce the specified output. A computer system that handles only batches of input in the manner just described is called a *batch* system.

Although *batches* of information may be input to a computer system in forms other than punched cards, when most people mention a batch computer or a batch terminal they are referring to one that accepts input from punched cards and produces output on a line printer. Figure 1.19

Figure 1.19   Batch terminal

shows a typical batch terminal. The computer system shown in Figure 1.10 could be used to support either a batch terminal or a timesharing terminal provided it had all necessary features.

Timesharing terminal.   Some computers are designed so that they can process the information from more than one input/output device simultaneously or at least what *appears* to be simultaneously. Recall that we previously mentioned that cpu's can perform operations at the rate of 10 million per second or even faster. No input/output device has been designed that can handle information at that rate, so back in the mid 1960's someone thought of connecting more than one input/output device to the same computer so as to share its computing time among the several input/output devices. Figure 1.20 shows a diagram of the logic used in designing such a system. Think of the cpu in that diagram as being connected successively for equal amounts of time to the smaller circles ①, ②, ③, ④, ⑤. When the connecting arrow is on small circle 1, terminal 1 has access to the cpu. Similarly, each of the other terminals has access to the cpu. If we assume that the connecting arrow rotates at a rate of one rotation every tenth of a second, then on each rotation any given terminal has the use of the cpu for .020 second. If the cpu can perform 10,000,000 operations per second, then during the time allotted to each terminal it can perform 200,000 operations. That's quite a few additions, subtractions, comparisons, and so forth! If we assume further that the small numbered circles represent an intermediate storage place into which the cpu can dump large volumes of information very quickly and from which that information can flow to and from the terminals at slow-

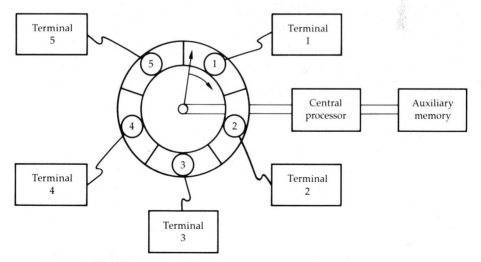

Figure 1.20   Diagram of a timesharing computer system

er rates, then it's not difficult to understand how each terminal can be receiving cpu information at its own full capacity while the same cpu serves terminal after terminal. Therefore, it seems that all terminals are being served simultaneously while in reality they are being served in consecutive spurts of time. Such an arrangement is called a *timesharing system* and has come to be widely used mainly because the cost of computing capability per user is less than if a single computer user had to pay the entire cost of the cpu.

Input/output devices connected to timesharing computer systems are called *timesharing terminals.* There are basically two kinds of timesharing terminals and they are shown in Figure 1.21 (video terminal) and Figure 1.22 (printer terminal). Most video terminals are about the size of a small portable television set, and desk-top models of printer terminals are nearly like electric typewriters (sometimes, if installed on their own floor stands, however, printer terminals seem larger than typewriters).

On a timesharing terminal, input is almost always accomplished by keying information on a keyboard not unlike a typewriter keyboard. Output on timesharing terminals occurs either as printed information (on a printer terminal) or as images on a TV-like screen (video terminal).

Real-time terminal.   There are many situations in which computer systems control or monitor a major activity, as, for example, the launching and flight of a space vehicle. In such a computer application it

Figure 1.21   Video computer terminal

Figure 1.22   Printer computer terminal

is necessary to have instantaneous access to information about the activity being controlled or monitored. It is also essential to have the capability of entering information into the computer so that the computer's controlling activities can be modified according to the judgement of the human being responsible for the project. A computer terminal used in a situation like the one here described is called a *real-time terminal* and the computer application is called a *real-time* application. The phrase "real-time" very likely derived from a process being controlled or monitored by a computer that could be modified while it was happening—that is, during the real time of its occurrence.

Real-time terminals, like timesharing terminals, may be printer terminals or video terminals like the ones in Figures 1.21 and 1.22. It is the *function* of terminals that dictates that they be called real-time terminals—not their physical appearance. In fact, it is entirely possible that the same model terminal could be functioning as a real-time terminal in one installation and as a timesharing terminal in another installation.

Memory

This function of a computer more than any other sets it apart from other calculating machines. Without memory a computer would really be just another calculator.

Computer memory consists of electronic components capable of storing information in a form that may be thought of as the 1's and 0's of a binary number. The computer science abbreviation for a binary digit is *bit*. In computer memory, eight (six in some computers) adjacent bits designed to store the binary code for a single character (letter, decimal digit, or other character) are referred to as a *byte*. In some computers, two or more bytes are designed to operate together to store a single unit of information that is called a *word*.

Regardless of the organization of computer memory into bits, bytes, and words, there are some other technical terms used to refer to the kind of memory used by a computer. The first of these is *random access memory* (RAM), also called *direct access memory*. This kind of memory is capable of having its contents *changed* or examined for output in some form. Thus, one says that such memory may be written into or read from. The segments (words or bytes) of random access memory are accessible independently and are therefore said to be *randomly* or *directly* accessible.

A second kind of memory is *serial-accessed memory* (SAM). The segments of this kind of memory must be accessed in sequential order, from beginning to end—not independently.

A third kind of memory is *read-only memory* (ROM). Once the contents of ROM are in place, they may not be changed by ordinary computer instructions. The contents may only be *read;* hence the name.

Another way of categorizing computer memory is in terms of its physical characteristics. Using this method, one may refer to several types, but we shall discuss two—namely, core memory and semiconductor memory.

Core memory.   A magnetic core is a tiny doughnut-shaped device made of ferrite material that is capable of being magnetized. Fine wires are threaded through the core and when electricity passes through both conducting wires (see Figure 1.23) in the same direction, the core becomes magnetized with one of its faces having south polarity and the

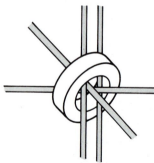

Figure 1.23   Magnetic core (greatly magnified) and core plane

other having north polarity. The third wire through the core is used to sense its magnetic condition. If electricity flows in opposite directions through the two conducting wires, the magnetic polarity of the core faces will be made the opposite of what it was previously. As you might suspect, these two magnetized conditions of the core can be used to represent 1 and 0, respectively. Thus, many cores attached to each other can be used to store a collection of 1's and 0's representing information. This information may actually be the binary form of a decimal number or it may be the binary coded version of a letter of the alphabet, or some other character of information.

Semiconductor memory.   More recent developments in storage devices make use of solid state materials called semiconductors that will retain either of two magnetic conditions like the magnetic core does. Semiconductor memory is becoming relatively inexpensive because mass production methods have been developed. As a consequence, semiconductor memory is replacing core memory as the most commonly used storage medium in computers built since 1976.

Other storage media are being developed as technology makes these possible. In every case, the new developments are intended to provide storage for large masses of information, storage that can be inexpensively produced and that will react ever more quickly to storage and retrieval of information.

Control unit

The control function is accomplished through devices consisting mainly of numerous electronic switches that are opened or closed in tiny fractions of a second. Such devices operate so quickly that new units of time have been defined specifically because of the computer industry. Nanosecond (one-billionth of a second) and picosecond (one-trillionth of a second) are now common terms when discussing the speed of computer components.

The main function of the control unit is to *fetch, interpret,* and *execute* program instructions. Electronic switches are connected together to form a place for storing the electronic equivalent of ones and zeros. Such storage places are called *registers* and are designed to temporarily hold instructions and information closely associated with instructions.

First an instruction must be retrieved (fetched) from memory and placed in the instruction register (IR). This is initially accomplished by first having placed in a special register called the program counter (PC) the memory address of the instruction about to be decoded. The contents of the PC are transferred to the memory address register (MAR) and a read operation is performed that transfers a copy of the instruction about to be decoded into the memory buffer register (MBR). From the MBR the instruction is transferred to the IR where it is decoded and executed. The process of retrieving the instruction from memory and placing it in the instruction register is called the *fetch phase* while the decoding and executing of it is called the *execute phase.* The control unit is designed so that each time an instruction is moved into the IR, the program counter is automatically incremented; thus, when the execution phase is completed, the PC will contain the memory address of the next instruction. In most computers, facilities are available to load the PC with a memory address other than the next one in sequence when that is desirable.

Arithmetic/logic unit (ALU)

This module of a computer contains the circuitry for performing arithmetic operations and logical operations. Like the control unit, it contains several registers, among which the most important is the *accumulator.* The accumulator is the register in which the results of arithmetic and logical operations are formed. Typical instructions will load the accumulator with a number from memory, add to it, store the result in memory again, clear the accumulator to zero, shift its contents left or right, or complement its contents. The ALU also contains circuitry called adders, multipliers, and other such circuitry associated with arithmetic and logical operations.

MICROCOMPUTERS

We have been discussing computer modules in terms of their functions, with little reference to the physical properties of the devices that perform these functions. The truth is that the physical size of *all* computers has been decreasing steadily. In 1965 the first arbitrary division relative to size and computing capability was made in identifying one category of computer from another. The smaller computers were called *minicomputers,* and the first one of these was Digital Equipment Corporation's PDP-8. There really is no clear-cut definition for distinguishing between computers and minicomputers. Minicomputers are generally the smaller of the two kinds and take more time to perform their tasks. Nevertheless, there surely are computers (not called minicomputers by their manufacturers) that are smaller than some minicomputers and take longer than some minicomputers to perform their tasks. So the categorization seems to be arbitrary.

In 1970, another arbitrary category was created when certain very small computers were called *microcomputers.* Their arrival on the scene was almost an accident. In 1969, Datapoint Corporation of San Antonio, Texas, a manufacturer of terminals capable of some memory in the terminal itself (called *intelligent terminals*), designed a very elementary computer that they planned to incorporate into their intelligent terminals. Datapoint contracted with Intel Corporation and Texas Instruments to fabricate the simple computer it had designed on a single chip of silicon less than an inch square. Intel succeeded in the task but its product performed instructions only one-tenth as fast as Datapoint had specified, so it was rejected. Datapoint went on to build their own computers for their terminals.

Meanwhile, Intel decided to market the single-chip computers rejected by Datapoint. Intel marketed these devices as 8008 "microcomputers," and with that action the microcomputer was born. These devices were sold primarily to function as mechanisms to control the operations of other machines. They processed information and produced signals that controlled the operations of these other machines. Hence microcomputers were (and are) also called microprocessors.

Microprocessors, then, are electronic devices made on a single chip of silicon capable of performing the functions of a computer. More recently, some distinction has come to be made between micro*processors* and micro*computers*. *Microprocessor* is generally used to refer to the chip that controls other machines. *Microcomputer* usually refers not only to the tiny computer-on-a-chip but also to the keyboard and screen that serve as the input/output devices for the processor chip. Since 1978, microcomputers have been sold in special retail stores at prices that large numbers of people find acceptable. The retailing term for a microcomputer is *personal computer*. Figure 1.24 shows a microprocessor and a microcomputer.

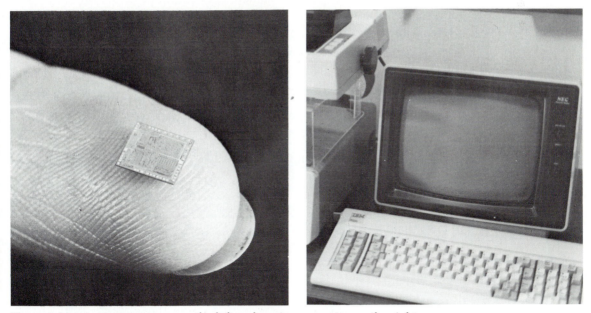

Figure 1.24   A microprocessor on the left and a microcomputer on the right

DATA ENTRY

Data commonly used by human beings are not in forms directly usable by computers. Such data must be prepared by special machines making them understandable to the computer *(machine readable)*. Machines used for this purpose are called *data entry* devices and are sometimes categorized into *indirect* data entry devices and *direct* data entry devices. Indirect data entry devices prepare data in a medium which is later processed by a computer (punched cards, magnetic tape, punched paper tape). Direct data entry devices feed their output directly into a computer. In this section we introduce you to a variety of data entry devices.

Punched card machines

The machine that has been used the longest for data entry is the keypunch machine. Figure 1.25 shows an early keypunch machine.

This machine places rectangular holes in any of 960 positions in the card. The card used is a standard data processing card measuring 7⅜ inches from left to right and 3¼ inches from top to bottom. (See Figure 1.26.)

Single punches in any of rows zero through nine represent the decimal digit for which the row is named. Letters and punctuation marks are represented by two or three punches in a given column. Figure 1.27 shows a card punched with 63 different characters (digits, letters, and other characters) so that you can see how the punched-card code is designed.

Figure 1.25   IBM 031 card punch machine, an early keypunch machine

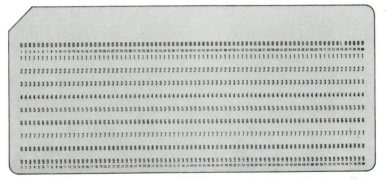

Figure 1.26 Standard data processing card showing lines and columns

Figure 1.27 Standard data processing card with punches representing characters as printed at the top of the card

The first machines designed and built to punch and detect information in punched cards were invented by Herman Hollerith; thus the name *Hollerith code* is sometimes used today for the standard punched-card code.

Magnetic tape encoders and key-to-disk machines

Even early central processors could accept data faster than card readers (machines for detecting information in punched cards) could process cards. This problem gave impetus to the invention of machines that could electronically transmit the information from punched cards onto long ribbons of plastic tape coated on one side with iron compounds, particles of which respond to magnetic fields. On such machines, these tapes were loaded with information from punched cards and then placed on tape drives that were connected to the central processor. These tape drives would then accomplish the input function at a rate greatly accelerated beyond what would be possible when going directly from punched

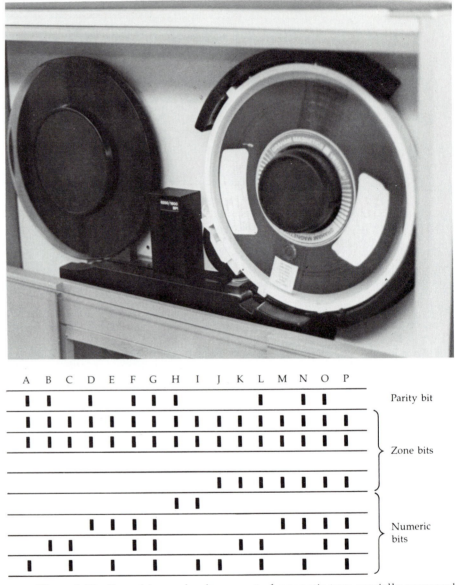

Figure 1.28 Tape drive and enlargement of magnetic tape specially processed to show magnetized spots

cards to the central processor. Figure 1.28 shows a typical tape drive and a portion of magnetic tape specially processed to show the magnetized particles.

On some computer systems, steel cylindrical devices were used instead of plastic tape. Such devices are called magnetic drums and Figure

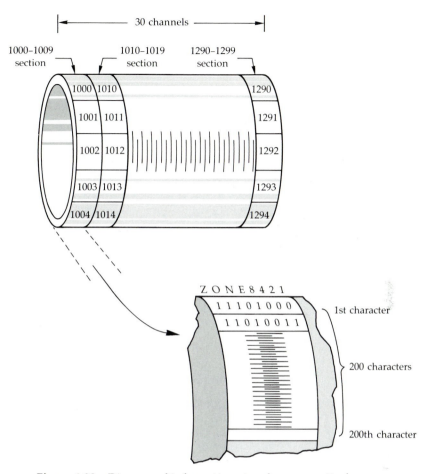

Figure 1.29   Diagram of information stored on magnetic drum

1.29 shows how information is stored on them in the form of magnetized particles. Such drums are still used on some of the very large-scale computer systems.

Because the amount of data waiting to be processed was ever growing, it became imperative to develop even faster means of presenting the data to the computer central processors. This led to the invention of magnetic disks, the medium whereby most modern computer systems input and store large masses of data. Magnetic disks are usually classified as *rigid* or *flexible* (also called *floppy*) and are the most widely accepted means for storing large masses of data. Figure 1.30 shows a diagram of a rigid disk including the read/write heads. Figure 1.31 shows a photograph of a disk pack and the machine (disk drive) that makes it possible to utilize the storage capabilities of a magnetic disk.

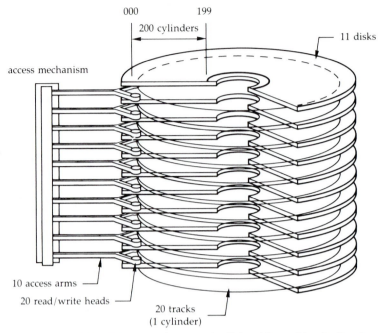

Figure 1.30   Diagram of magnetic disk with read/write heads

Figure 1.31   Disk pack and disk drives

The availability of smaller, less expensive computer systems has made floppy disks very popular. Figure 1.32 shows such a floppy disk and the device that makes it possible for a computer central processor to place information on the disk and detect information stored on it. These machines are much smaller than the ones in Figure 1.31 and can be produced much less expensively. Thus, floppy disk drives are widely used with smaller computers (minicomputers) and personal computers (microcomputers). Figure 1.33 shows a microcomputer with an attached floppy disk drive.

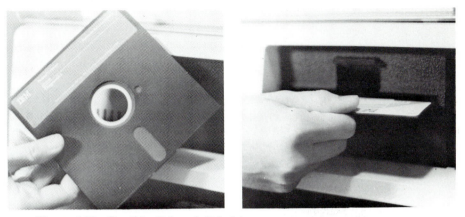

Figure 1.32   Flexible disk and disk drive

Figure 1.33   Microcomputer with video monitor and floppy disk unit

Since some of these magnetic methods for storing information generated by computer central processors have been around for a while, it is natural that someone should have thought of designing a machine that could take information directly from a typewriter-like keyboard and store it on either magnetic tape or magnetic disk without going through the intermediate step of punched cards. The former is called a magnetic tape encoder and the latter a key-to-disk machine. Although magnetic tape encoder and the latter a key-to-disk machine. Although magnetic tape encoders are still used, they are gradually being replaced by key-to-disk machines because disks provide a faster medium of input to computers. Even key-to-disk machines are losing out to on-line terminals.

On-line terminals

Since the early 1970's, an increasing number of computer installations have been designed to use on-line terminals for data entry. These devices are about the size of a small television set with an attached keyboard. Figure 1.34 shows an on-line terminal. Such a terminal is connected directly to a computer central processor (hence the name on-line) and when information is keyed at the terminal, the central processor responds by either storing the information away for later processing or by processing it immediately, producing appropriate results on the terminal's video screen.

Although there are other types of data entry devices in use, the ones here described are the most common.

Figure 1.34   On-line computer terminal

SUMMARY

In this chapter we have considered some history of computer development and introduced a few of the people who played significant roles in that development. Of particular importance in this regard is the development of UNIVAC I, the first computer designed specifically to be marketed. We have also introduced the five fundamental computer functions and learned something about the components of a computer that carry out those functions. Three kinds of computer systems were presented—namely, batch, timesharing, and real-time computers. The characteristic differences among these machines were discussed.

Finally, we concluded the chapter with a description of data entry and some of the machines used to perform data entry.

EXERCISES

1. What major computational device did Blaise Pascal invent? List some capabilities of the machine.
2. Charles Babbage is sometimes called the "father of the computer." Why is he given this honor?
3. What were some of Babbage's problems in the developmental work that he did?
4. What major features set the Analytical Engine apart from the Difference Engine?
5. For what purpose did Herman Hollerith develop punched-card processing machines?
6. What company currently in business had its beginnings in Hollerith's Computing Tabulating Recording Company?
7. Why is James Powers remembered in most histories of computing machines?
8. List in chronological order and with approximate dates the inventions of Aiken, Atanasoff, Babbage, Eckert and Mauchly, Hollerith, Pascal, and Powers.
9. What major company provided support to Howard Aiken in his development work of the MARK I?
10. Compare the speeds of operation in terms of additions per second for these machines: the Analytical Engine, MARK I, ENIAC, and UNIVAC I.
11. What is the name of the first computer designed and built for the commercial market?
12. What was the model number assigned by IBM to its first commercial computer?

13. Compare the three computer generations in terms of (a) basic switching device, (b) speed of operation, (c) energy consumption, and (d) approximate years of operation.
14. Identify the five basic computer hardware functions.
15. Distinguish between main memory and auxiliary memory.
16. What is meant by the cpu of a computer system?
17. List some different input/output media used for computer systems.
18. Describe a memory core.
19. What is meant by a peripheral device?
20. Define the following terms: bit, address, byte, and word.
21. Find out what you can about the physical size of today's computers compared with first-generation computers. Also compare costs.
22. Give definitions for the following terms: data entry, floppy disk, sequential storage, random storage, computer terminal, batch terminal, RJE terminal, timesharing computer system, and real-time terminal.
23. List at least three media used for data entry.
24. What is another name for punched-card code?
25. How many rows and how many columns are there in the layout of the standard data processing card?
26. Describe how information is stored on magnetic tape and magnetic disks.
27. Why is magnetic disk as a form of data entry replacing punched cards and magnetic tape?
28. Compare the speeds of line printers and matrix printers.
29. Explain how a timesharing computer system works.
30. Give an example of a computer input or output device that does not utilize standard input/output media.
31. List advantages and disadvantages of batch computer systems and timesharing computer systems.
32. What is meant by a real-time computer application? Give an example.
33. List advantages and disadvantages of microcomputers compared to larger computer systems.

CHAPTER 2
Introduction to the ForTran Language

By now you know something about how computers came into existence and that a modern computer system comes in quite a variety of shapes and sizes. A functioning computer system is made up of both equipment (hardware) and stored instructions (software). In this chapter we begin the major purpose of this book: the teaching of a computer programming language called ForTran.

COMPUTER PROGRAMMING LANGUAGES

Before we go into some specifics of the ForTran language, let's consider some concepts about computer programming languages in general. When a complete computer hardware system leaves the factory and is installed for a customer, it typically has the capability of performing the functions identified in Figure 1.9. However, all instructions to such a system given through the input device must be provided in machine language form. Nevertheless, almost all computer systems sold today are sold with software that makes it possible for them to accept instructions and data in forms other than machine language.

Machine languages

Machine languages are, by far, the most difficult and time-consuming means of communicating with computers. Each make and model of computer system has its own unique machine language and that language is always numeric in form. Furthermore, most computers are designed so that their machine languages must be input to them in binary numbers (1's and 0's) or some closely related number system, like the hexadecimal system.

To help you understand how a machine language functions, let's invent a very simple machine language for a simple hypothetical computer and then use it to solve a trivial problem. Note that the language we develop is not one associated with any known computer. It is a hypothetical language for a hypothetical computer intended only to serve as a learning aid in this book.

Suppose the computer on which we are to execute this machine language has memory locations (*words*) each of which can store sixteen binary digits (*bits*) of information. Suppose further that when the contents of a word are intended as an instruction, the leftmost six bits contain the code for the operation to be performed, and the rightmost ten bits contain the address of the memory location from which data are to be used or the numeric code for the device to be used in the execution of the instruction. Figure 2.1 shows a diagram of what has just been stated.

Assume that our hypothetical computer system has a card reader, a printer, and a console keyboard as input/output devices. Assume also that the only operations that can be performed are the following:

Store in memory,   Add,   Input,   Output.

Now we design the operation codes. Table 2.1 shows the codes we shall use. Table 2.2 shows the codes we shall use for the three input/output devices.

Now let's use the machine language we have defined to solve the problem of accepting three numbers, finding their sum, and producing the sum as output on the printer. We assume that the accumulator is the device into which the input operation places data and from which the output operation extracts data. The "Store in memory" operation extracts data from the accumulator and places it in the specified memory location. The "Add" operation requires that the accumulator must previously have had a number stored in it and that the contents of the specified memory location is then added to the number in the accumulator. The sum is

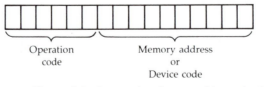

Operation
code

Memory address
or
Device code

Figure 2.1   Instruction format of hypothetical machine language

TABLE 2.1   Machine language operation codes

Operation	Code
Store	000010
Add	000011
Input	000100
Output	000101

TABLE 2.2   Machine language device codes

Device	Code
Card reader	0000001000
Printer	0000001100
Console keyboard	0000001110

stored back in the accumulator, all other numbers temporarily there having been erased.

Here is a machine language program that solves the problem:

Instruction	Comments
0001000000001000	Input first number from card reader
0000100001000000	Store first number in location 0001000000
0001000000001000	Input second number from card reader
0000100001000001	Store second number in location 0001000001
0001000000001000	Input third number from card reader
0000110001000001	Add second number to third number
0000110001000000	Add first number to sum of the other two
0001010000001100	Output sum of all three numbers on printer

Notice what a tedious task it is to write a machine-language program to solve a really trivial problem! Most of the tedium is caused by the need to get all the 1's and 0's correctly written so that the right instruction is executed at the appropriate time and so that data are stored in and extracted from the correct memory locations. It should be clear why developers of computer systems began early to think about ways to make the task of communicating with computers easier than it was when done in machine languages.

Assembly languages

As pointed out in the previous section, the worst aspect of writing programs in machine language is the care that must be taken to use the right sequence of 1's and 0's to get the desired operations correctly accomplished. Suppose we were to modify Tables 2.1 and 2.2 as shown in Tables 2.3 and 2.4 on the next page.

TABLE 2.3   Assembly language operation codes

Operation	Code
Store	STO
Add	ADD
Input	INP
Output	OUT

TABLE 2.4   Assembly language device codes

Device	Code
Card reader	CR
Printer	PR
Console keyboard	CK

A computer language made up of abbreviations of the names of the operations and devices is commonly called an *assembly language*. When designating memory locations in an assembly language, it is customarily possible to do so by using a meaningful name made up of some combination of letters and numbers not to exceed a specified number like, say, six.

Now let's see what an assembly-language program would look like for solving the same problem as described earlier in this chapter. Here it is:

Instruction	Comments
INP,CR	*Input first number from card reader*
STO,N1	*Store first number in location N1*
INP,CR	*Input second number from card reader*
STO,N2	*Store second number in location N2*
INP,CR	*Input third number from card reader*
ADD,N2	*Add second number to third number*
ADD,N1	*Add first number to the sum of the other two*
OUT,PR	*Output sum of three numbers on printer*

One thing you'll notice is that there is no reduction in the number of instructions needed to solve the problem. However, it is possible to look at the assembly language program and tell what is happening, at least if one knows the assembly language operation and device codes. These codes are easy to learn because they remind us of the words for the code itself. Such codes are called *mnemonics*.

Thus, writing a computer program in an assembly language is definitely easier than doing so in a machine language primarily because we can use mnemonics, codes which are significantly easier to remember than the binary numeric codes. However, assembly languages like machine languages are machine dependent. That is, for a given make and model of computer system, the assembly language for it is generally different from that for any other computer. Therefore, programs written in assembly language for one computer system would generally require rewriting before they could be executed on a different computer system.

High-level languages

It occurred to early programmers that it would be useful to have computer languages that were not machine dependent and that were easier to use, in the sense that they would be more like human languages. But human languages make it easy to be ambiguous and it is extremely difficult to be precise enough to make it plain to a machine exactly what operations are to be performed when using a human language.

As some sort of compromise between human languages and machine languages, computer languages have been developed that use much of the terminology and symbolism of a given area of activity. For example, the programming language COBOL (COmmon Business-Oriented Language) has been developed for computer users in the field of business and commerce. ForTran was developed primarily for scientific and mathematical applications.

Most authors designate ForTran as FORTRAN. When a word is written entirely in capital letters, however, it usually means that it is an acronym—each letter is the first letter of a word in a series of words. For example, BASIC is the acronym for Beginner's All-purpose Symbolic Instruction Code. Since ForTran is not an acronym, but is coined from *For*mula *Tran*slation, we have chosen the form ForTran.

There have been many such programming languages developed in the past twenty years. Besides COBOL and ForTran there are PL/I, ALGOL, APL, LISP, Pascal, BASIC, SNOBOL, and others. Since the vocabulary and structure of these languages are closer to those of human languages, such languages are referred to as *high-level* languages. ForTran is properly called a high-level language.

The problem of accepting three numbers, finding their sum, and printing that sum is accomplished in ForTran as follows:

```
REAL N1,N2,N3
READ*,N1,N2,N3
SUM=N1+N2+N3
PRINT*,SUM
END
```

Not only is the ForTran program shorter than its assembly language counterpart discussed earlier but it is also much more readable. Even though you have learned very little yet in this course about ForTran you could probably tell what most of the lines in the program accomplish.

Executing ForTran programs

Recall that computer hardware systems are generally designed so that instructions to them must be given in the form of machine language. Obviously, programs written in high-level languages or in assembly languages are *not* in the form of machine language. A valid question is, "How are programs ever executed that have been presented to a computer in a language other than machine language?" The answer is that such programs must first be translated, then executed. The translation is done by a sophisticated translating program readily available to the computer. However, the concepts needed to understand such translating programs are beyond the scope of this book, so no further attention will be given to them.

Most computer systems are marketed with one or more translating programs. Such programs are designed to function in conjunction with a master control program called an *operating system*. It is the job of the operating system to call into use the appropriate translating program as well as to constantly monitor the usage of the various hardware components of the computer system. A well-designed operating system makes efficient use of all aspects of the computer system, both hardware and software. Typically, the computer user precedes the input of a program with appropriate instructions to the operating system describing how the program is to be handled and in what language it is written. Such instructions to the operating system are called *system commands* or *job control language (JCL)*. Any further discussion of system commands or JCL will not be given in this book. However, the reader must be alert to the fact that such commands are essential to having a ForTran program executed, but are highly dependent on the computer system on which the execution is to take place. Therefore, local procedures for system commands or JCL must be learned.

Before we proceed to the ForTran language itself, we shall consider a brief history of its development.

HISTORY OF FORTRAN

According to Jean Sammet in *Programming Languages: History and Fundamentals,* the first document referring to ForTran is dated November 10, 1954 and was issued by the Programming Research Group in the Applied Science Division of IBM. In this document it was stated that

"The IBM Mathematical Formula Translation System or briefly, ForTran, will comprise a large set of programs to enable the IBM 704 to accept a concise formulation of a problem in terms of a mathematical notation and to produce automatically a high-speed 704 program for the solution of the problem."[1]

It took two and a half years to complete the project and in early 1957 the promised ForTran for the IBM 704 was released. It is interesting to note that records from those days show that customers were not willing to quickly accept this boon to programmers. A major objection of early customers seemed to be that the ForTran translating program could not produce as good a machine-language code as the customers' best programmers could—that is, programs resulting from ForTran translating programs took more computer time. Nothing was said about the greatly reduced time needed to develop useful programs.

In June 1958 a new version of ForTran was released again for the IBM 704. This version was dubbed FORTRAN II. Later in 1958 ForTran systems were issued for the IBM 709 and 650 computers, and in 1960 for the IBM 1620 and 7070. There is no record of a FORTRAN III ever having been released by IBM or any other manufacturer but in 1962 FORTRAN IV was issued for the IBM 7030, a very large-scale computer used for scientific research.

It appears that the first non-IBM introduction of ForTran was in January 1961 by UNIVAC for its Solid State 80 computer. Records show that by 1963 virtually every computer manufacturer had promised to provide a ForTran translating program for use with its equipment. How remarkable that in less than six years, this programming language had made such a significant impact on the world of computing! ForTran had made it practical for engineers and scientists to actually program their own problems without having to work through an assembly-language programmer. The similarity between the normal language of scientists and engineers and ForTran made it significantly easier for many such professionals to learn ForTran.

Throughout the 1960s a myriad of ForTran compilers appeared with language differences not only among the various manufacturers but also among the models of any given manufacturer. It became apparent that some effort must be made to standardize the language. The first move in that direction was the calling of a meeting of Working Group X3.4.3 of the American Standards Association in August 1962. After many meetings and considerable reactions from the computing community, two sets

[1] Sammet, J. *Programming Languages: History and Fundamentals.* Englewood Cliffs, N.J.: Prentice-Hall, 1969.

of ForTran standards were approved in March 1966. The two standards were named FORTRAN and Basic FORTRAN and were the industry standards for the language until early 1978. In April of that year the same organization, now named the American National Standards Institute, adopted specifications for ForTran, called ForTran 77. This book will adhere to those standards unless otherwise stated.

STRUCTURE OF A FORTRAN PROGRAM

The solution to a problem in the form of a ForTran program consists of three types of structures, which we now discuss: *sequential, selective,* and *iterative.*

Sequential structures

Statements that do not change the natural flow of executing statements are called sequential structures. There are five types of ForTran statements that fall in the category of sequential structures.

1. **Declaration statements.** These are statements that define the type and form of data, and all are non-executable statements. Non-executable means that they do not result in action by the computer but do affect the action resulting from executable statements.

2. **Structural statements.** These statements define the beginning or ending of a program or program segment. Some statements in this group are used to pass information back and forth among program segments.

3. **Input and output statements.** These statements make it possible to provide a program with data for processing and to produce program results in a form usable by human beings.

4. **Initialization statements.** These are statements that define initial values for variables at the beginning of a program.

5. **Computational statements.** These are statements that cause arithmetic operations to be performed.

In this chapter we will discuss most of the sequential structures of ForTran, though some are left until later chapters because of their special applications.

Selective structures

Statements that specifically alter the normal sequential execution of one statement after the other are called selective structures. Several types of selective structures will be discussed in Chapter 3.

Iterative structures

Statements that result in the repetitive execution of a segment of a program are called iterative structures. These statements are discussed in Chapter 5.

DATA TYPES IN FORTRAN

ForTran 77 provides for three data types: numeric, character, and logical. Numeric data are further subdivided into integer, real, and complex. We proceed now to a discussion of these.

Numeric data

Integer data.   Recall from mathematics that any whole number (also called a counting number) or its negative is called an integer. For example, 2, −5, 7, −36, 1278961, and 0 are all examples of integers. In ForTran any of these integer values could be the name of an integer constant whose contents are whatever the name is. To denote a ForTran integer constant, we therefore simply write the integer value, being careful *not* to include a decimal point or any commas, which one might normally use in large numbers. Following are examples of integer constants that could appear in ForTran programs:

```
5       − 15629
−8      89542
725     − 16
```

Real data.   Real data in ForTran may be in either of two forms: (1) standard decimal form or (2) exponential form. The reference "real" probably comes from the mathematics term "real number" and generally refers to a number that may have a fractional part.

Standard decimal form. In ForTran the standard decimal form of real data is simply the number written with a decimal point. Following are some examples:

```
3.056       0.0165
− 12.576    −46654.9
375.4       15.
```

In the preceding list, the last entry is 15. Note that 15 is an integer, but in ForTran, when a decimal point is written as a part of the constant, as in 15., it is stored as a real datum, not as an *integer* datum. This is a very important distinction because it means that 15. (with a decimal point) is

TABLE 2.5   Comparison of scientific notation and ForTran exponential form

Standard notation	Scientific notation	ForTran exponential form
0.0000000667	6.67×10^{-8}	6.67E$-$8
30000000000	3.0×10^{10}	3.0E10
0.0000589	5.89×10^{-5}	5.89E$-$5
983500000	9.835×10^{8}	9.835E8
-0.0000025	-2.5×10^{-6}	-2.5E$-$6

stored in computer memory in an entirely different form than 15 (without the decimal point) would be stored. For further information on this topic, see Appendix A.

Exponential form. When very small or very large numbers are involved in computations done by hand, it is often more convenient to use a form of such numbers called scientific notation, or exponential form. For example, the small number 0.000000067 in scientific notation would be 6.7×10^{-8}. Similarly, the large number 30000000000 in scientific notation would be 3.0×10^{10}. Numbers of these magnitudes appear often in several areas of scientific endeavor and are much more easily handled in the exponential form.

ForTran provides for an exponential form of real data, though it is not exactly the same as scientific notation. There are three parts to the exponential form of ForTran data: (1) the coefficient part, which looks exactly like the standard decimal form of a real constant; (2) the letter E; and (3) the exponent part which has the same form as an integer constant. Table 2.5 shows ForTran exponential form compared with standard decimal notation and scientific notation. Although in scientific notation it is standard to write the coefficient portion as a number between 1 and 10 (adjusting the exponent accordingly), in ForTran exponential form that standard does not hold. Therefore 6.67E$-$8, 66.7E$-$9, .667E$-$7, and 667.E$-$10 are all acceptable forms of the same real datum in ForTran exponential form (namely, the datum 0.0000000667). Note that, in both scientific notation and ForTran exponential form, unsigned numbers are positive numbers.

In summary, ForTran exponential form is related to scientific notation as follows:

$$(\text{coefficient})\text{E}(\text{integer}) = (\text{coefficient}) \times 10^{(\text{integer})}$$

To obtain the ForTran exponential form from standard decimal form, carry out these steps:

1. Decide on the coefficient to be used in exponential form, including the placement of the decimal point.
2. The integer exponent is equal in magnitude to the number of positions through which the decimal point has been moved to form the coefficient, and its sign is positive if the movement from the original number has been to the left in forming the coefficient and negative if it has been to the right.

Suppose we apply the above rules to the number 37900000000. First, we decide that the coefficient is to be 37.9. Then the integer exponent is 9, because in forming 37.9 from 37900000000., the decimal point was moved 9 positions to the left. Therefore, the exponential form of 37900000000. is 37.9E9 with the coefficient we decided upon.

Complex data.   Complex numbers are used mostly in engineering and physics problems. Mathematicians have defined complex numbers as numbers of the form $x + yi$ where x and y are real numbers and i is the square root of -1. Examples of complex numbers are

$$2 + 3.21i$$
$$2.7 + 0i$$
$$0 + 2i$$

In ForTran, complex numbers are written as a pair of real numbers enclosed in a set of parentheses. The above three complex numbers would be written as follows:

```
(2,3.21)
(2.7,0)
(0,2)
```

Character data

Any set of acceptable characters (letters, digits, punctuation marks) not intended for computation is called a *character datum, string constant, Hollerith constant,* or *literal.* The use of Hollerith in this context is in honor of Herman Hollerith's punched-card code as mentioned in Chapter 1.

There are two ways to denote Hollerith constants in ForTran. One of these makes special use of the letter H in that immediately to the left of H an integer is placed to specify the number of characters in the constant, and immediately to the right of H are placed the characters that form the constant. This method of defining character constants has been available for a long time, so although preferred methods are now available, we still include it. For example,

```
9HJOHN KNOX
```

In this example, 9 specifies the number of characters in the Hollerith constant, while the constant itself is made up of the characters

```
JOHN KNOX
```

where the space between the first and last names is included in the count of 9.

The second method, probably preferred by most computer professionals, denotes a string constant by enclosing the string in a pair of single quotation marks. Consider these examples:

```
'ABC'
'MARY AMES'
'516282716'
```

Here the quotation marks are used to specify both the extent of the string constant and the characters of which it is to be composed.

Logical data

In mathematical logic there are exactly two logical values—namely, "true" and "false." In ForTran these are written as

```
.TRUE.
.FALSE.
```

NAMING MEMORY LOCATIONS

Any information to be processed by a computer must first be in the main memory of the computer. This implies that there must be rules whereby individual memory locations can be identified. In order for a memory location to be referenced in a computer program it must have a name (or identifier) such as SALARY, SUM, N1, N2, and so forth.

ForTran constant names

In the case of data that never change, called constants, such as 25, −10.6, and 'ROBERT BOE', the name is the same as the constant to be stored. Thus a memory location named 25 in a ForTran program would contain the number 25. Similarly, a memory location named 'ROBERT BOE' would contain the string ROBERT BOE.

Names used to identify memory locations must be unique. In the case of constant data that requirement is not a problem because the contents of a given memory location are the same as its name. For variables, however, this rule must be carefully followed. Later in this book you will learn of a method provided by ForTran to assign two or more names to a single memory location.

ForTran variable names

We turn next to the identifying of memory locations whose contents may change either during a given program execution or from one execution to another. A name used to identify such a memory location is called a *variable name* or simply *variable*. The rules for naming a variable are:

1. The only characters permitted are the letters A through Z and the digits 0 through 9.
2. The first character of the name must be a letter.
3. The number of characters in the name should not exceed 6 (ForTran 77 standard).

Here are some examples of valid ForTran variable names:

```
JOHN
NAM E
R    7
J76
FORM21
```

It should be noted that blanks are ignored in ForTran (as in NAM E and R   7 above), except when they occur in string constants. This means that NAM E and NAME are interpreted exactly the same, as are R   7 and R7. Here are some examples of *in*valid ForTran variable names and the reasons why they are invalid:

```
FIRSTNAME    More than 6 characters
3JOHNS       First character not a letter
K.ART        Can't use decimal point
R(           Can't use parenthesis
F—3          Can't use minus
```

Variable-name declaration

As previously mentioned, in ForTran there are three major types of data and one of those is further subdivided into three subtypes, making a total of five kinds of data. A variable name is an identifier for the data stored at a given memory location. The name should correspond to the type of data stored. ForTran makes such correspondence possible through *type declaration* statements, which we discuss next.

INTEGER statement.   This statement is used to identify variables to be used in a ForTran program for storing integer data. The general form of this statement is

```
INTEGER VAR1,VAR2, . . . ,VARN
```

where VAR1,VAR2,...,VARN are any acceptable ForTran variable (or function) names. This statement has the effect of causing all variables listed to be of type *integer.*

REAL statement.   This statement identifies variables as those to be used in a ForTran program for storing real data. Its general form is

```
REAL VAR1,VAR2, . . . ,VARN
```

where VAR1,VAR2,...,VARN are any acceptable ForTran variable (or function) names. It has the effect of causing all variables listed to be of type *real.* Thus the statement

```
REAL NETPAY
```

appearing in a program would declare NETPAY as a real variable and would allow us to use it, for example, as the location for storing the net salary, where fractions of dollars are significant information.

COMPLEX statement.   Any variables appearing in a COMPLEX statement are identified as memory locations for storing complex data. The general form of this statement is

```
COMPLEX VAR1,VAR2, . . . ,VARN
```

where VAR1,VAR2,...,VARN are any acceptable ForTran variable (or function) names. Recall that complex data consist of a real part and an imaginary part (as defined in mathematics). A complex constant is written in ForTran as two numbers enclosed in parentheses. The first number in the parentheses is the *real* part of the complex number while the second number is the *imaginary* part.

Here are some correct examples of the three kinds of type-declaration statements discussed so far:

```
INTEGER ADDRES,ZIP,NAME
REAL HOURS,NET,GROSS
COMPLEX ROTATN,DIRECN
```

CHARACTER statement.   This statement identifies variables to be used in a ForTran program for storing data of type *character*. Recall that character data are any acceptable characters available on the given computer system. Character data are not to be used for arithmetic computations even though they could consist entirely of numbers. (An example of such data would be a Social Security number.)

There are three acceptable general forms of the CHARACTER statement.

1. `CHARACTER V1,V2, . . . ,VN`
   . . . where V1,V2,..., VN are acceptable ForTran variable (or function) names.
2. `CHARACTER V1*N1,V2*N2, . . . ,VN*NN`
   . . . where V1,V2,...,VN are as described above and N1,N2,..., NN are positive integer data.
3. `CHARACTER *K,V1,V2, . . . ,VN`
   . . . where V1,...,VN are as described in 1 above and K is any positive integer datum. Note that the comma following K is optional in ForTran 77 but is forbidden in some versions of ForTran.

In the first form, the variable names are listed without any length specification. In this case the system assumes a length of one character. That is, any variable in the list of such a CHARACTER statement stores exactly one character. Here are two examples of CHARACTER statements of the first form:

```
CHARACTER ID
CHARACTER INIT1,INIT2,INIT3
```

In the first of the preceding examples, ID is the name of a variable capable of storing one character. In the second example, there are three variables, each of which can store a single character.

In the second form, the length of each character variable is specified by the integer following the asterisk that separates the variable name and the integer. Here is a specific example:

```
CHARACTER ID,ID2*10,FNAME*8
```

ID will store one character (because no length is specified, by default a length of 1 applies), ID2 will store ten characters, and FNAME will store eight characters.

In the third form, one is able to specify character variables of identical lengths by using only a single integer indicating that length. Here is an example:

```
CHARACTER*15 FNAME,LNAME,CITY
```

Each of the three variables FNAME, LNAME, and CITY will store fifteen characters.

One of the most important reasons for using CHARACTER statements is to make the use of character data independent of the hardware system on which the program is being processed. For example, some computers store six characters per word, others store four, and some store two. It is necessary for the programmer to remember the situation on the local computer since it affects whether one may use six or four or two as the maximum length when storing character data. When CHARACTER statements are used to specify lengths, it causes such matters to be under software control rather than being dependent on hardware and, therefore, frees the programmer from having to remember these details.

LOGICAL statement.   There are times when it is essential to include in a program variables whose values can be either "true" or "false." For example, if one process is to be done for odd numbers and another for even numbers, a variable called FLAG might be used to determine which process to execute. If a test of FLAG indicates a value of "false," this could signal an odd number and the execution of one process, whereas a value of "true" could signal an even number and call for the other process to be performed. Whenever decisions are to be made in a ForTran program, variables like FLAG are very useful. Such variables, whose only values can be "true" or "false," are called *logical* variables. The name of a logical variable is formulated according to the same rules that apply to any ForTran variable. When a logical variable is assigned one of the two acceptable logical values "true" or "false," those values are written

```
.TRUE. or .FALSE.
```

It is the LOGICAL statement that is used to declare a variable for storing the logical values ".TRUE." or ".FALSE." Its general form is

```
LOGICAL VAR1,VAR2, . . . ,VARN
```

where VAR1,VAR2, . . . ,VARN are any acceptable ForTran variable (or function) names.

There are situations in programming where it would be convenient to declare a whole group of variables as a certain type without specifying every variable name in a type-declaration statement. This is accomplished by using the IMPLICIT statement.

IMPLICIT statement. This statement makes it possible for the programmer to identify a group of variable names as being of a certain type according to the letter of the alphabet that starts the variable name. Here is the general form of the IMPLICIT statement:

```
IMPLICIT type (L1-L2)
```

where "type" refers to one of variable types REAL, INTEGER, COMPLEX, CHARACTER, LOGICAL, or DOUBLE PRECISION. (For a discussion of DOUBLE PRECISION type of variable, see Chapter 8.) L1 and L2 refer to letters of the alphabet (in alphabetical order) with which the names of variables are to begin. Thus, the variable type is implied to be the one specified by "type." There need not be *two* letters inside the parentheses, in which case all variable names beginning with the single letter that appears are implicitly assumed to specify variables of the type indicated.

Consider these examples:

```
IMPLICIT INTEGER(A-D)
```

This statement would specify that any variable or function names that begin with A, B, C, or D would implicitly identify *integer* variables.

Here is another example:

```
IMPLICIT REAL (A-Z)
```

The appearance of this statement in a program would cause *all* variables to be of type *real*.

One last example:

```
IMPLICIT CHARACTER (A)
```

This statement identifies all variable and function names that begin with A as being character variables of length 1 (the default length).

Undeclared variables.   If a variable is not declared by using one of the declaration statements previously discussed, the default condition in ForTran provides that the variable is of type integer if its name begins with I, J, K, L, M, or N. Otherwise it is of type real. Only types integer and real occur if the default rule is allowed to apply.

Although the default rule makes it possible to avoid declaring your variables, we urge you to always declare all program variables. The only exception might be an occasional variable used to store temporary data or a variable used as an index for a program loop. (A program loop is a set of statements repeated a finite number of times. This concept is discussed at length in a later chapter.) The benefit of declaring all major variables is that it helps make the program more understandable, a goal well worth the extra effort of declaring variables.

Also important to writing an understandable program is the choosing of meaningful variable names. We consider this topic next.

Using ForTran variables

Although a standard translating program for ForTran 77 assumes that the previously stated rules for variable names are adhered to, these rules allow for a great amount of flexibility on the part of the programmer. You are, therefore, urged to make variable names as meaningful as possible, especially toward the goal of writing computer programs that are understandable even long after they were written. To help you see the contrast between using variable names that add nothing to the understandability of a program and using variable names that tell much about what a program accomplishes, consider the following problem and two different procedures for solving the problem.

Problem: For a given employee, input the number of hours worked and the hourly rate of pay. Calculate gross wages, FICA tax, Federal tax, and State tax. Prepare an output report to show net earnings, gross earnings, the three tax amounts, and the sum of all three taxes. Assume the tax rates are .06 for FICA, .20 for Federal, and .08 for State.

Solution 1:   Here is one procedure for solving the problem:

Input X and Y
Compute Z = product of X and Y
Compute A = product of .06 and Z
Compute B = product of .20 and Z
Compute C = product of .08 and Z
Compute D = sum of A, B, and C
Compute E = difference of Z and D
Output E, Z, A, B, C, and D
End

If you follow through each step of the preceding procedure you will find it does everything the problem requires. However, to recognize that D is really the total tax deduction you must trace the steps to its computation.

Solution 2:   The following procedure is exactly like the previous one except that variable names are chosen to be more meaningful.

Input HOURS and RATE
Compute GROSS = product of HOURS and RATE
Compute FICA = product of .06 and GROSS
Compute FEDTAX = product of .2 and GROSS
Compute STATAX = product of .08 and GROSS
Compute TOTDED = sum of FICA, FEDTAX, and STATAX
Compute NETSAL = difference of GROSS and TOTDED
Output NETSAL, GROSS, FICA, FEDTAX, STATAX, and TOTDED
End

Even a quick reading of the second procedure makes it apparent what is happening at each step, because the variable names mean something in the context of the problem being solved.

Although we have not yet discussed all the concepts necessary to understand a complete ForTran program, we believe it is worthwhile to present to you such a program corresponding to the above procedure. One reason for doing this is to demonstrate the close relationship between a good solution procedure and the corresponding ForTran program. The following program corresponds to the above procedure.

```
PROGRAM PAYROL
REAL HOURS,RATE,GROSS,FICA,FEDTAX
REAL STATAX,TOTDED,NETSAL
READ*,HOURS,RATE
GROSS = HOURS * RATE
FICA = GROSS * .06
FEDTAX = GROSS * .20
STATAX = GROSS * .08
TOTDED = FICA + FEDTAX + STATAX
NETSAL = GROSS - TOTDED
PRINT*,NETSAL,GROSS,FICA,FEDTAX,STATAX, TOTDED
END
```

The preceding ForTran program could be entered into a computer and if two input numbers are provided (the number of hours worked and the rate of pay per hour) the program would produce the output requested in the problem as originally stated.

The remainder of this chapter defines the rules of word usage and punctuation marks for some of the ForTran statements essential to the writing of ForTran programs. This aspect of program statements is called *syntax*.

INPUT/OUTPUT IN FORTRAN

Recall from Chapter 1 that when the input for a computer system is a card reader and the output is a printer, the system is called a batch processing system. Also discussed in that chapter was an interactive computer system characterized by the use of a conveniently located computer terminal, on which input is keyed on-line to the cpu; output is printed on paper if the terminal is a printer terminal or appears on a video screen if the terminal is a video terminal. We discuss next the processes of input and output in ForTran when the computer is a batch system.

Program and data preparation for a batch system

You learned in Chapter 1 that a standard data processing card has eighty columns in it. ForTran translating programs (also called compilers) make use only of the information punched in columns 1 through 72. Following are the special uses made of information in certain columns.

Column 1:   Reserved for the letter "C" to indicate that the card is a *comment* card. Comment cards are used by the programmer to insert information among the program statements as a means of describing the function of groups of statements when this may not be so apparent by reading the program statements themselves. Comment cards are not translated by the translating program and are, therefore, not really a part of the program; they do not cause any action by the computer. They are, however, listed (printed) as lines in the program whenever such a program listing is requested and should add significantly to the understandability of the program. We recommend that column 1 be left blank if the card is not a comment card.

Columns 2–5:   Reserved for a reference number also called a statement number. Many times the logic of a program requires that program control be transferred to a statement other than the next one in sequence. When this need arises, a means must be available whereby a specific statement can be referenced. This is accomplished with the statement number, or reference number, placed in columns 2–5 of the card. If the program contains so many statements that it is necessary or convenient to use more than four digits for the statement number, it is permissible to use column 1 for that purpose also. However, we recommend that this *not* be done unless really necessary, since the practice makes programs less readable. Note that statement numbers are unsigned integers from 1 to 99999.

Column 6:   Reserved to indicate that a card is a continuation of the information in the previous card. That is, if a program statement is longer than can be contained on one card, the statement may be continued in columns 7–72 of the next card provided any character other than a 0 is punched in column 6 of that next card. More than one card may be a continuation card as long as some character appears in column 6 of that card. In this book, we will use a plus sign (+) as the continuation symbol, but *any* character is acceptable except 0.

Columns 7–72:   Used for the ForTran statement, the special words and symbols about which we will soon learn.

Columns 73 through 80 are ignored by ForTran translators and may, therefore, contain any information useful to the programmer or may be left blank. Among professional programmers, these columns are often used to provide sequence numbers for all cards in the deck, thus providing a means of sorting the cards should they ever accidentally be dropped or otherwise have their order mixed up. Figure 2.2 shows a standard ForTran coding form with some lines of code.

Unformatted READ statement

The ForTran statement easiest to use for entering data into computer memory for later processing is the unformatted READ statement. There are two forms of this statement. The first form we discuss is the form

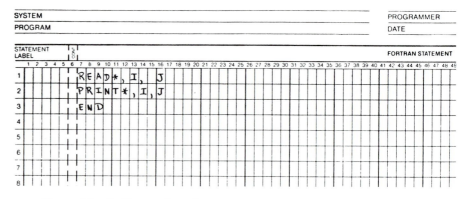

Figure 2.2   ForTran coding form

used when the data are to be read from a card reader or terminal. The general forms of this READ statement are as follows:

(a) READ*, Var-list

(b) READ(5,*) Var-list

where "Var-list" represents any list of acceptable ForTran variables separated by commas. In form (b), the integer 5 is system-dependent but is the most common one used to designate that a card reader or terminal is to be used for input. The asterisk in both forms indicates unformatted input and is required in ForTran 77. Recall that this statement must be keyed in positions 7–72 of the card or line. Sometimes the variables in Var-list are referred to as the *list* of the READ statement because they constitute the list of variables used in that statement.

Whenever a READ statement appears in a ForTran program, corresponding data must also be provided on a separate card or line keyed to match the variable list of the READ statement. (Later in this book we will learn about other methods of providing the data.) Thus, for the READ statement

        READ*,VAR1,VAR2,VAR3,NAME

there must be an associated card similar to the following:

In order to comply with our policy of declaring all major variables, we add two statements along with the READ statement:

```
REAL VAR1,VAR2,VAR3
CHARACTER NAME*6
```

When the preceding READ statement has been executed in conjunction with the given data card, the value stored in location VAR1 would be the real numbers 5.0, in VAR2 the real number 10.0, in VAR3 the real number 15.0, and in NAME the character data ROBERT. Notice that the order of variables in the READ statement corresponds to the order of the data on the data card. Also note that on the card the data are separated by a blank space. This is characteristic of data read by unformatted READ statements. There must be exactly as many data items provided as there are variables in the list of the READ statement. In the previous example, four variables appeared in the READ statement and four values were provided in the data card. Each of the four values could have been punched on a separate card if there were some reason to do so.

When using either form of the unformatted READ statement, if the data being input are *character* data, such data should be enclosed within a pair of single quotation marks (apostrophes) as in the previous example. Normally the type of each datum entered on the data card (or line) must match the type of the variable corresponding to it in the associated READ statement. It is possible to enter integer data into a real variable (the data are stored as real data) but not real data into an integer variable. Thus, in the previous example, 5 and 10 were shown as integer data (no decimal point) but would be stored as real data because VAR1 and VAR2 are declared as real variables. The variable NAME was declared to be of type character. Therefore, we enclose the datum ROBERT in a pair of single quotation marks to designate it as type character.

Let us examine another example of an unformatted READ statement:

```
INTEGER NUMBER, ZIP
REAL VALUE
READ* , NUMBER, VALUE, ZIP
```

2nd data card | 5 | 6 | 5 | 6 | 0 | | |

1st data card | 2 | 9 | 6 | | 2 | 3 | 6 | . | 4 | 1 |

Here the READ has three variables and there are three numbers provided in two data cards. All three numbers *could* have been punched in the first data card but we want to demonstrate that as long as three numbers are provided it doesn't matter how many data cards are used for that purpose. Note that in this example the first and third variable names are integer type so that the first and third numbers in the data cards are punched without decimal points. The second variable, VALUE, is real type so that the second data item, 236.41, contains a decimal point. It is very important to note this correspondence between type of variable and type of data provided.

Nothing has been said about the physical placement of the ForTran READ statements and the corresponding data cards. Without going into further details at this time, suffice it to say that all data cards come after all program statements. Data cards are not interspersed with program-statement cards.

Unformatted PRINT and WRITE statements

These statements are used to produce a printed copy of data electronically stored in computer memory. In the case of a batch system—the type of system we are presently considering—the printed output is produced on a computer printer, typically a line printer as shown in Chapter 1. The general forms of these statements are as follows:

(a) `PRINT*,` Output-list

(b) `WRITE(6,*)` Output-list

where the asterisk is used to indicate the absence of a format specification and "Output-list" refers to any collection of constants, variables, or arithmetic expressions whose values are to be printed or displayed and which are separated by commas. The number 6 in form (b) is the integer used by many computer systems for designating the line printer of the system or the terminal from which the program is being executed.

Suppose that a READ statement has already been executed so that real variables VAR1, VAR2, and VAR3 contain the values 2.5, 10.0, and −3.1, respectively. We now consider some examples to illustrate the use of these statements.

The first example is

```
PRINT*,VAR1,VAR2
```

When executed, this statement would cause the values stored in VAR1 and VAR2 to be printed like this:

```
2.5    10.0
```

Here is another example:

```
WRITE(6,*)VAR3,VAR1+VAR2
```

This statement causes the printing of two numbers. The first number is a copy of what is stored in location VAR3 while the second number is the sum of the numbers stored at VAR1 and VAR2. The line of output looks like this:

```
-3.1    12.5
```

Consider this third example:

```
PRINT*,VAR1, '+',VAR2,'=',VAR1+VAR2
```

There are five items in the output-list (separated by commas) thus causing five pieces of information to be printed as follows:

```
2.5    +    10.0    =    12.5
```

Note that each of the five items printed appears to be in its own field, spaced across the line. The first field contains the value of VAR1; the second field has the string constant "+"; the third field has the value of VAR2; the fourth field the string constant "="; and the fifth field has the number equal to the sum of VAR1 and VAR2. If you go back and reread the PRINT statement you will see that each item in the output-list of that statement corresponds to one item of output as shown.

Finally, a fourth example of an unformatted PRINT statement:

```
PRINT*, 'THE SUM OF THE VARIABLES IS', VAR1+VAR2+VAR3
```

The output produced is

```
THE SUM OF THE VARIABLES IS    9.4
```

An analysis of this PRINT statement makes it clear that there are two items specified for output, the string enclosed in single quotation marks, and the sum of numbers stored in locations VAR1, VAR2, and VAR3. The string constant is twenty-seven characters in length and, therefore, requires a field of at least that width while the sum, 9.4, is printed in its own field to the right of the string.

Notice that the PRINT and the WRITE statements as just discussed cause identical effects if their output-lists are identical. The value of the WRITE statement over the PRINT statement will become apparent when we discuss the occurrence of output on devices other than printers or terminals.

Program and data preparation for an interactive system

When an interactive computer system is used, the lines of ForTran code and the input data are not submitted to the computer on punched cards as we have been discussing in the preceding portion of this chapter. Instead, such information is keyed at a computer terminal of the kind mentioned in Chapter 1. The rules and procedures for entering programs and data on a terminal that is part of an interactive system vary from system to system. It is not just the hardware configuration that determines these procedures but the system software as well. Therefore, anyone using this book in connection with an interactive system should consult those responsible for providing local computing facilities in order to determine the proper procedures for entering programs and data and for making changes in stored programs. No attempt will be made to describe any single interactive system because of the lack of standards among such systems.

Our attention is now turned to three ForTran statements that are used to start or conclude a program, or to conclude a segment of a program. These we shall call *structural statements*.

STRUCTURAL STATEMENTS

We use this name for those ForTran statements that result in identifying the beginning and end of a program or subprogram, or specifying the end of the action. Thus they serve in a way to indicate certain program structure. Three of this type of statement used in most ForTran programs are discussed here. Others will be presented in later chapters.

PROGRAM statement

This identifies the beginning of the program (only comment statements may precede it) and assigns a name to the program. Its form is

```
PROGRAM NAME
```

where NAME represents any valid ForTran variable name. The use of this statement is not required in a program under the rules of ForTran 77 but we recommend it as a convenient method for assigning an identifying name to your program. If the PROGRAM statement appears in a program it must precede all statements with the exception of comment statements. Although the PROGRAM statement as described here is a part of ForTran 77, some translating programs (compilers) do not accept it in this form.

END statement

Every ForTran program and subprogram must conclude with a statement that signals the compiler that the end of the program has been reached. This statement is the END statement and has this form:

```
END
```

This must be the last card (or line) of the program.

STOP statement

The third structural statement is the STOP statement, whose form is simply

```
STOP
```

It may appear anywhere and as often in a program as makes logical sense. When this statement is executed it brings an immediate halt to any further program execution.

The major difference between an END and a STOP statement is that a STOP statement may be used anywhere in a program and can, therefore, stop program execution at any point. An END statement is used only at the physical end of a program, hence it only terminates a program at its natural termination point. A second difference is that STOP statements may include a reference number, while the END statement, which is a nonexecutable statement, can never contain anything but the word END. Note that the use of a STOP statement in ForTran 77 is optional, but there are other ForTran translating programs that require this statement.

A SAMPLE PROGRAM

We have now presented enough information about ForTran that a complete program can be given for your consideration. Before we do that, however, we comment on another matter important to getting a ForTran program executed on a given computer system.

Job control language (JCL)

All computers of the 1980s include as a part of their software an operating system. The most important function of an operating system is to simplify human interaction with the computer system. This is accomplished by requiring that human beings provide certain instructions to the operating system. All such instructions are referred to as the *Job Control Language (JCL)* for that computer system. We shall refer to such instructions as *JCL commands,* or *system commands.* The JCL commands are not a part of ForTran or any other programming language but must be used in order to get the computer system to process your program. Since JCL is dependent on the given computer system, we shall not include any reference to specific JCL commands in this book. However, it is important that you learn what JCL commands are needed to execute your ForTran programs properly on your local computer system.

Sample program

This program inputs three numbers, then prints each number with an identifying message on a separate line. Also, a part of the output is a line of digits to help determine the horizontal spacing of the rest of the output. Some sums involving the three original numbers are also included in the output.

```
10      PROGRAM SAMPLE
11      REAL NUM1,NUM2,NUM3
12      READ*,NUM1,NUM2,NUM3
13      PRINT*,'012345678901234567890123456789 0123456789'
14      PRINT*,'THE VALUE OF THE FIRST VARIABLE IS',NUM1
15      PRINT*,'THE VALUE OF THE SECOND VARIABLE IS',NUM2
16      PRINT*,'THE VALUE OF THE THIRD VARIABLE IS',NUM3
17      PRINT*,'NUM1 + NUM2=',NUM1 + NUM2
18      PRINT*,'NUM2 + NUM3=',NUM2 + NUM3
19      WRITE(6,*) 'NUM1 + NUM3=',NUM1 + NUM3
20      WRITE(6,*) 'THE SUM OF ALL 3 VARIABLES IS',
21     +NUM1 + NUM2 + NUM3
22      END
```

The numbers at the left of each program line are not a part of the program but are included so as to be able to refer to individual lines later in our discussion.

If this program were executed with input data of 5, 50, and 500, the output would be as follows:

```
0123456789012345678901234567890123456789
THE VALUE OF THE FIRST VARIABLE IS 5.0
THE VALUE OF THE SECOND VARIABLE IS 50.0
THE VALUE OF THE THIRD VARIABLE IS 500.0
NUM1 + NUM2 = 55.0
NUM2 + NUM3 = 550.0
NUM1 + NUM3 = 505.0
THE SUM OF ALL 3 VARIABLES IS 555.0
```

Line 10 assigns the name SAMPLE to this program and results in no other action. Line 11 declares three variables, NUM1, NUM2, and NUM3, as real variables. Line 12 provides for the input of three numbers to be stored in NUM1, NUM2, and NUM3. As a result of the data provided, 5 is stored at NUM1, 50 at NUM2, and 500 at NUM3. Line 13 produces the output line of forty digits intended to help count the characters in each line of output. Line 14 produces as output the string bounded by the pair of single quotation marks and the value stored in NUM1. Similarly, lines 15 and 16 produce both string and numeric output. Lines 17 and 18 also each produce a line of output in which there is a string (bounded by quotation marks in the associated PRINT statement) followed by a number that is the result of the addition called for by the last item in the PRINT statement. Lines 19 and 20 are unformatted WRITE statements that function in exactly the same way as the preceding PRINT statements. Line 21 is a continuation of line 20 and, therefore, has a plus sign in the space corresponding to column 6 of a punched card.

COMPUTATIONAL STATEMENTS

We consider next some methods used in ForTran to accomplish arithmetic calculations. There are six arithmetic operations available in ForTran: addition, subtraction, multiplication, division, exponentiation, and negation.

Addition

The symbol used for addition is + just as it is in mathematics. Thus, to indicate the sum of two variables NUMBR1 and NUMBR2, we write

```
NUMBR1+NUMBR2
```

Subtraction

Again the symbol is a familiar one from mathematics, the minus sign, −. The difference of two variables NUMBR1 and NUMBR2 is designated as

```
NUMBR1-NUMBR2
```

Multiplication

In mathematics a common symbol used is the × or "times" sign. Unfortunately, most people think of × and x as interchangeable. As you know from the discussion earlier in this chapter on naming variables, x is a legitimate name for a ForTran variable. When communicating with a computer there must be no ambiguity, so to avoid that problem, the asterisk, *, is used to indicate multiplication. For instance, the product of the variables RATE and HOURS is specified by

```
RATE * HOURS
```

Division

In mathematics a common method of indicating division is to place the dividend over the divisor and draw a short horizontal line between them, as

$$\frac{15}{22}$$

Also in mathematics, a method of indicating division similar to the above method is the use of the slash, as in 15/22. This second method is the one used in ForTran, probably because it lends itself more easily to the limitations imposed by computer input devices. Thus, the general form for designating division in ForTran is

```
VAR1/VAR2
```

where VAR1 and VAR2 represent any valid ForTran constants or variables. If at least one of the operands, VAR1 and VAR2, is of type real, then the quotient is real. If both operands are of type integer, then the quotient is of type integer. For example, if the real number 12.6 is stored in variable SUM, then the quotient SUM/3 is the real number 4.2. If the integer value 195 is stored in variable N, then the quotient N/50 is 3, the largest integer less than or equal to the true quotient. Notice the loss of any fractional portion in the quotient when division of integers occurs. This chopping off of the fractional part is called *truncation.*

Note that this matter of the *type* of the result carries over into addition, subtraction, and multiplication. That is, if either of the operands is of type real, then the result is of type real. Only if *both* operands are of type integer is the result of type integer.

Exponentiation

The designation of the value obtained by raising a variable to some power is done in mathematics by writing the power as a superscript (a smaller number a little above and to the right of the variable). For example, Y^2 means the value obtained by raising Y to the second power. In mathematics, no special symbol is used to specify this operation commonly called exponentiation. However, in ForTran the symbol used is the double asterisk. Thus Y**2 means Y raised to the second power. Of course, it is possible to raise a variable to a variable power as follows:

```
VAR1**VAR2
```

This expression has the value resulting from raising the value of variable VAR1 to the power specified by the value of variable VAR2. So, if "5." were stored at VAR1 and "3." stored at VAR2, then the above expression has the value 125.

Negation

The preceding arithmetic operations are all binary operations in the sense that *two* operands are required. There are certain mathematical operations in which only one operand is required. Such operations are called *unary* operations and one of these available in ForTran is negation. The symbol for negation is the minus sign. To indicate the negation of a variable, B, we write

−B or (−B)

Whatever value is stored in variable B will be negated, so that (−B) will have a value equal to the contents of B with opposite sign.

Arithmetic expressions

When referring to a combination of variables and constants together with arithmetic operation symbols, we use the phrase *arithmetic expression*. In order to make this phrase more general people have taken it to include a single variable or a single constant. That is, a numeric variable by itself, a numeric constant by itself, or any combination of these together with arithmetic symbols, are all called arithmetic expressions. Following are some examples of arithmetic expressions:

```
A
5
A*5
(A*5)+B/6
X**2−4*B
X+Y/3*T
```

Since arithmetic expressions without parentheses may often be interpreted as having more than one meaning, it is important to know what interpretation is used by the ForTran compiler.

Hierarchy of arithmetic operations. Recall from mathematics that if the five fundamental arithmetic operations occur in the same expression, the order in which they are to be done is as follows: first, exponentiation; second, multiplication and division in the order of appearance from left to right; and third, addition and subtraction also in order from left to right. In mathematics, the method for specifying clearly what the order of operations is involves the appropriate use of parentheses. If parentheses are used in an arithmetic expression, that portion within parentheses is computed first. In the case of parentheses within parentheses the expression within the innermost parentheses is computed first, then the expression in the next-innermost parentheses, and so on.

Fortunately, all of these hierarchical rules apply in ForTran also. To shed further light on the situation we consider four examples. Where applicable in these examples, assume that the values stored in A and B are as follows:

A=5   B=3

Example 1:   7 + 3 *A—B**2

Applying the above rules tells us that first the square of B is computed as an intermediate value. Following this the expression to be further evaluated is

7 + 3 *A—9

since the square of B is 9. In this latter expression, the rules indicate that the product 3 *A is calculated next. Since A is 5, 3 *A is 15, and now the original expression is reduced to 7 +15 −9. In this latest expression, the addition and subtraction operations have equal priority but are performed from left to right. Therefore, 7 +15 is computed to give 22, after which 9 is subtracted to yield the final value of 13.

Example 2:   7 + (3 *(A—B))**2

This expression has the same variables, constants, and operation symbols as Example 1 but now parentheses have been inserted. Let's apply the hierarchical rules to evaluate this expression. Starting in the innermost parentheses means that A—B must be calculated first, yielding a result of 2. Now the original expression becomes

7+(3*2)**2

Next we find the product of 3 and 2 inside the parentheses, which is 6. The expression to be evaluated now is

7 + 6 **2

The exponentiation operator has priority, so 6 **2 is computed, giving 36. Finally we evaluate 7 +36 to obtain the value of the original expression as 43. Obviously, the use of parentheses changed the value of the expression, a fact that should alert the reader to careful consideration of arithmetic expressions that involve more than one operation.

Example 3:   2 *A/B*7 +B**2**3

In this expression the new concept is the appearance of the exponentiation operator twice in succession. In mathematics this situation would appear like this:

$$B^{2^3}$$

The interpretation would be $(B)^{2^3}$ or B^8. In order to comply with this interpretation from mathematics, ForTran provides that multiple appearances of the exponentiation operator are performed from right to left. Therefore, an evaluation of the original expression produces the following steps:

2*A/B*7+B**2**3	*2**3 is computed first*
2*A/B*7+B**8	*B**8 is computed next*
2*A/B*7+6561	*2*A is computed next*
10/B*7+6561	*10/B is computed next*
3.33333*7+6561	*3.33333*7 is computed next*
23.3331 + 6561	*23.3331+6561 is computed next*
6584.33	

Example 4:   −A*B+100*(90/100)

The use of parentheses in this expression requires the division operation to be done first. Here are the steps in the evaluation process:

−A*B+100*(90/100)	*90/100 is computed first, resulting in zero because this is integer division*
−A*B+100*0	*A*B is computed next*
−15+100*0	*100*0 is computed next*
−15+0	*−15+0 is computed next*
−15	

Note that when the negation operation is not enclosed in parentheses, its precedence is the same as for subtraction.

Assignment statement

Now that we know how arithmetic expressions are evaluated in ForTran, we are ready to discuss the process of storing or *assigning* the results to a variable. The ForTran statement to accomplish this is called the assignment statement and has the following form:

RESULT = Any acceptable arithmetic expression or string constant

The "equals" symbol must not be interpreted in the same sense as it is in mathematics but must be read as "is assigned." Consider this example:

RESULT = 3 * B

This must be interpreted: "variable RESULT is assigned the product of 3 and B." The actions that occur in the computer are (1) the multiplication of 3 by the contents of location B; (2) the erasing of whatever was stored in location RESULT; and (3) the assigning of the product of 3 and B to location RESULT.

Following are more examples of assignment statements with brief explanations of what actions take place.

VAR=126.5	*Assigns the constant 126.5 to variable VAR.*
VAR=VAR*2 +1	*The current value of variable VAR is multiplied by 2 and 1 is added to that product. Then this result replaces the previous contents of VAR.*
NAME='ALI'	*The string ALI replaces whatever was stored in variable NAME.*
J=7.3*A—23	*The expression on the right is evaluated (result is 13.5 if A is 5.) and truncated because the integer variable J is to contain the final result. Therefore, the integer 13 is stored in J.*
X=(X+1)*(3—8/3)+2**3	*Assume the current value of X is 2.5. Then the steps in evaluating the arithmetic expression on the right are*
	(2.5+1)(3—8/3)+8*
	(3.5)(3—2)+8*
	*3.5*1+8*
	3.5+8
	11.5
	Then X is assigned the new value 11.5. Note that the result is a real number because one of the variables on the right—X—is a real variable, in spite of the fact that all constants are integers.

Mixed-mode arithmetic expressions. Recall the importance of distinguishing between integer and real values as stored in a computer. Since the representation in computer memory is so different for these two kinds of information, we need to be well aware of what happens if an arithmetic expression contains *both* kinds. Some mention was made of this in our previous treatment of arithmetic operations, where it was stated that if one or more operands in an arithmetic expression are real then the value of the whole expression is real. Only if all variables and constants in an arithmetic expression are of integer type will the entire expression be of integer type. This is true but needs to be elaborated on just a little. Suppose that the integer 5 has been stored in integer variable N, the integer 6 in integer variable M, and the real value 4.0 in real variable A. Now consider this expression:

    (3 +A) * (N/M + 1)

The fact that parentheses appear in the expression causes the ForTran compiler to handle each parenthesized part separately in terms of integers and reals. Note that the rightmost parenthesized group consists only of integer type values, causing integer arithmetic to be performed within that group. The result is 1 since N/M = 0 with N = 5 and M = 6. The leftmost parenthesized group has the real value 7.0, so that when the result of the left group is multiplied by the result of the right group, the entire expression has the real value 7.0. Therefore, *portions* of arithmetic expressions may result in integer values, though the entire expression will be of type real as long as at least one operand is real.

When an arithmetic expression contains both real and integer values, like the preceding example did, it is called a mixed-mode expression. Because it is so important to know just what results are obtained from mixed-mode expressions, we discuss two examples, both a little more complex than the previous one. *In both examples assume 2.5 is stored at A, 10.2 at B, 5 at N, 6 at M, and 10 at L.* Here's the first expression to be evaluated:

    M/L*A+N/M*B

Here are the steps in evaluating it:

    M/L=6/10=0
    N/M=5/6=0
    (0*2.5)+(0*10.2)=0.0+0.0=0.0

Does it surprise you that an arithmetic expression with no zero values initially and no subtraction operation should ultimately have the value zero? It does most people. The reason, of course, is that integer arithmetic results are sometimes unexpectedly zero. Note that the final value is of type real because of A and B in the original expression.

Consider this expression:

    A*M/L+B*N/M

The steps in evaluating it are

```
A*M=2.5*6=15.0
(A*M)/L=15.0/10=1.5
B*N=10.2*5=51.0
(B*N)/M=51.0/6=8.5
```

Therefore,

```
A*M/L+B*N/M=1.5+8.5=10.0
```

You probably noticed that these two examples involve the same arithmetic operations, though the order of some of the variables is different. The results are not the same at all. The reason, of course, is that the rearranged order resulted in no integer arithmetic, hence no unexpected zeros.

Thus far, any references to mixed-mode have likely given the impression that this is a concern only when real and integer operands are involved. This is not true since mixed-mode arithmetic operations can occur among any types of *numeric* data. (Character type data do not cause mixed-mode errors but may cause syntax errors.) In order to present a more general discussion of mixed-mode, consider this general form of an arithmetic expression:

```
OPRND1 OP OPRND2
```

where OPRND1 and OPRND2 are two operands in a ForTran expression and OP is any of the binary operations +, -, *, /, and **. When OPRND1 and OPRND2 are of the same type there is no mixed-mode expression. But if OPRND1 and OPRND2 are of unlike types the expression is a mixed-mode expression, for which the following rules apply in determining the type of the result:

1. If one of the two operands is of type real and the other one is of type integer the result is of type real.
2. If one of the two operands is of type complex (or double precision) the result is of type complex (or double precision).

Therefore, if an arithmetic expression with several operations in it has at least one complex operand, the final result of all operations is of type complex. If there is no operand of type complex but at least one of type real, then the result is of type real. Otherwise the result is of type integer.

Consider this program as we give some examples of mixed-mode expressions:

```
1     PROGRAM MIX
2     INTEGER A
3     REAL B
4     COMPLEX C
5     C = (2.1, 7.0)
6     B = 3.6
7     A = 2
8     C = A * B + C + (2,3)
9     PRINT*, C
10    C = C * B
11    PRINT*, C
12    A = C
13    B = C
14    PRINT *, A, B
15    STOP
16    END
```

If the above program were executed, the output would be

```
11.3000    10.0000
40.6800    36.0000
40         40.6800
```

Notice that line 5 in the program assigns a complex value to C; therefore, when both C and the constant (2,3) are involved in the arithmetic expression, as they are in line 8, because both are complex numbers the result is also complex and is stored at variable C. Line 9 calls for the printing of the contents of variable C, causing the first line of output.

Line 10 in the above program causes the multiplication of a complex number by a real number, meaning that the real number multiplies both parts of the complex number, producing another complex number, which is stored at variable C. Line 11 causes the output of C as the second line of output.

Line 12 causes the integer portion of the real part of the complex number stored at C to be stored in the integer variable A. Line 13 causes the real part of the complex number at C to be stored in the real variable B. Finally, line 14 causes the output of the integer at A and the real number at B to be output as the third line of output.

ROUNDING AND TRUNCATING

Suppose you purchase a dozen oranges for $1.29. The price per orange can be obtained, of course, by dividing 1.29 by 12, for which the quotient is $.1075, or 10.75 cents per orange. If you were asked how much you had paid for an orange, you would probably say 11 cents, not 10.75 cents. In other words, you would have *rounded* 10.75 to 11.

In mathematics, generally, when we have a real number containing *n* digits (integer or fractional) that we wish to round to *m* digits (*m* less than *n*), we proceed as follows. If the digit in position *m*+1 from the left is less than 5, we drop all digits to the right of the *m*th digit if they are also to the right of the decimal point, or change them to zeros if they are to the left of the decimal point. If the digit in position *m*+1 is 5 or more, we add 1 to the *m*th digit and treat those digits to the right of the *m*th digit as in the previous case.

Here are some examples:

1. Rounding 32.68717 to four significant digits gives the result 32.69 because the digit in position five is 7, a number greater than 5, so 1 is added to the fourth digit to get the indicated result.
2. Rounding 2.8998 to three significant digits gives the result 2.90 because the digit in position four is 9, a number greater than 5, so 1 is added to the third digit. But 1 added to 9 gives 0 plus a carry digit, which then makes the 8 in position two 9. Thus the result is given.
3. Rounding 21.398 to two significant digits gives the result 21, since the 3 in position three is less than 5; hence the three digits to the right of the second position are dropped.
4. Rounding 238,691 to four significant digits gives the result 238,700 because the fifth digit is 9, so we increase the fourth digit from 6 to 7 and change the fifth and sixth digits to zeros.

Recall in the earlier section on division when integer type quantities were involved, we mentioned the term *truncation*. In that context, truncation meant "dropping any fractional digits," leaving an integer result. Truncation in ForTran usually means exactly that. However, in a more general sense, one can speak of truncating to some number of digits, similar to rounding. When truncation occurs, all digits to the right of the specified number are dropped or changed to zero, regardless of their size. For example, if 20 is divided by 3 and the result is truncated to two digits, the result is 6.6, though the true quotient is 6.666666 . . . .

In ForTran 77, when a number is being output, its value is rounded to the greatest number of digits allowed for output by the given computer system. Internally, the number is stored to as many digits as the size of the computer word will allow, then rounded.

In order to impress you with the importance of the computer's not being able to store all fractions in their exact representation we present an example. Suppose the variable X is assigned the value resulting from dividing 1.0 by 300.0. If the size of the computer word allows for storing the equivalent of twelve significant digits, then X is assigned the value .003333333333. Suppose, next, that Y is initially assigned the value 0. Now suppose the value at X is added repeatedly 300 times to the value at Y. (This seems like it should have the same effect as multiplying X by 300.) What value do you think is stored at Y after such repeated additions?

Now assume a variable A is defined by

```
A=(1-Y)*10**14
```

where Y is the value just discussed as the result of 300 additions of X. What value do you think is stored at A? The common answers given to the two questions just posed are 1 and 0, respectively. However, those answers are incorrect, as you can discover for yourself by running this ForTran program:

```
      PROGRAM TRUNC
      REAL X,Y,A
      X=1.0/300
      Y=0
      DO 10 I=1,300
   10 Y=Y+X
      A=(1-Y)*10**14
      PRINT*, 'A=',A,'Y=', Y,'X=',X
      STOP
      END
```

The output of this program is

```
A= 230.686  Y= 0.999999  X= 3.33333E-3
```

In the preceding program, we have included the lines

```
      DO 10 I=1,300
   10 Y=Y+X
```

for the purpose of adding the value of X to the variable Y 300 times. This ForTran structure will be discussed thoroughly in a later chapter. We use it now simply as a convenient means of accomplishing 300 additions.

Now suppose that instead of performing 300 additions of X to Y we simply assign to Y the result of multiplying X by 300. The previous program would be altered to appear like this:

```
PROGRAM TRUNC
REAL X,Y,A
X=1.0/300.0
Y=300*X
A=(1-Y)*10**14
PRINT*, 'A=',A,'Y=',Y,'X=',X
STOP
END
```

This program has the output we would expect:

```
A= 0  Y= 1.00000  X=3.33333E-3
```

In both of the preceding programs the value of X is exactly the same. The difference lies in the method used to compute the value of Y. In the first program, repeated additions result in repeated roundings, which cause the unexpected results. In the second program, no such repeated operations occur. Only one multiplication of 300 and X produces the final value of Y. The student should be constantly alert to the possibility of incorrect results because of faulty computation techniques.

We turn our attention now toward a ForTran statement that causes no computer action whatever but that may be used effectively to improve the readability and understandability of a ForTran program. That statement is the COMMENT statement.

COMMENT STATEMENT

No matter how careful we are in developing the structure and organization of a program so that its logic is easily followed and understood, it is not possible to make it as understandable as prose in one's native language. A ForTran statement that virtually makes this possible is the COMMENT statement. This statement requires that the letter C appear in the first space of the card (or line) and any other useful information in the rest of the card (line). In other words, the COMMENT statement has no effect as far as accomplishing any useful steps in the solution of the problem for which the program was written. Its only purpose is to allow for readable information to be inserted here and there within the program, but only so that a human being reading a listing of the program will be informed about the program's logic. We repeat: no effect on steps in the problem solution result from COMMENT statements. Their usefulness is to increase the readability and understandability of a program listing.

Now we introduce you to some guidelines to help you develop skill in using COMMENT statements effectively.

Dennis VanTassel in *Program Style, Design, Efficiency, Debugging, and Testing*[2] has stated that COMMENT statements can be considered in two categories: (1) prologue comments and (2) explanatory comments.

Prologue comments

The use of COMMENT statements as a prologue results in their appearing at the beginning of a program (or a subprogram) or a group of adjacent statements. Following are some things that prologue comments can do:

1. Describe briefly what the program (subprogram) or group of statements accomplishes.
2. Provide a list of the important variables together with a one-line explanation about each variable.
3. Provide a list of all array variables (to be described in a later chapter) together with brief descriptions of their use.
4. Provide a detailed description of input data required by the program.
5. Provide a detailed description of all possible output generated by the program.
6. List any special instructions needed to correctly execute the program.
7. List the program author, date of completion, company or institution of affiliation, and any other such related information.
8. List any subprograms within the program and describe their functions.
9. Provide a summary description of any special scientific method or procedure used in the program. (A reference should be included for those who want to do further study.)
10. Provide an estimate of computer memory and cpu requirements for executing the program.
11. Explain any special operating instructions.
12. Describe any special cases that the program cannot process.
13. List all data and problem-formulation limitations of the program.

---

[2] VanTassel, D. *Program Style, Design, Efficiency, Debugging, and Testing.* Englewood Cliffs, N.J.: Prentice-Hall, 1974.

Example of prologue COMMENT statements

```
C*************************************************
C*  PROGRAM NAME:  EXAMPLE                        *
C*  AUTHORS:  A. BEHFOROOZ, M. HOLOIEN.           *
C*  DATE: FEBRUARY, 1983.                         *
C*  PURPOSE: TO PROVIDE AN EXAMPLE OF             *
C*      PLACEMENT OF COMMENT LINES IN A           *
C*      PROGRAM OR SUBPROGRAM.                    *
C*  VARIABLE DEFINITIONS:                         *
C*      NUM=NUMBER OF CASES.                      *
C*      OBSV=OBSERVED VALUE.                      *
C*      ESTV=ESTIMATED VALUE.                     *
C*  ARRAY DEFINITIONS:                            *
C*      A=ONE-DIMENSIONAL ARRAY OF SIZE 100       *
C*        FOR STORING OBSERVED VALUES.            *
C*      B=TWO-DIMENSIONAL ARRAY OF SIZE 200       *
C*        X 2 FOR STORING THE DIFFERENCES         *
C*        AND THEIR SQUARES OF THE                *
C*        OBSERVED AND ESTIMATED VALUES.          *
C*  INPUT:                                        *
C*      NUM=NUMBER OF CASES.                      *
C*      NUM IS FOLLOWED BY THAT SPECIFIED         *
C*      NUMBER OF OBSERVED VALUES.                *
C*                                               *
C*  OUTPUT:                                       *
C*      VALUE OF CHI-SQUARED AND THE              *
C*      ASSOCIATED DEGREES OF FREEDOM.            *
C*                                               *
C*  SCIENTIFIC METHOD USED:                       *
C*      THE CHI-SQUARED GOODNESS OF FIT           *
C*      IS USED.  FOR DETAILS SEE ANY             *
C*      APPLIED STATISTICS TEXTBOOK.              *
C*  USING THE PROGRAM.                            *
C*      BEFORE EXECUTION, USER MUST ENTER         *
C*      LINE 39 OF THE PROGRAM AS THE DEFINING    *
C*      EQUATION BY WHICH THE ESTIMATED VALUE     *
C*      OF X IS COMPUTED.  LINE 39 IS             *
C*      ENTERED AS FOLLOWS:                       *
C*        EST(X)=ANY FORTRAN EXPRESSION           *
C*          INVOLVING X.                          *
C*  SUBPROGRAM USED: NONE                         *
C*  MEMORY/TIME REQUIRED: AVERAGE                 *
C*  SPECIAL OPERATING INSTRUCTIONS: NONE          *
C*  SPECIAL CASE: THE PROGRAM WILL NOT WORK       *
C*      FOR NEGATIVE INPUT VALUES.                *
C*  LIMITATIONS: NO MORE THAN 100 OBSERVED        *
C*      VALUES CAN BE PROCESSED.                  *
C*************************************************
```

Explanatory comments

The use of COMMENT statements for explanations is intended to clarify any program logic that would otherwise be difficult to follow. When using these kinds of comments it may be assumed that the reader is familiar with the language in which the program is written. Here are some examples of explanatory comments. It is not intended that the examples given relate to each other.

```
C    GO TO STOP IF END-OF-DATA.
C    BRANCH TO ERROR ROUTINE IF NEGATIVE ACCOUNT NUMBER.
C    RE-ENTER DATA IF BLANK CARD IS READ.
C    REJECT OFF-LIMIT DATA.
C    INPUT DATA FROM FILE XTRA.
C    PRINT HEADINGS FOR OUTPUT.
```

Placement of comments

Although there are no standardized rules for the use of COMMENT statements, it is useful to follow certain guidelines. We suggest the following:

1. Separate comments from program statements by one blank line. This means that there must be a COMMENT statement consisting only of the C in column 1 with all other columns blanks.
2. Comment lines and program lines should be easily distinguished. This may be done by using comment lines consisting of asterisks appropriately placed so that comments are enclosed in boxes. It may also be done through proper use of indentation. Consider the following example:

```
C***********************************************
C*    THE NEXT FIVE PROGRAM STATEMENTS COMPUTE *
C*    CHI-SQUARE.                              *
C***********************************************
C
 }    (PROGRAM LINES)
C
C***********************************************
C* THIS LOOP INITIALIZES ARRAY A.             *
C***********************************************
      ⎫
      ⎬   (PROGRAM LINES)
      ⎭
```

We conclude this section with three problems and show their complete solutions including the ForTran programs.

Problem 1.   Write a ForTran program that accepts as input a number that represents degrees Fahrenheit, converts that number to its equivalent in degrees centigrade, and outputs the converted number.

Solution steps

1. Input degrees Fahrenheit into location F.
2. Recall that degrees centigrade equals (5/9) (degrees Fahrenheit-32) or
   C = (5/9) (F-32).
3. Output the result, C.

These solution steps translate into the following ForTran statements.

ForTran program

```
PROGRAM EXCHNG
REAL F,C
READ*,F
C=(5.0/9.0)*(F-32)
PRINT*,C
END
```

Note that in order for this program to run successfully, input data must be provided either on a punched card if it is run on a batch processing system or on a keyboard if run on an interactive system.

Although the preceding problem statement and solution are a satisfactory disposition of the problem, if a user of the program had available only the program listing of six statements, it probably would not be completely clear what the program accomplishes. This would be especially true if the user knew nothing about the relationship between centigrade and Fahrenheit degrees. It is our goal to provide examples of correct, understandable programs. Much more will be said of this in Chapter 4, but as a step in that direction consider this program as a replacement for the previous one.

```
      PROGRAM EXCHNG
C***************************************************
C*      PRBLEM DEFINITION                         *
C*      THIS PROGRAM INPUTS FAHRENHEIT DEGREES,   *
C*      CONVERTS IT TO CENTIGRADE DEGREES AND     *
C*      OUTPUTS THE RESULT.                       *
C***************************************************
```

```
REAL F,C
READ*,F
C=(5.0/9.0) * (F-32)
PRINT*,C
STOP
END
```

It is clear that the inclusion of comment statements does a great deal to elucidate the program and its functions. We know from our current discussion that the number that is output is the number of centigrade degrees corresponding to the input Fahrenheit degrees. But a user without the benefit of this discussion and without a listing of the program would be hard pressed to know what either the input or output to the program is. In response to this problem we urge beginning programmers to develop early the habit of including statements in programs that result in understandable output so that the user is made well aware of what the program accomplishes. For example, suppose the PRINT statement in program EXCHNG were replaced by the following statement:

```
PRINT*,F,'DEGREES FAHRENHEIT EQUALS',C,'DEGREES CENTIGRADE'
```

If the input were 98.6, the output would be as follows:

```
98.6 DEGREES FAHRENHEIT EQUALS 37.0 DEGREES CENTIGRADE
```

The user would feel much more comfortable with the output of this latter program than the one that gives no clues as to what the output of the program represents.

Here is another problem with complete solution.

Problem 2. Input the radius of a circle and compute and output (a) the area of the circle; (b) the area of the largest square contained within the circle; and (c) the ratio of the result of (a) to the result of (b). Use for the constant, pi, the value 3.1416.

Solution steps

1. Input radius and store in RADIUS.
2. Compute the area of the circle using the equation

```
CAREA=RADIUS*RADIUS*3.1416
```

3. Compute the area of the largest square contained within the circle according to the formula

```
SAREA=RADIUS*RADIUS*2
```

4. Compute the ratio, `RATIO=CAREA/SAREA`.
5. Output all results, then stop.

ForTran program

```
      PROGRAM  AREA
C****************************************************************
C*          PROGRAM DEFINITION                                 *
C*              PROGRAM TO READ RADIUS OF A CIRCLE,            *
C*              COMPUTE AREA OF THE CIRCLE,                    *
C*              COMPUTE AREA OF THE LARGEST SQUARE             *
C*              CONTAINED IN THE CIRCLE,                       *
C*              AND COMPUTE THE RATIO OF CIRCLE                *
C*              AREA TO SQUARE AREA.                           *
C*              IT ALSO PRINTS THE THREE                      *
C*              COMPUTED VALUES.                               *
C****************************************************************
C*          VARIABLE DEFINITION                                *
C*              RADIUS IS THE GIVEN RADIUS OF A CIRCLE.*
C*              CAREA  IS AREA OF CIRCLE.                      *
C*              SAREA  IS AREA OF SQUARE.                      *
C*              RATIO  IS THE RATIO OF THE TWO AREAS.  *
C*              PI IS 3.1416.                                  *
C****************************************************************
      REAL RADIUS, CAREA, SAREA, RATIO, PI
      PI = 3.1416
      READ*, RADIUS
      CAREA = RADIUS * RADIUS * PI
      SAREA = RADIUS * RADIUS * 2
      RATIO = CAREA / SAREA
C*
      PRINT*, ' THE RADIUS GIVEN IS ', RADIUS
      PRINT*, ' THE AREA OF THE CIRCLE IS ',CAREA
      PRINT*, ' THE AREA OF THE SQUARE IS ',SAREA
      PRINT*, ' THE RATIO OF THE TWO AREAS IS ',RATIO
      STOP
      END
```

Problem 3.   Input a person's birthdate in the form of three numbers: year, month, and day. Input today's date similarly. Determine the person's age.

Solution steps

1. Input today's year, month, and day as TY, TM, and TD, respectively.
2. Input year, month, and day of birthdate as BY, BM, and BD, respectively.
3. If TM is greater than BM, then `AGE=TY-BY`.
4. If TM equals BM and if TD is greater than or equal to BD, then `AGE=TY-BY`.
5. In any other case, `AGE=TY-BY-1`.

Although the preceding solution steps will yield the desired results, translating them into ForTran statements goes beyond the discussion so far in this book. To avoid this complication, we suggest an alternate solution based on the assumption that every month in the year has 30 days, making a year of 360 days.

Alternate solution steps

1. Input today's year, month, and day as TY, TM, and TD, respectively.
2. Input year, month, and day of birthdate as BY, BM, and BD, respectively.
3. Convert today's date to days as follows:

   `TDAYS=360*TY+30*TM+TD`

4. Convert birthdate to days as follows:

   `BDAYS=360*BY+30*BM+BD`

5. Then `AGE=(TDAYS-BDAYS)/360`

Although the alternate solution is less accurate, it can be written as a ForTran program with the statements previously discussed in this book.

ForTran program

```
        PROGRAM  AGES
C***********************************************************
C*          PROGRAM DEFINITION                             *
C*             APPROPRIATE DEFINITION SHOULD BE PROVIDED   *
C*             BY THE READER.                              *
C***********************************************************
C*          VARIABLE DEFINITION
C*             TY, TM, TD ARE TODAY'S YEAR, MONTH AND DAY, *
C*             RESPECTIVELY.                               *
C*             BY, BM, BD ARE BIRTH YEAR, MONTH AND DAY,   *
C*             RESPECTIVELY.                               *
C*             TDAYS  IS TODAY'S DATE IN DAYS.             *
C*             BDAYS  IS BIRTHDATE IN DAYS.                *
C*             AGE  IS AGE IN YEARS.                       *
C*             NAME  IS THE NAME OF THE USER.              *
C***********************************************************
C*          INPUT SPECIFICATION                            *
C*             TY AND BY ARE 4-DIGIT NUMBERS, TD, BD, TM,  *
C*             BM ARE 2-DIGIT NUMBERS AND NAME IS 1 TO 10  *
C*             CHARACTERS.                                 *
C***********************************************************
C*
        INTEGER  TY,TM,TD,BY,BM,BD,TDAYS,BDAYS,AGE
        CHARACTER  NAME*10
C*
        READ*, TY, TM, TD
        READ*, NAME, BY, BM, BD
C*
        TDAYS = 360 * TY + 30 * TM + TD
        BDAYS = 360 * BY + 30 * BM + BD
        AGE = (TDAYS - BDAYS)/360
C*
        PRINT*, NAME, ' YOU ARE ', AGE, ' YEARS OLD.'
        STOP
        END
```

As you know from information presented earlier in this book, besides numeric information, a computer is able to process information consisting of letters and other characters. We have called such data *string data* or *character data*. As there are arithmetic operations for processing numeric data, there are certain operations reserved for processing character data. We proceed next to a discussion of two such string data operations.

STRING DATA OPERATIONS

We begin by introducing you to the operation called *concatenation*.

Concatenation

Suppose variable A is a character variable of size *m* characters and B is a character variable of size *n* characters. By the *concatenation* of A and B is meant the string consisting of the *m* characters of A followed immediately by the *n* characters of B. The symbol in ForTran for the operation of concatenation is // (double slash). Thus if we want to indicate the concatenation of variable A and variable B, we do so as

```
A // B
```

Consider the following program segment:

```
CHARACTER A*8,B*11,C*19
A ='AS GOOD'
B =' AS THIS ONE'
C =A // B
```

In this program segment, after the last statement has been executed, variable C contains the characters

```
AS GOOD AS THIS ONE
```

Note that blank spaces are included as legitimate characters in a string. As an example illustrating this, here is another program segment:

```
CHARACTER A*8,B*11,C*19
B='  THIS       '
A='OR THAT'
C=B // A
```

After the execution of these statements the contents of C is

```
  THIS     OR THAT
```

where the two spaces before THIS and the five spaces after it are preserved in the concatenation of B and A.

Here is another example.

```
CHARACTER A*7,B*3,C*12
A='THIS'
B=' OR'
C=' THAT'
A=A // B
C=A // C
```

After the execution of these statements the contents of C is

```
THIS OR THAT
```

The reader is urged to verify this result by examining the contents of variables A, B, and C at each line of the above program segment.

If a character variable is too small to contain all the characters assigned to it, the leftmost characters (as many as can be stored) of the assigned string will be stored while the excess characters to the right are lost. Consider this program:

```
PROGRAM  CHAR
CHARACTER *10 A, B
CHARACTER C*15
READ*,A, B
C = A // B
PRINT*,A, B, C
STOP
END
```

Suppose this program is run four times with these sets of input data:

1. 'ALI' 'BEHFOROOZ'
2. 'MARTIN O. HOLOIEN' 'ALI BEHFOROOZ'
3. 'A' 'B'
4. '1234567890' '1234567890'

The four resulting sets of output are

Case 1.	ALI	BEHFOROOZ	ALI	BEHFO
Case 2.	MARTIN O.	ALI BEHFOR	MARTIN O.	ALI B
Case 3.	A	B	A	B
Case 4.	1234567890	1234567890	123456789012345	

A careful analysis of these outputs gives us some insights into the way character data are handled.

Case 1.   Notice that variables A and B are each specified as having length 10, and C as having length 15. The input provided for A is only three characters and that for B is nine characters. A look at the spacing of the ouput seems to indicate that for A the output consists of the letters A, L,I, and seven spaces. Similarly, B is output as the nine letters B,E,H, F,O,R,O,O,Z and one space. Finally, to the right in Case 1 output, the concatenation of A and B, stored in variable C, is output as the fifteen characters consisting of A,L,I, seven spaces, B,E,H,F, and O. All of this indicates that if the input provided for a character variable is not enough to fill the entire length of the variable, then the unfilled rightmost characters become blank spaces.

Case 2.   In this case, more characters are provided as input for both variables A and B than the ten-character length at which they are both declared. By looking at the output for A,B, and C (the concatenation of A and B at the right of Case 2 output) it is evident that A consists of the ten characters M,A,R,T,I,N,space,O,period, and space. Similarly, B consists of the ten characters A,L,I,space,B,E,H,F,O,and R. Variable C consists of the ten characters of A plus the first five characters of B to make up the fifteen-character length it is specified as.

Therefore, it appears that we can conclude that if the length of the character string provided as input for a character variable is not as long as the length specified for the variable, then blank spaces are automatically filled into the rightmost part of the variable. If *more* characters are provided as input than the length specified for the variable, then only as many characters as are needed are taken from the left part of the input string to fill the length of the variable.

You are urged to analyze the output for Cases 3 and 4 above to determine if the preceding statements are valid.

The substring operation

A second string operation available in ForTran 77 is the *substring* operation. This operation makes it possible to get at certain specified characters of those that make up the string. This operation differs from concatenation in that it is a unary operation requiring only one operand. The symbol for the substring operation on the string STRING is

STRING(i:j)

where i and j are integer constants or variables such that i is less than or equal to j and both i and j are not greater than the length of STRING. The default value for i is one and for j is the length of the string STRING. For example, STRING(:5) means the first five characters of string STRING and STRING(5:) means the last N-4 characters of STRING where N is the length of STRING.

Let's consider an example to illustrate this operation. Suppose ST has been declared as a character variable of length 15 and that variable B has been declared as a character variable of length 5. Then the statement

```
B = ST(3:5)
```

will cause the third through the fifth characters of ST to be stored in the first three positions of variable B. Note that the substring symbol appears immediately to the right of the variable on which it is to operate.

Here is a complete program to illustrate further uses of the substring operation.

```
      PROGRAM STRING
      CHARACTER ST1*20,ST2*15,ST*15
      READ*, ST1, ST2
C***********************************************************
C*              RECALL THAT IF THE 2ND INTEGER IN THE    *
C*              SUBSTRING SYMBOL IS OMITTED, THEN WE      *
C*              INTEND TO SPECIFY THE RIGHTMOST           *
C*              CHARACTER OF THE STRING IN THE SUB-       *
C*              STRING SYMBOL. THUS IN THIS PROGRAM       *
C*                   ST2(6:15) AND ST2(6:)               *
C*              ARE EQUIVALENT.                           *
C***********************************************************
      ST = ST1(1:5) // ST2(6:)
      PRINT*,ST1,ST2,ST
      STOP
      END
```

Suppose we run the above program with this input:

```
'ALI' 'BEHFOROOZ'
```

The three letters of the name ALI are stored in the leftmost three positions of variable ST1 and the next seventeen positions of ST1 are filled in with blank spaces. The nine letters in BEHFOROOZ are stored in the leftmost nine positions of variable ST2 and the next six positions of ST2 are filled in with blank spaces. In ST are stored the first five characters of

ST1 (three letters and two blanks) followed by the last ten characters of ST2 (since only ten positions remain in ST and those ten positions will be filled from the rightmost characters of ST2, namely, the letters R,O,O, and Z, and six blank spaces). The output of this program consists of the contents of the three variables ST1, ST2, and ST, and would appear as follows:

```
ALI                  BEHFOROOZ    ALI  ROOZ
```

Suppose we run the program again with this new input:

```
'ALI BEHFOROOZ OR'    'MARTIN O. HOLOIEN'
```

The output this time would be

```
ALI BEHFOROOZ OR    MARTIN O. HOLOI    ALI BN O. HOLOI
```

Note in the preceding case that ST2, with length 15, is capable of storing only the first fifteen characters of the input

```
MARTIN O. HOLOIEN
```

We present one more set of input to be used with the previous program STRING:

```
'FFIVE NEXT FIFTEEN'    'LAST NINE CHARACTERS'
```

The output produced by the program this time is

```
FFIVE NEXT FIFTEEN    LAST NINE CHARA    FFIVENINE CHARA
```

Note that in this last case, ST1, with specified length of 20, was large enough to store all of the input provided. However, ST2, with specified length of 15, was *not* large enough to contain the twenty characters of input data provided. Therefore, ST2 contained only the first fifteen characters of input data provided. The substring operations caused characters 1 through 5 of ST1 and characters 6 through 15 of ST2 to be used in the concatenation operation, resulting in the output shown at the right, above.

With arithmetic operations, the same variable could appear on both sides of the equals sign in an assignment statement. For example,

A = A + B

is a valid and sometimes useful statement. However, a similar statement involving string operations would not produce results that you might at first think should be produced. Therefore, the statement

```
ST = ST1 (1:3) // ST (1:9)
```

(with variable ST on both sides of the equals sign) would not produce the results in ST that you would most likely expect. Let's examine a program to help us understand what happens:

```
PROGRAM TEST
CHARACTER ST*12, ST1*3
ST = ' EXAMPLES'
ST1 = 'TWO'
ST = ST1(1:3) // ST(1:9)
PRINT*,ST
STOP
END
```

When the ForTran compiler executes this program, it follows these steps:

1. Determines ST1 (1:3).
2. Assigns the results of step 1 to ST.
3. From ST the characters in positions 1 through 9 are assigned *one at a time* to positions 4 through 12 of ST. That is, a copy of the first character of ST is stored in the fourth position of ST; a copy of the second character of ST is stored in the fifth position of ST; a copy of the third character is stored in position 6; a copy of the fourth character (the one currently occupying that position) is stored in position 7; and so on until the twelfth position is filled from the character last occupying position 9. Thus, there are nine times that the contents of ST are modified after the first three characters of ST are filled with the three characters of ST1.

To make it clear what actually happens when the program line

```
ST = ST1 (1:3) // ST (1:9)
```

is executed, we show the contents of ST at ten different times. Time 0 shows the contents of ST after having positions 1–3 filled with the characters of ST1. Subsequent times show the position-by-position filling described in step 3 above.

TIME	CONTENTS OF ST
0	TWO
1	TWOT
2	TWOTW
3	TWOTWO
4	TWOTWOT
5	TWOTWOTW
6	TWOTWOTWO
7	TWOTWOTWOT
8	TWOTWOTWOTW
9	TWOTWOTWOTWO

Once more, let us remind you that the preceding example is intended to illustrate the potential *problem* that arises when the same variable appears on both sides of the equals sign in an assignment statement.

To obtain the *intended* results, PROGRAM TEST should be written as follows:

```
PROGRAM TEST
CHARACTER ST*12, ST1*3, TEMP*9
ST = ' EXAMPLES'
ST1 = 'TWO'
TEMP = ST(1:9)
ST = ST1(1:3) // TEMP
PRINT*,ST
STOP
END
```

The output of this program is

TWO EXAMPLES

Note that it is *not* incorrect to use the substring operation within an assignment statement. It is simply a matter of understanding exactly what happens. For example, the statement

```
ST (1:5) = ST(7:11)
```

calls for positions 1 through 5 of ST to be filled with a copy of the characters in positions 7 through 11 of the same variable. Thus, if the contents of ST were originally

```
ABCDEFGHIJKL
```

then, after the execution of the above statement, the contents would be

```
GHIJKFGHIJKL
```

The last concept to be introduced in this chapter is that of a system library function.

FORTRAN SYSTEM LIBRARY FUNCTIONS

There are a number of computational processes that require more than one step to obtain the desired results and are used again and again by many different people. For example, the process of finding the square root of a positive real number is fairly complicated to describe step by step no matter which of the several available methods one chooses. Once constructed in the form of a computer program, the process would be readily available for anyone to use. Many other processes, especially in the fields of mathematics and programming, could also be made accessible. The ForTran system makes a number of such stored programs available to any user of the system and these are called *system library functions* or *ForTran library functions*. Appendix D provides a complete list of all the system library functions available in ForTran 77. We discuss here the use of such library functions and list some of the ones you will likely be wanting to use early in your programming.

Using a ForTran library function

To use a library function we need only provide the correct *name* of the function followed by a pair of parentheses in which is identified the object on which the function is to operate. We call the object the *argument* or *parameter* of the function. For example, suppose we have a positive number stored at real variable X and we want to compute the square root of X, storing the result at real variable Y. The following statement accomplishes this:

```
Y = SQRT(X)
```

Note that we use an expected ForTran name for the square-root function, which is SQRT. To the right of that name is a pair of parentheses in which we specify the object of the process performed by the function SQRT. As we said above, this is called the argument or parameter of the function.

So *using* a ForTran library function is no more difficult than providing its correct name and an argument. As with simple ForTran variables, library functions may be used in assignment statements (as above), in arithmetic or string expressions (depending on the library function being used), in PRINT statements, and as arguments of other library functions. Thus, all of the following are valid examples of using the ForTran system library function SQRT:

```
A = B *(R+SQRT(S))
PRINT *, A, B, SQRT(T)
Z = SQRT (X+SQRT(B))
```

As you can tell from these examples, system library functions are used in ForTran statements in the same manner as simple ForTran variables.

At this point we introduce you to some of the more important ForTran system library functions.

Some important ForTran system library functions

With each of the following library functions we show its correct name, an argument name reminding you of the type of the argument, and a brief description of what the library function does.

1. SQRT(R)
   This function computes the square root of a real value stored at R, the result also being real.
2. INT(R)
   This function produces an integer value, which is the greatest integer less than or equal to the real value at R.

3. ABS(R)

This function produces a real value that is the same as the real value at R if R is zero or positive but is the negative of R if R is negative. This result is called the absolute value of R.

4. MAX0(I1,I2, . . . ,IK)

This function determines the greatest of the K integers I1, I2, . . . ,IK and makes that integer value available in the program statement where it appears.

5. MIN0(I1, I2, . . . ,IK)

This function is like the MAX function except that the *smallest* of the K integers is returned to the program.

6. FLOAT(I)

This function produces a real value that is the real representation of the integer value at I.

7. MOD(I1,I2)

This function computes the integer remainder when the integer value at I1 is divided by the integer value at I2. This is called the *modular quotient* of I1 and I2.

8. EXP(R)

This function computes the real result when e (e = 2.178 . . .) is raised to the real power R.

9. LEN(STRING)

This function computes the length (number of characters) of the string STRING. This result is an integer value.

10. ICHAR(C)

This function produces the integer value code for the single character represented by the argument C.

11. CHAR(I)

This function produces the single character value that has the code equal to the integer value I.

12. INDEX (ST1,ST2)

This function searches the character string ST1 to see if substring ST2 is contained in it. If substring ST2 is not contained in ST1, the integer value zero is returned to the program. If ST2 does occur in ST1, then this function returns the integer value equal to the leftmost position in ST1, where the substring ST2 first occurs.

This is just a partial list of the ForTran system library functions available. As stated earlier, the complete list is given in Appendix D.

Examples using system library functions

Earlier we gave three sample statements using the library function SQRT. Now we present some examples using some of the system library functions we have just described. In these examples, CST is a character variable of length 4, I is an integer variable, and all other variables are of type real.

1. `Y = INT(X/4)`
The real value at X is divided by 4, and the greatest integer less than or equal to that quotient is stored at Y.
2. `X = ABS (3 − SQRT(X))`
First the square root of the value at X is computed and the result subtracted from 3. Then the absolute value of the difference is computed and stored at X.
3. `Z = FLOAT(MOD(I,20)) / 15`
The modular quotient of I and 20 is computed and the resulting integer value is changed to an equal real value which is then divided by 15 and the real result is then stored at Z.
4. `CST = 'ABCD'`
   `Y = LEN(CST) + LEN('QRSTUV')`
Here the length of CST, which is 4, is added to the length of the string QRSTUV, which is 6, and the result, which is 10, is stored as a real number at Y.
5. `I = ICHAR('A')`
The integer code for the letter A used by the computer system on which this statement is executed is assigned to I.
6. `CST = CHAR(22) // 'BC'`
The character whose code on the given computer system is the integer 22 is concatenated with the string BC and the resulting character string is stored in CST.

The random number function

There is one last function that we want to discuss even though it is not one of the system library functions available in ForTran 77. It is available on most computer systems that have ForTran translating programs and is used in almost the same manner as system library functions. Therefore, we discuss its use here.

This function is called the *random number generator* and is used to produce random numbers. The range of random numbers generated by this function extends from 0 to 1, second endpoint excluded. Although this function requires an argument, it is a dummy argument in the sense that it makes no difference what it is as long as there is one. We give the name RANDOM to this function. It is used in ForTran statements in the same way as the ForTran system library functions. Here are some examples:

1. X = RANDOM(1)

   A random number between 0 and 1 is produced and stored in real variable X.

2. Y = RANDOM(1)*99 + 1

   A random number between 0 and 1 is multiplied by 99 and 1 is added to the product. The result is that the number stored in real variable Y is a real number between 1 and 101, endpoints excluded.

3. D = INT(RANDOM(X)*9 + 1)

   A random number between 0 and 1 is multiplied by 9 and 1 is added to the product. Then the INT function computes the greatest integer less than or equal to the computed results just described. Therefore, the integer finally stored in integer variable D is a single digit from 1 to 9, inclusive.

SUMMARY

In this chapter we have considered a variety of concepts related to the software portion of a computer system. We learned about the three levels of programming languages and were introduced to the idea of system commands or job control language.

Then we turned our attention to the programming language, ForTran, considering a little of its historical development first, followed by an introduction to concepts that must be mastered if one is to write programs in ForTran. The use of constants and variables was discussed.

Attention was also given to data declaration statements by which we define variables as real, integer, complex, logical, or character. The general forms of these statements are

1. REAL var-list
2. INTEGER var-list
3. LOGICAL var-list
4. COMPLEX var-list
5. CHARACTER Vl*nl,V2*n2, . . . ,VN*nn
6. CHARACTER *n,var-list

In the above statements, "var-list" represents a list of one or more variable names separated by commas. In statement 5, n1,n2, . . . ,nn represent integers specifying the maximum number of characters in each of the associated character variables. In statement 6, n represents an integer specifying the maximum number of characters in each of the variables in "var-list."

Statement 1 declares all variables in "var-list" as *real* variables. Statement 2 declares all variables in "var-list" as *integer* variables. Statement 3 declares all variables in "var-list" as *logical* variables. Statement 4 declares all variables appearing in "var-list" to be *complex* variables. Each complex variable stores two real values corresponding to the real part and the imaginary part of the complex number stored.

The IMPLICIT statement was introduced as the one that enables the programmer to specify a whole set of variables according to the first letter of their names as being of a certain type. For example, consider these:

1. `IMPLICIT INTEGER (A-D)`

2. `IMPLICIT REAL (X-Z)`

3. `IMPLICIT CHARACTER (M)`

Statement 1 causes all variables whose names begin with A, B, C, or D to be considered of type integer. Statement 2 causes all variables whose names begin with X, Y, or Z to be considered of type real. Statement 3 causes all variables whose names begin with M to be of type character, each capable of storing one character.

Input and output statements discussed in this chapter are:

1. `READ*,var-list`

2. `PRINT*,out-list`

In the above statements, "var-list" refers to zero or more variables separated by commas and "out-list" refers to zero or more variables, constants, or arithmetic expressions.

When "var-list" is missing in a READ statement, it results in one data card (line) being skipped. When "var-list" is missing in a PRINT statement, the result is one blank line of output.

Also discussed in this chapter were the comment card (line), the STOP, END, and PROGRAM statements. Comment cards are used to insert clarifying comments in the program. The STOP statement is used to halt further program execution. The END statement must be the last card (line) in any program, and the PROGRAM statement (if it is used) must be the first card (line) except that comment cards (lines) may precede it. The PROGRAM statement assigns a name to the program if it is used.

Arithmetic operations, mixed-mode expressions, and the hierarchy of arithmetic operations were discussed. It was pointed out that

** has highest priority,
/ and * have second priority, and
+ and − (subtraction) have lowest priority.

When two or more operations of the same priority appear in the same arithmetic expression, they are executed in order from left to right, except for exponentiation (double asterisks), which is executed in order from right to left. As in mathematics, parentheses may be used to specify order of operations. When this is done, expressions inside the innermost parentheses are evaluated first, then within the next-innermost pair, and so on until all operations have been performed.

This chapter has included a discussion of two operations whose operands are character strings. These operations are concatenation (whose symbol is the double slash, //) and the substring operation (whose symbol consists of a pair of parentheses in which are placed two integer values separated by a colon, as in (i:j), where i and j can be integer constants or variables). The concatenation symbol is placed between the two character strings being concatenated (which means essentially the joining together of two character strings to form a new character string). Thus if ST1, ST2, and ST3 are all character variables, then the statement

```
ST3=ST1 // ST2
```

causes the character string made up of the string at ST1 followed by the string at ST2 to be stored at ST3. Of course, the length of ST3 must be large enough to accommodate the combined string made up of ST1 and ST2.

The substring operation works on only one operand, so that the statement

```
ST2=ST1(3:9)
```

causes the substring consisting of characters 3 through 9 of ST1 to be stored in ST2. We assume both ST1 and ST2 have been properly specified to accommodate strings of the lengths given.

The last topic discussed in this chapter is ForTran system library functions. A definition was given for a system library function and twelve important such functions were described with several examples to illustrate them.

EXERCISES

1. Give a correct response for each of the following:
   (a) Why is machine language rarely used by the average programmer?
   (b) What are some advantages of assembly language over machine language?
   (c) Give two examples of machine-dependent languages.
   (d) Give three examples of machine-independent languages.
   (e) Give the meaning of these words: bit, byte, word.
   (f) What category of languages is referred to as high-level languages? Low-level?
   (g) List some advantages for standardizing a programming language.
   (h) What is meant by a *constant* in ForTran?
   (i) What is meant by a *variable* in ForTran?
   (j) How do *integer* and *real* constants differ?
   (k) How do *integer* and *real* variables differ?

2. Identify all incorrect *variable* names in the following list and state why each is incorrect.

(a) VARIABLE	(p) NUMBR	(ee) KOOL	(tt) XX1
(b) SQUARE	(q) 2X	(ff) KOLD	(uu) IXX2
(c) DON'T	(r) X−1	(gg) A 15 B	(vv) JXY
(d) GAS-PUMP	(s) NEW	(hh) 'ABC'	(ww) NEWX
(e) PRIME	(t) OLD/NEW	(ii) KOUNT	(xx) OLDX
(f) I/O	(u) 25	(jj) COUNT	(yy) INPUT
(g) IN OUT	(v) A25	(kk) FOOT*	(zz) OUTPUT
(h) ENTER	(w) NO.	(ll) $100	(a1) GRADE
(i) IN	(x) PRIMNO	(mm) D100	(b1) TESTNO
(j) INOUT	(y) FACTOR	(nn) LAST	(c1) SCORE
(k) INDOOR	(z) RATIO	(oo) LEAST	(d1) NTEST
(l) OUTDOOR	(aa) RATE	(pp) LENGTH	(e1) NOFTST
(m) SUM	(bb) WAGE	(qq) WARM	
(n) MAX	(cc) NET	(rr) DEGREE	
(o) MIN	(dd) JOB1	(ss) X1	

3. When preparing ForTran cards to be processed on a batch computer system, what special use is made of each of the following designated card fields?
   (a) Column 1
   (b) Columns 2–5
   (c) Column 6
   (d) Columns 7–72
   (e) Columns 73–80

4. If yours is a timesharing computing environment, answer question 3 with respect to that environment.

5. Examine each of the following ForTran statements individually and independently of each other. Those that are incorrect are to be so identified and a reason given for calling them incorrect. Where that is important, assume the variables are properly defined.

(a) READ NUM1,NUM2,NUM3

(b) CHARACTER 10*NUM1,NUM2,NUM3

(c) NUM1=NUM2/−NUM3

(d) X*X=X**2

(e) READ*,10, 20

(f) READ*,A, A+B

(g) PRINT*,A, A+B

(h) A(1:5)=B(2:7)

(i) A(1:5)=B(1:3) // A(1:6)

(j) CHARACTER THIS, OR, THAT

(k) COMPLEX END, READ, WRITE

(l) STOP = END

(m) END = START

(n) LAST = END OR 0

(o) 200A = 100A * 2

(p) A+2 = A+4−2

(q) TWO = 2.5

(r) FIVE = 0

(s) TEN = 2*FIVE

(t) LOGICAL A NAME

(u) PRINT*,'A+B', = , A+B

(v) PRINT*,'A+B =' , A+B

(w) PRINT*,SUM, ' = TOTAL'

(x) A = B * AX(AX+1)

(y) B = B * (B+1)/A TIME B

(z) X = (X+Y) / ZERO

(aa) READ*

(bb) PRINT*

(cc) IMPLICIT (A TO Z)

(dd) A=A**−B

(ee) X OR Y = Y OR X

(ff) X2 = X*2

(gg) A(1:4)=A(5:8)

(hh) X=X // Y

(ii) IFA = THEN = ELSE

(jj) INTEGER A OR B

(kk) REAL I,J,X

(ll) CHARACTER 2*A*3B

(mm) CHARACTER A*2,B*3

(nn) CHARACTER 2*A,B,C*3

(oo) IMPLICIT REAL(A,B)

(pp) IMPLICIT INTEGER(I-N)

(qq) IMPLICIT CHARACTER(A-Z)

(rr) IMPLICIT REAL(X)

(ss) LOGICAL A,B

(tt) IMPLICIT LOGICAL(A)

(uu) LOGICAL REAL,INTEGER

(vv) LOGICAL REAL,VAR

(ww) A = B // C(1:10)

(xx) X = A(1:10)

(yy) AB=A(1:10) // B(10:)

(zz) A = A(1..10)

6. Write a single ForTran statement equivalent to each of these mathematical equations. Each different letter represents a different variable. When two or more letters are written together, we assume the convention of mathematics as meaning their product.

(a) $$x = \frac{a^2+b^2+ab}{1 + \dfrac{a}{a+b}}$$

(b) $v = (a-2b)^3 (3a+b)^2$

(c) $w = \dfrac{-b + \sqrt{b^2 - 4ac}}{2a}$

(d) $t = (a^n)^m + a^n a^m$

(e) $p = 37 \left(\sqrt[8]{a^2} + \sqrt{a^3} \right) / \left(\sqrt{a\sqrt{b}} + a \right)$

(f) $y = a + \dfrac{1}{1 + \dfrac{1}{1+a}}$

(g) $z = \left(a^{b^2} + b^{a^2} \right) \left(a + b + \dfrac{ab}{a+b} \right)$

(h) $u = abc + \dfrac{1}{ab} + \dfrac{bc}{ab+bc} + \dfrac{ab+bc+ac}{\dfrac{a+b+c}{abc}}$

7. Determine the value that would be stored in real variable X or integer variable IX if each of the following ForTran statements were executed given that A=200., B=10., and C=.5.

  (a) X=A*C + (B*C)**2
  (b) X=(A+B)*C/B*C
  (c) X=B**2**3/A**2
  (d) IX = (A+B)**(B+C)*(5/9)
  (e) IX = (9/5)*(A-32)
  (f) IX = (A+B+30*C) * 1/3
  (g) IX = A*(10/200) * C
  (h) X = A * C * (10/200)
  (i) IX = C/B + C**2*A
  (j) X = C**2*A + C/B
  (k) IX = (A+B+30*C)*(1/3)
  (l) X = (B/A)*(A/B)**C
  (m) X = (A/B*B/A)**C

8. The following programs should run and produce output on any computer system that has a ForTran 77 translating program. Give the output generated by each one if it were executed with indicated input data.

(a)
```
PROGRAM A
INTEGER NUM
PRINT*, 'WHAT IS YOUR STUDENT NUMBER?'
READ*, NUM
PRINT*, 'YOUR STUDENT NUMBER IS', NUM
STOP
END
```

Assume 810001 is input for NUM.

(b)
```
PROGRAM B
REAL AVE,SUM
INTEGER NUM1,NUM2
READ*,NUM1,NUM2
SUM=NUM1+NUM2
AVE=SUM/2
PRINT*, 'THE TWO NUMBERS ARE:',NUM1,NUM2
PRINT*,NUM1, '+',NUM2,'=',SUM
PRINT*,'THE AVERAGE OF THE 2 NUMBERS IS',AVE
STOP
END
```

Assume 8 and 12 are input for NUM1 and NUM2, respectively.

(c)
```
PROGRAM C
READ*,CENTIG
FAHR=CENTIG*9/5+32.
PRINT*,'CENTIGRADE','FAHRENHEIT'
PRINT*,CENTIG,FAHR
END
```

Assume 40. is input for CENTIG.

(d)
```
PROGRAM D
READ*,X,Y
AREA=X*Y
PER=(X+Y)*2
PRINT*,'SIDE ONE OF A RECTANGLE',X
PRINT*,'SIDE TWO OF A RECTANGLE',Y
PRINT*,'AREA OF THAT RECTANGLE IS',AREA
PRINT*,'PERIMETER IS',PER
END
```

Assume 5 and 8 are input for X and Y, respectively.

9. The following ForTran programs each have at least one fatal error (meaning an error that makes it impossible for the program to be executed). Identify the errors and provide corrected statements.

(a)
```
READ*,A,B
X=A+B
PRINT*,X,A*B
READ*,N+M
PRINT*,N+M
END
```

(b)
```
PRINT*'HEADING LINE'
PRINT*,'************'
READ*,ONE,TWO
THREE=ONE+TWO
PRINT*,ONE+TWO,EQUALS,THREE
END
```

(c)
```
READ*,F
C=5/9*(F-32)
PRINT*,'F DEGREES', 'C DEGREES'
PRINT*,F,C
PRINT*,A,A*2
END
```

10. For each of the following situations write a simple ForTran program to produce the required results.
    (a) Input number of miles, output equivalent number of kilometers.
    (b) Input pounds, output equivalent kilograms.
    (c) Input degrees Fahrenheit, output equivalent degrees centigrade.
    (d) Input yards, output equivalent meters.
    (e) Input acres, output equivalent square kilometers.
    (f) Input cubic feet, output equivalent cubic meters.
    (g) Input gallons, output equivalent liters.

11. Write a ForTran program to input two numbers, compute what percent the first number is of the second, and output both numbers input as well as the computed percent.

12. Write a ForTran program to input three numbers, X1, X2, and X3, and find their average, $\overline{X}$, where

$$\overline{X} = \frac{X1+X2+X3}{3}$$

Also find the standard deviation, SD, where $SD=(1/3)(X1^2+X2^2+X3^2-3\overline{X}^2)$, and the relative percentage, RP, for each number where

$$RP1 = \frac{X1}{X1+X2+X3} \cdot 1000$$

and similarly for RP2 and RP3. Output the standard deviation and the three relative percentages.

13. Write a ForTran program that inputs the necessary dimensions for each of the following geometric figures and outputs the perimeter.
    (a) Square (one dimension)
    (b) Rectangle (two dimensions)
    (c) Diamond (one dimension)
    (d) Circle (one dimension)
    (e) Right triangle (lengths of two legs from which the hypotenuse can be computed)

14. Write a ForTran program to solve the following problem where the only input is the thickness of the paper and the output is the number of times the paper must be folded.

    Problem: Suppose you are given a piece of paper as large as needed and suppose you are always able to fold the paper no matter how many thicknesses there are. If the paper is 1/64 inch thick, how many times must the paper be folded so that the height of the folded layers is at least one yard. (Note that every time the paper is folded, the resulting layers form a thickness twice the thickness before it was folded.)

15. Write a ForTran program that accepts three names each having a maximum length of twenty characters and produces the following output:
    (a) The first letter of each of the three names all on the same line.
    (b) A single character string of length 3 made up of the first letters of the three names.
    (c) A line of output on which appear three character strings, each of them made up of the ninth, tenth, eleventh, and twelfth letters of one of the three names.
    (d) A single string of length 21 such that it contains the first seven characters of the first name, characters 8 through 14 of the second name, and characters 14 through 20 of the third name.

CHAPTER 3

Selective Structures in ForTran

In Chapter 2 we introduced the three major structures used in most high-level programming languages and discussed in detail the first of these—namely, sequential structures.

Now we shall proceed to *selective structures*, those structures that involve decisions being made to determine which program segment will be executed next in moving toward the solution of the problem. We will discuss selective structures under two major categories: two-way and *k*-way, where *k* is a positive integer greater than 2.

TWO-WAY SELECTIVE STRUCTURE

The general form of the two-way selective structure is diagrammed in Figure 3.1. In abbreviated English (or pseudocode), we can represent the two-way selective structure as follows:

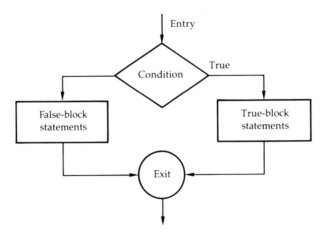

Figure 3.1   Diagram of the two-way selective structure

```
IF   (Condition) THEN
          Statement 1 ⎫
          Statement 2 ⎪
                      ⎬   True-block statements
              .       ⎪   (also called an IF block)
              .       ⎪
              .       ⎪
          Statement m ⎭
     ELSE
          Statement m+1 ⎫
          Statement m+2 ⎪
                        ⎬   False-block statements
              .         ⎪   (also called an ELSE block)
              .         ⎪
              .         ⎪
          Statement n   ⎭
     ENDIF
```

In this form, statements 1, 2, . . . , n are any executable program statements.

In the discussion to follow you may refer either to Figure 3.1 or the pseudocode form just presented. At the entry to a two-way selective structure, the condition given is tested for being either true or false. If it is true, then statements 1, 2, . . . ,m (the true-block statements) are executed, followed by whatever statement comes immediately after the ENDIF.

If the condition is false, statements 1 through m are ignored and statements $m+1$ through n (the false-block statements) are executed, followed by the statement immediately after the ENDIF. Note that in either the *true* case or the *false* case the normal exit from the two-way selective structure is at the ENDIF statement.

To illustrate the two-way selective structure let us study some examples using ordinary English statements.

1. IF the value of the variable is positive, THEN
        Add A to B;
        Report the result;
   ELSE
        Add X to B;
        Report both X and the sum;
   ENDIF.
2. IF it is raining, THEN
        Call friends Jane, Pete, and Mary;
        Prepare lunch for the four of us;
        Watch the football game on television;
   ELSE

Call friends Mike and Lee;
Prepare for a camping trip;
Go camping;
ENDIF.

3. IF a test is scheduled for tomorrow, THEN
Get extra materials at library;
Spend the evening in my room studying;
Get a decent night's rest;
ELSE
Plan a party;
Arrange for finances to pay for refreshments;
Invite friends;
Purchase refreshments;
Ask two friends to help serve refreshments;
Enjoy the party;
Eventually get to bed;
ENDIF

In each of the preceding examples, between the words IF and THEN the first line contains a condition that may be judged true or false. This line is followed by statements of action, as is the ELSE line.

The two-way selective structure in ForTran is called by either of two names: the *IF-THEN-ELSE structure* or the *Block-IF structure.* Note that "Block-IF" and "IF block" are not the same. An IF block is one of the parts of a two-way selective structure and, therefore, a part of a Block-IF structure.

We consider now two special cases of the two-way selective structure.

Special Case 1 (IF-THEN)

If it happens that the ELSE block (false block) contains *no* statements, the structure takes a somewhat different form, as follows:

IF (Condition) THEN
Statement 1
Statement 2
.
.
.
Statement *m*
ENDIF

Here, if the condition is true, then statements 1 through *m* are executed, followed by the statement after the ENDIF. If the condition is false, the next statement executed is the one after the ENDIF. This special case of the two-way selective structure is called the *IF-THEN structure* in For-Tran and will be discussed fully later in this chapter.

Special Case 2 (logical IF)

If it happens that the IF block (true block) consists of only *one* statement, and there is no statement in the ELSE block (false block), that statement is included in the IF line, and the word THEN is omitted, as follows:

IF (Condition) Statement

As you can see, the entire structure consists of one line, so that the ENDIF statement (the normal exit) is not needed. If the condition is true, the statement in the IF line is executed, followed by the next line after the IF. Should the condition be false, the statement in the IF line is ignored and the next line executed is the one following the IF line. This special case of the two-way structure in ForTran is called the *logical IF structure.*

RELATIONAL EXPRESSIONS

A relational expression is a combination of arithmetic or string expressions and relational operators. There are six relational operators in ForTran, comparable to six symbols of comparison in mathematics. Table 3.1 shows these relational operators.

TABLE 3.1   Relational operators in ForTran

Relational operator	Its application in ForTran	Meaning
.EQ.	A.EQ.B	A is equal to B.
.NE.	A.NE.B	A is not equal to B.
.LT.	A.LT.B	A is less than B.
.GT.	A.GT.B	A is greater than B.
.LE.	A.LE.B	A is less than or equal to B.
.GE.	A.GE.B	A is greater than or equal to B.

The operators specified in Table 3.1 are all binary operators because they require *two* operands. The general form of a relational expression is

Operand   RELOPR   Operand

where Operand is any acceptable ForTran expression and RELOPR is one of the relational operators in Table 3.1. The result of a relational expression is always one of the logical values "true" or "false."

Let's examine four relational expressions and determine their values. We will assume, in the process of evaluating these relational expressions, that the real variables A=10, B=2, and the integer variable N=5.

1. `A+1 .GT. N*B`
   Observe that `A+1=11` and `N*B=5*2=10`; therefore, the value of this relational expression is "true."

2. `A/B .EQ. N`
   Observe that `A/B=10.0/2.0=5.0`; `N` is the integer `5`; therefore, the value of this relational expression is "true." Recall that the ForTran compiler takes care of the mixed data types in the expression.

3. `3*(A+B)/4 .LE. 2*A+5*B−21.01`
   First we evaluate the left operand: `A+B=12.0`, and therefore `3*(A+B)/4=3*12.0/4=36.0/4=9.0`. Now we evaluate the right operand: `2*A=20.0` and `5*B=10.0`. Therefore, `2*A+5*B−21.01=20.0+10.−21.01=30.0−21.01=8.99`. Since 9 is not less than or equal to 8.99, the relational expression has the value "false."

As you can tell from these examples, the rules for evaluating arithmetic expressions are applied to each of the two operands in the relational expression.

4. `ST(1:5) .LE. ST1(1:2) // 'END'`
   The action resulting from this statement begins with the concatenation of the first two characters stored at ST1 with the word END. This result is automatically stored in a temporary location whose contents are then compared to the first five characters stored at ST. If the value of the numeric code for the first five characters of ST is less than or equal to the numeric code for the characters in the temporary storage location, the relational expression has value "true." Otherwise it is "false."

Note that when a relational expression involves the comparison of two character strings, as in Example 4, just given, the comparison occurs character by character from left to right in both strings. The question of one string being less than or equal to the other is answered on the basis of the numeric code for a single character. For example, suppose that the numeric codes for letters is such that

$$A < B < C < \cdots < Z$$

and that ST1='ABC' and ST2='ABD'. Then the relational expression

```
ST1 .LE. ST2
```

has value "true" because the code for C is less than the code for D.

Two character strings are considered equal if and only if they have an equal number of characters such that identical characters are in corresponding positions in both strings. If strings have unequal length, the shorter one is expanded with blank space(s).

Although it is not incorrect to do so, one should avoid comparing character string expressions with arithmetic expressions because it is so easy to be unaware of the process used for comparing such expressions. There might be some situations where such comparisons are useful but one should do so only when one knows exactly what happens in such cases.

The next concept we shall introduce in our development of selective structures is that of logical expressions.

LOGICAL EXPRESSIONS

A *logical expression* is a combination of two or more relational expressions together with logical operators, which we discuss next.

Logical operators

The three logical operators available to us are .AND., .OR., and .NOT. The functions of these operators correspond closely to the functions performed by the English words "and," "or," and "not." Let's begin our examination of them by reviewing some concepts about these three logical conjunctions as used in simple English statements. Suppose we have two statements, A and B, as follows:

Statement A: It is warm.
Statement B: It is raining.

When statements A and B are connected with the conjunction "and" we get the statement

It is warm *and* it is raining.

This is called a compound statement and it is possible to consider its truth or falsity. If statements A and B were both true, then the compound statement obtained by connecting the two statements with "and" would also be true. However, if it were false to say "It is warm," then to say "It is warm *and* it is raining" would also be false. Similarly, if statement B, "It is raining," were false, the compound statement would be false. Thus, to summarize, only if statements A and B are both true is the compound statement A *and* B true.

Now let's use the same statements A and B as above, but this time we connect them with the conjunction "or" to get the compound statement

It is warm *or* it is raining.

Although the grammar seems a bit awkward, we can discuss the truth or falsity of this compound statement. If it is false to say "It is warm," but the statement "It is raining" is true, then it is true to say "It is warm or it is raining." Similarly, if statement A is true, the compound statement using "or" is true regardless of the condition of statement B. Only if both statements A and B are false is the "or" of A and B false. To summarize, then, we can say that given two statements A and B, the compound statement A *or* B is false only if both A and B are false.

The conjunctions "and" and "or" are used to connect *two* simple statements to form a compound statement. These words can therefore be thought of as performing a binary operation; "and" and "or" can be called *binary logical operators* because they operate on two statements and because logical considerations (truth or falsity) are the important consequences.

The word "not" used in connection with English statements has important implications for their truth or falsity, also. "Not" changes the truth or falsity of a statement in which it is inserted. Again consider

Statement A: It is warm.

By inserting "not" as follows,

It is *not* warm

we obtain a new statement that is the negative of Statement A. Therefore, if Statement A were true, the "not" of A (or the *negation* of A) would be false, and if A were false, the negation of A would be true. Sometimes the negation of a statement is called the *complement* of the statement. Note that the use of "not" involves a single statement, hence "not" can be thought of as a *unary logical operator.*

Now that we have discussed the three logical operations "and," "or," and "not" in the context of English statements, we are ready to return to ForTran to consider the use of these operations in that setting. The form of these logical operators is similar to that of the six relational operators explained earlier in this chapter. Here are the three operators as they appear in ForTran:

```
.AND. .OR. .NOT.
```

Next we shall see how they are used in expressions called compound relational expressions.

Compound relational expressions

Recall earlier in this chapter our discussion of relational expressions like A .GT. B and A*B .LE. C. When we connect two or more such relational expressions with the "and" or "or" operator, or precede one such expression by the "not" operator, we get what we shall call a *compound relational expression.* Here are some examples using the two relational expressions just given:

1. (A .GT. B) .AND. (A*B .LE. C)
2. A .GT. B .OR. A*B .LE. C
3. .NOT. (A .GT. B)
4. .NOT. (A .GT. B) .OR. A*B .LE. C

Let's examine these compound relational expressions with the goal of evaluating them. In doing so we shall assume that the values 2, 3, and 6 are stored at locations A, B, and C, respectively. We now evaluate each of the above examples.

1. A .GT. B has the value "false." A*B .LE. C has the value "true." The .AND. of a false expression and a true expression is "false." (Recall our discussion of logical operators in the setting of simple English statements.) Therefore, the compound relational expression in example 1 has the value "false."

2. In this example we have the .OR. of a false expression and a true expression. Thus the value of the compound expression of this example is "true."
3. The expression A .GT. B has the value "false"; therefore, .NOT. (A .GT. B) has the value "true."
4. In this example .NOT. (A .GT. B) has the value "true" and A*B .LE. C has the value "true." Therefore, the compound relational expression, which is the .AND. of two true statements, also has the value "true."

Consider one more example of a compound relational expression in ForTran in which we assume the same values mentioned previously stored in variables A, B, and C.

```
A .GT. B .AND. (A .LE. C .OR. B .NE. C) .OR. .NOT. (A .EQ. B)
```

We begin evaluation inside the parentheses. A .LE. C has value "true" because A = 2 and C = 6; B .NE. C has value "true" because B = 3 and C = 6; (A .LE. C .OR. B .NE. C) thus has the value "true."A .GT. B has the value "false"; (A .GT. B) .AND. (A .LE. C .OR. B .NE. C) thus has the value "false"; (A .EQ. B) has the value "false"; .NOT. (A .EQ. B) has the value "true." Therefore, the final .OR. of a false expression with a true expression has the value "true."

Priority of relational and logical operations

A word should be said here about the priority of operations. Recall from Chapter 2 that there is a priority of execution among the arithmetic operations, with the unary operations—minus and exponentiation—being executed first, followed by multiplication and division, and that the lowest priority is given to addition and subtraction. With operations of equal priority, execution occurs from left to right, except for exponentiation, which is done from right to left. It was also pointed out that parentheses could be used to interrupt this priority, in which case innermost parentheses are processed first.

The relational operations (.EQ.,.NE.,.GT.,.LT.,.GE.,.LE.) all have equal priority and are executed before the binary logical operations .AND. and .OR.. Following is a listing of the various operations we have discussed so far in this book, together with the priority assigned to them if no parentheses are used.

**	*Highest priority*
* , /	*Next priority*
+ , -	*Next priority*
All relational operations	*Next priority*
.NOT.	*Next priority*
.AND.	*Next priority*
.OR.	*Lowest priority*

Before considering the second special case of the two-way selective structure, we introduce the GO TO statement. Its form is

GO TO n

where n represents the reference number of the next statement to be executed.

Logical IF statement

Notice that the GO TO statement *always* causes an interruption in the normal sequential order of executing program statements. This fact has resulted in its being called an *unconditional* selection statement, in contrast to the statement we shall now discuss—namely, the logical IF statement—in which selection depends on a condition. We begin with the kind that has only one relational operator and shall refer to this kind of statement as a *simple logical IF* statement.

Simple logical IF statement. The general form of this statement is:

IF (Relational expression) Statement

where "Statement" is any executable ForTran statement except for DO, ELSE, ELSE IF, block IF, or ENDIF, all of which will be discussed later in this chapter. The END statement and another logical IF also cannot appear as the "Statement" in a logical IF statement.

When a logical IF is executed, the value of the relational expression is determined and if it is "true," the "Statement" part of the logical IF is executed next. If "Statement" does not transfer control to another point in the program, when it has been executed, the statement following the logical IF is executed next. If the relational expression has the value "false," then "Statement" is not executed and the next statement to be executed is the one following the logical IF.

Here are some examples of simple logical IF statements:

1. `IF (A .GT. B+X) X=25.2`
   The value of B+X is computed and compared with the value of A. If A is greater than B+X, then X is set to 25.2; otherwise, X remains the value it was. In both cases the next statement executed is the one following the logical IF.
2. `IF(FIRST .EQ. 'SECOND') LAST=FIRST`
3. `IF(A*B-1 .NE. A*(B-1)) B=0`
4. `IF(A*A-1) .GE. 6.0) STOP`
5. `IF(A .LE. 3) A=A*5./B-1`
6. `IF((X+Y)*(X-Y) .GT. X*Y+25) GO TO 25`
7. `IF(R-S .LT. 0.6) GO TO 100`

Compound logical IF statement.   Now we are ready to use compound relational expressions together with the logical IF statement to form what is called the *compound logical IF statement.* We do so by considering some examples.

1. `IF(B .EQ. 5 .AND. A*A .LT. 100)STOP`
   The value stored at variable B is compared with 5 and if they are equal, the value of B .EQ. 5 is "true." Otherwise it is "false." Then the value at A is multiplied by itself and the product is compared with 100. If the product is less than 100, the value of A*A .LT. 100 is "true"; otherwise, it is "false." Finally, the .AND. of the values of the two relational expressions is determined. Recall that if both relational expressions have the value "true," then the .AND. of them has the value "true"; otherwise, the .AND. has the value "false." If the .AND. value is "true," the STOP at the right of the IF statement is executed, thus causing a halt to any further program execution. If the .AND. value is "false," the program statement immediately following the IF statement is executed next and further program execution depends on each subsequent statement.

2. `IF(A+B .GT. C .OR. A-B .LT. C) GO TO 200`
   The sum of the values at A and B is compared to the value at C. If the sum is greater, then the value of this first relational expression is "true"; otherwise, it is "false." Next, the difference of the values at A and B is compared to the value at C. If the difference between A and B is less than C, then the value of this second relational expression is "true"; otherwise, the value is "false." Finally, the .OR. of the values of the two relational expressions is determined. Recall that the .OR. has value "true" if either of the operands has value "true"; otherwise, the .OR. has value

"false." If the .OR. has the value "true," then the statement GO TO 200 is executed, which, of course, transfers program control to the statement whose reference number is 200. If the .OR. has the value "false," GO TO 200 is *not* executed but, instead, the statement that next follows the IF statement is executed.

3. IF(A .GT. B .AND. (B .LE. C .OR. A .GE. B-C))STOP

Starting inside the innermost parentheses means that we first compare the value at B with the value at C. If the B-value is less than or equal to the C-value, the value of B .LE. C is "true"; otherwise, its value is "false." Then the value at A is compared with the difference of the values at B and C. If the A-value is greater than or equal to the difference, the value of the relational expression A .GE. B—C is "true"; otherwise, its value is "false." Then the .OR. of the values of these two relational expressions is determined. The .OR. has value "true" if either of the two relational expressions has value "true." Otherwise, the .OR.-value is "false." Now the value at A is compared to the value at B. If the A-value is greater than the B-value, the value of the relational expression A .GT. B is "true." Otherwise, its value is "false." Finally, the .AND. of the value of (A .GT. B) and the .OR.-value discussed previously is determined. If the .AND.-value is "true," the STOP statement is executed. Otherwise, the statement following the IF statement is executed next.

4. IF(.NOT.(A .EQ. B) .OR. B+A .GT. -C) X=50

Because it is enclosed within a set of parentheses, the first operation to be performed is the relational operation .EQ.. If the contents of A and B are the same, then A .EQ. B has the value "true"; otherwise, its value is "false." Next the negative of the value at location C is compared to the sum of the values at locations B and A, and if the sum is greater than the negative of the value at C, then the relational expression has the value "true." Otherwise, its value is "false." The next operation executed is the .NOT. of the value of A .EQ. B. Finally, the .OR. of the negation of (A .EQ. B) and the value of (B+A .GT. −C) is executed and if the result of the .OR. operation is "true," then X is set to 50 and program execution continues with the statement following the IF statement. If the .OR.-value is "false," then the value of X is left as it is (it is *not* set to 50) and program execution continues with the statement following the IF statement.

PROGRAM LOOPS

One of the most significant capabilities of any programming language is the ability to repeat a given set of program statements many times while having that set of statements appear only once in the program. This property is what is called a *program loop* or *iterative structure* of the programming language. The iterative structure is the major topic of Chapter 5. However, any meaningful programming in ForTran would benefit from the use of a loop, so we introduce here the simplest manually controlled loop structure. This structure involves the use of the logical IF and the GO TO statements.

We can introduce this concept through a simple example. Suppose there are some number—say fifty—of data cards to be processed, where each card contains two numbers, the first an integer that is a student identification number and the second a test score for that student. Each card is to be input to the program whose output for each card is the two numbers input. Obviously this is a trivial problem of input, output; input, output; and so on until all fifty cards have been processed. If there were only one card the program might look like this:

```
       PROGRAM RDONE
C*****************************************************
C*           PROGRAM DEFINITION                     *
C*              THIS PROGRAM READS ONE DATA CARD*
C*              AND PRINTS THE DATA READ.          *
C*****************************************************
C*           VARIABLE DEFINITION                    *
C*              NUMBER IS STUDENT ID NUMBER.        *
C*              TEST IS TEST SCORE.                 *
C*****************************************************
C*
       INTEGER NUMBER , TEST
       READ*,NUMBER,TEST
       PRINT*,NUMBER,TEST
       STOP
       END
```

The program just shown does not handle the problem of reading all fifty cards and generating appropriate output for each, as the example requires. One way to do this would be to modify the program so that there were fifty pairs of READ and PRINT statements. Although this modification would make the program correctly process fifty cards, intuitively it seems that the new program would be unnecessarily lengthy. As a correction of this weakness, we show how a logical IF statement can be used to introduce a loop into the program, to process all fifty cards with a short program.

This can be done with the addition of a fifty-first card on which is punched a student identification number of −1 and a test score of −1. The data on this card could not possibly be valid data for a real student (how could a score of −1 be attained?) but would be carefully selected so that they could be used to inform the computer that all data had been processed. To utilize this newly proposed data card, we therefore modify the previous program to appear as follows:

```
      PROGRAM RDMANY
C****************************************************************
C*          PROGRAM DEFINITION                                 *
C*              THIS PROGRAM READS A NUMBER OF                  *
C*              DATA CARDS AND PRINTS THE DATA READ.*
C*              IT ENDS WHEN A NEGATIVE ID NUMBER IS*
C*              READ.                                           *
C****************************************************************
C*          VARIABLE DEFINITION                                *
C*              NUMBER IS STUDENT ID NUMBER.                    *
C*              TEST IS TEST SCORE.                             *
C****************************************************************
C*
      INTEGER NUMBER , TEST
      READ*,NUMBER,TEST
   10 IF(NUMBER.LT.0)STOP
      PRINT*,NUMBER,TEST
      READ*,NUMBER,TEST
      GOTO 10
      END
```

Note that in this program the IF statement has reference number 10 and that the GO TO statement near the end of the program transfers control back to the IF statement. The sequence of four executable statements that are repeated fifty times constitutes the loop in this program. We know that, because the fifty-first card has −1 in the first field, when the logical IF is executed the fifty-first time, the result will be the execution of the STOP statement that is a part of the logical IF. Hence the original problem has been solved.

Data that are added (like the fifty-first card in our example) to signal certain action within the program are called *sentinel data,* or *trailer data,* or *dummy data.* Such data must be carefully selected so that there is no possibility of their being confused with actual data. Furthermore, such sentinel data must not be processed in the same way as real data. Note in our example there was no output produced from the fifty-first card.

If you review the second program just discussed you will see that it contains only two more executable statements (the logical IF and the GO TO) than the first program. And yet the second program processes correctly fifty data cards while the first program processes only one card. Therein lies the strength of program loops: they make it possible to repeat certain actions again and again without significantly lengthening the program. In fact, a little more thought will convince you that the second program above could process *any* number of cards as long as the last card contains −1 in its first field and any number in the second field.

Since we are discussing the concept of loops, we shall present a second commonly used method for controlling the number of times a loop is executed. Recall that the first method involved the use of sentinel data. The second method involves the use of a *counter* whose value is changed by 1 each time the loop is executed. Then, when the counter attains a predetermined value, the loop is no longer executed. This process reminds us of primitive methods of tallying the count of items as, for example, when taking inventory in a store. Many of us have used the counting aid of making short vertical marks as shown below.

ℕ ℕ ℕ ℕ ℕ ℕ ℕ |||

By this method, four such marks are made and then the fifth mark is made diagonally across the first four. This procedure is used to keep track of the number of times some procedure has been repeated (such as counting items in a store).

Of course, the computer will not be used to reproduce this method of making marks, but the principle is the same. Every time a procedure (a set of steps) is completed a counter will be incremented, much like the adding of another mark to record the completion of the counting procedure just described.

Before we present a program that makes use of a counter to control the execution of a loop, we shall discuss a ForTran statement, a form of which is commonly used in such programs:

```
COUNT = COUNT + 1
```

Applying our knowledge of how the assignment statement functions, we know that when this statement is executed, the current value stored in location COUNT is added to 1 and the sum is stored back in location COUNT. Thus if 0 were stored in location COUNT before the first execution of the given statement, then *after* the first execution the value stored in COUNT would be 1. If the given statement were executed a second time, the value stored in COUNT would be 2. If it were executed a third time, the value 3 would be stored in location COUNT. This happens because each time the statement is executed, the latest value of COUNT is increased by 1. Therefore, location COUNT maintains a continuously changing record of the number of times the given statement has been executed, in much the same way that the tally marks keep a continuously changing record of the count of items as described in a previous paragraph.

The discussion in the immediately preceding paragraph assumed that location COUNT began with a stored value of 0 and was then successively *increased*. It is also possible to initially set variable COUNT to a predetermined value and then *decrease* it by 1 each time. Consider this statement:

```
COUNT = COUNT - 1
```

If COUNT initially had the value 10 stored in it, then, after the first execution of the preceding statement, 9 would be stored in location COUNT. If the statement were executed a second time, the value stored in COUNT would be 8. If this process were continued it should be clear that when the value stored in COUNT has become 0, the statement would have been executed ten times. This technique is presented here because there are situations where it is more appropriate to monitor the loop with a decreasing counter rather than one that is increasing.

Now let's return to the problem of reading fifty cards each having two numbers on them; this time the program contains a loop controlled by a counter instead of by a test for sentinel data.

```
                    PROGRAM RFIFTY
C*********************************************************
C*           PROGRAM DEFINITION                          *
C*                THIS PROGRAM READS A TOTAL OF 50       *
C*                DATA CARDS AND PRINTS THE DATA READ.*
C*********************************************************
C*           VARIABLE DEFINITION                         *
C*                NUMBER IS STUDENT ID NUMBER.           *
C*                TEST IS TEST SCORE.                    *
C*                COUNT IS TO COUNT NUMBER OF CARDS      *
C*                READ.                                  *
C*********************************************************
C*
            INTEGER NUMBER ,COUNT, TEST
            COUNT=0
   10       IF(COUNT.GE.50)STOP
              READ*, NUMBER, TEST
              PRINT*,NUMBER,TEST
              COUNT=COUNT+1
            GOTO 10
            END
```

This third program solves the problem no better than the second one did. It simply demonstrates another method of controlling program loops. As an exercise, try to modify this third program so that it will process *any* number of cards instead of exactly fifty. Obviously, this will make the program more general and, therefore, useful to more people.

Now we proceed to the other two cases of the two-way selective structure. We mentioned previously that the general two-way selective structure is usually called the IF-THEN-ELSE structure and that the special case where the ELSE block (false block) contains no statements is usually called the IF-THEN structure. Some authors refer to both of these structures in ForTran by the single name Block-IF structure.

IMPLEMENTING THE TWO-WAY SELECTIVE STRUCTURE (BLOCK-IF)

We begin this topic by considering a simple example.

Example 1

Problem: Input a number, HOURS. If HOURS is 40 or less compute WAGES=HOURS*5., then output WAGES and stop. If HOURS is greater than 40 separate the number into two parts, one part being 40 and the other part HOURS—40. Call this difference OVERTM and compute WAGES=200 + (OVERTM*7.5), then output WAGES and OVERTM, then stop.

A Programmed Solution

```
        PROGRAM  WAGE1
C*******************************************************
C*         PROGRAM DEFINITION                          *
C*             TO READ HOURS, COMPUTE WAGES AND REPORT  *
C*             THE RESULTS.                             *
C*******************************************************
C*         VARIABLE DEFINITION                         *
C*             HOURS  IS TOTAL HOURS WORKED.            *
C*             WAGES  IS TOTAL WAGES EARNED.            *
C*             OVERTM IS OVERTIME HOURS WORKED.         *
C*******************************************************
C*
        REAL HOURS, WAGES, OVERTM
C*
        READ*, HOURS
C*
        IF(HOURS .GT. 40) GOTO 20
        WAGES = HOURS * 5
        PRINT*, WAGES
        STOP
   20   OVERTM = HOURS - 40
        WAGES = 200 + OVERTM * 7.5
        PRINT*, WAGES, OVERTM
C*
        STOP
        END
```

Alternative solution. Now we introduce the block-IF structure and you will see how more than one program statement may be executed with respect to the value of the relational expression in the IF statement. It is also true that the block IF makes the program flow smoother and reduces or eliminates the use of the often troublesome GO TO statement. Recall the format of the IF-THEN-ELSE structure:

```
IF (Relational expression) THEN
          } IF block (True block)
ELSE
          } ELSE block (False block)
ENDIF
```

You will notice that there are three new statements, the IF-THEN, the ELSE, and the ENDIF. Each of these must appear as separate statements. Between these three statements appear two sets of executable statements. The first set is called the IF block, which is executed if the relational expression has a value of "true." The second set of statements is called the ELSE block, which is executed if the relational expression has a value of "false." In both cases, once the block of statements has been executed, the next statement to be executed is the first executable statement following the ENDIF.

To illustrate the use of the IF-THEN-ELSE structure, we show a second program for solving the preceding example.

```
      PROGRAM  WAGE2
C***************************************************************
C*          PROGRAM DEFINITION                                *
C*              TO READ HOURS, COMPUTE WAGES AND REPORT *
C*              THE RESULTS.                                  *
C***************************************************************
C*          VARIABLE DEFINITION                              *
C*              HOURS  IS TOTAL HOURS WORKED.               *
C*              WAGES  IS TOTAL WAGES EARNED.               *
C*              OVERTM IS OVERTIME HOURS WORKED.            *
C***************************************************************
C*
      REAL HOURS, WAGES, OVERTM
C*
      READ*, HOURS
C*
      IF(HOURS .LE. 40) THEN
        WAGES = HOURS * 5
        PRINT*, WAGES
      ELSE
        OVERTM = HOURS - 40
        WAGES = 200 + OVERTM * 7.5
        PRINT*, WAGES, OVERTM
      ENDIF
C*
      STOP
      END
```

Although Program WAGE 2 is no shorter than Program WAGE 1, its logic is somewhat easier to follow. And in that Program WAGE 2 has no GO TO statements, it should be a more understandable program. The problem solved in this example is so trivial that it does not provide a fair test of the power of the block IF. Examples will be given later that illustrate this power more clearly.

We present now some program segments and complete programs that make use of the IF-THEN-ELSE (block-IF) structure.

Example 2

Problem: The reader should examine the following program output to try to discover what problem is solved by this program.

```
      PROGRAM EXAMP2
C***********************************************************
C*         PROGRAM DEFINITION                              *
C*             READER SHOULD EXAMINE THE PROGRAM TO        *
C*             DISCOVER THE PROBLEM SOLVED BY IT.          *
C***********************************************************
C*         VARIABLE DEFINITION                             *
C*             SINCE THIS WOULD PARTIALLY SOLVE THE        *
C*             EXERCISE, WE DO NOT GIVE DEFINITIONS FOR    *
C*             THE VARIABLES.                              *
C***********************************************************
C*
      INTEGER M, N, LARGE, SMALL, Q, R
C*
      READ*, M, N
C*
      IF(M.GT.N)THEN
        LARGE = M
        SMALL = N
      ELSE
        LARGE = N
        SMALL = M
      ENDIF
C*
      Q = LARGE/SMALL
      R = LARGE - SMALL * Q
      PRINT*, LARGE, SMALL, Q, R
      STOP
      END
```

Example 3

Detail of the general two-way selective structure

```
        .
        .
        .
IF (CODE .LE. 10)THEN
  A=A*B
  B=2*B
  PRINT*,A,B
ELSE
  A=A+B
  B=B**2
  WRITE (6,*)A,B
ENDIF
        .
        .
        .
```

The foregoing is a *segment* of a program. If the relational expression in the IF-THEN has value "true," the block of statements following the IF-THEN is executed, after which the first executable statement following the END IF is performed. If the relational expression has value "false," the statements following the ELSE statement are executed.

Example 4

Another application of the IF-THEN-ELSE structure

```
        PROGRAM   EXAMP4
C**************************************************************
C*        PROGRAM DEFINITION                                 *
C*            THE PROGRAM READS A POSITIVE INTEGER AND       *
C*            DETERMINES IF IT IS EVEN OR ODD.               *
C**************************************************************
C*        VARIABLE DEFINITION                                *
C*            N   IS THE POSITIVE INTEGER READ.              *
C*            K   IS THE REMAINDER OF N/2.                   *
C**************************************************************
        INTEGER N,K
```

```
C*
        READ*,N
        K = (N/2) * 2 - N
C*
        IF(K .EQ. 0)THEN
           PRINT*,N, ' IS EVEN'
        ELSE
           PRINT*,N, ' IS ODD'
        ENDIF
C*
        STOP
        END
```

IF-THEN SELECTIVE STRUCTURE (BLOCK-IF)

As the last of our two-way selective structures we consider the special case of the IF-THEN-ELSE structure where the ELSE block (false block) is empty. In this case the ELSE statement is omitted and the general form of the structure is

IF (Relational expression) THEN

$\left.\rule{0cm}{1.2cm}\right\}$ IF block

ENDIF

When the relational expression (the condition) has the value "true," the statements in the IF block are executed. Otherwise all of these statements are ignored. In either case the statement following the END IF is the next statement executed. The IF-THEN structure is illustrated in the next example.

Example 5

Application of IF-THEN-ELSE structure with empty ELSE block

```
            PROGRAM  WAGE3
C**********************************************************
C*           PROGRAM DEFINITION                          *
C*              TO READ HOURS, COMPUTE WAGES AND REPORT   *
C*              THE RESULTS.                              *
C**********************************************************
C*           VARIABLE DEFINITION                         *
C*              HOURS  IS TOTAL HOURS WORKED.            *
C*              WAGES  IS TOTAL WAGES EARNED.            *
C*              OVERTM IS OVERTIME HOURS WORKED.         *
C*              RHOURS  IS REGULAR HOURS WORKED.         *
C**********************************************************
C*
         REAL HOURS, OVERTM, RHOURS, WAGES
C*
         READ*,HOURS
         OVERTM = 0
         RHOURS = HOURS
C*
         IF(HOURS .GT. 40)THEN
           OVERTM = HOURS - 40
           RHOURS = 40
         ENDIF
C*
         WAGES = RHOURS * 5 + OVERTM * 7.5
C*
         IF(OVERTM .EQ. 0)THEN
           PRINT*,WAGES
         ELSE
           PRINT*,WAGES, OVERTM
         ENDIF
C*
         STOP
         END
```

Analysis: In this program we assume overtime hours are 0. If more than 40 hours are worked, the hours are separated into regular (RHOURS) set at 40 and overtime hours (OVERTM) computed as the difference between total hours (HOURS) and 40. We assume the rate of pay is $5 per hour for regular hours and $7.50 per hour for overtime.

Example 6

Problem: Same as in Example 1.

```
      PROGRAM  WAGE4
C*****************************************************************
C*        PROGRAM DEFINITION                                    *
C*              TO READ HOURS, COMPUTE WAGES AND REPORT         *
C*              THE RESULTS.                                    *
C*****************************************************************
C*        VARIABLE DEFINITION                                   *
C*              HOURS  IS TOTAL HOURS WORKED.                   *
C*              WAGES  IS TOTAL WAGES EARNED.                   *
C*              OVERTM IS OVERTIME HOURS WORKED.                *
C*****************************************************************
C*
      REAL HOURS, OVERTM, WAGES
C*
      READ*,HOURS
      OVERTM = 0
      IF(HOURS .GT. 40) OVERTM = HOURS - 40
      WAGES = HOURS * 5 + OVERTM * 2.5
C*
      IF(OVERTM .EQ. 0)THEN
         PRINT*,WAGES
      ELSE
         PRINT*,WAGES, OVERTM
      ENDIF
C*
      STOP
      END
```

Of all the four programs for solving the problem of Example 1 (Programs WAGE 1, WAGE 2, WAGE 3, and WAGE 4), Program WAGE 2 has the best structure and is most understandable. These examples should make it clear that there are usually many ways to write a program to solve a given problem. Some of these may be efficient (in terms of using computer resources) and structurally sound but others are not. It is important to learn to write programs that are easy to read and understand. Efficiency is also important, although understandability is more important for beginning programmers. In Chapter 4 we discuss in considerable detail problem solving and preparation of structurally sound programs.

Now we present a complete solution to a problem together with a ForTran program corresponding to the solution. The most important aspect of this example is its handling of data verification.

PROBLEM WITH COMPLETE SOLUTION

Problem

Input a data card on which are punched four numbers: (1) the number, N, of employee data cards to follow, $(0<N<100)$; (2) the FICA rate, FICAR, $(0<FICAR<.10)$; (3) the state income tax rate, STATER, $(0<STATER<.15)$; and (4) federal income tax rate, FEDR, $(0<FEDR<.25)$. The four input data all have permissible ranges and should be carefully verified, and any errors should be reported. If any of the data is invalid, processing should terminate as soon as such an invalid datum is encountered.

From each of the N employee data cards three items are to be input: (1) Social Security Number, SSN, $(111111111<SSN<999999999)$; (2) hours worked, HOURS, $(0<HOURS<80)$; and (3) rate per hour, RATE, $(3.00<RATE<50.00)$.

All input data are to be verified for validity. If any is incorrect, an appropriate message is to be printed. For each employee data card that has no data errors, compute the following:

OVERTM = overtime hours = HOURS−40
GROSS = gross wages=HOURS*RATE for first 40 hours plus
   1.5*OVERTM* RATE for hours over 40
FICA = FICA tax = FICAR*gross wages
STAX = state tax = STATER*gross wages
FTAX = federal tax = FEDR*gross wages
TTAX = total tax = FICA tax + state tax + federal tax
NET = net pay = gross wages − total taxes

Output

Output for each correct data card should be as follows:

SOCIAL SECURITY NO. _____

TOTAL HOURS _____   OVERTIME _____

GROSS WAGES _____

FICA TAX   _____

STATE TAX   _____

FEDERAL TAX _____

TOTAL TAXES _____

NET PAY   _____

A major function of this program is verifying input data—that is, checking to determine if the input data are valid. Any good computer program should include editing of input data so as to improve the validity of the program. The solution to this problem in English statements follows.

English statement solution to "Procedure PAYROLL"

1. Input N, FICAR, STATER, and FEDR from first data card. Check each for validity, printing appropriate error messages where necessary.
2. If all data just input are valid, proceed to next step; otherwise, stop any further processing.
3. If there are more employee data cards, input SSN, HOURS, and RATE from data card. Check each for validity, printing appropriate error messages where necessary. If there are no more employee cards, stop further processing.
4. If any of the employee data just input are invalid, ignore this employee and repeat the process from step 3.
5. Compute:
   OVERTM = HOURS−40
   If no overtime, set OVERTM=0
   GROSS = HOURS*RATE+OVERTM*.5*RATE
   FICA = GROSS*FICAR
   STAX = GROSS*STATER
   FTAX = GROSS*FEDR
   TTAX = FICA+STAX+FTAX
   NET = GROSS−TTAX
6. Output the results and repeat the process from step 3.

Solution to "Procedure PAYROLL" (second-level steps)

Now we present Procedure PAYROLL in more detailed steps, in preparation for writing the ForTran program that solves this problem. The numbering scheme can be compared with the preceding version in that all detailed steps derived from a given step in that version are assigned numbers whose integer parts are the same.

1.1 Input N, FICAR, STATER, and FEDR.
1.2 Initialize a switch, call it STOPCD, at 0.
1.3 If N is invalid, set STOPCD to 1 and report an appropriate message.
1.4 If FICAR is invalid, set STOPCD to 1 and report an appropriate message.
1.5 If STATER is invalid, set STOPCD to 1 and report an appropriate message.

1.6 If FEDR is invalid, set STOPCD to 1 and report an appropriate message.

2.1 If STOPCD is 1, then stop any further processing.

3.1 If N is zero, then terminate the process.

3.2 Input SSN, HOURS, and RATE.

3.3 Decrease N by 1.

3.4 Initialize an error switch, say SKIPCD, to 0.

3.5 If SSN or HOURS or RATE is invalid, set SKIPCD to 1 and report error in employee input data.

4.1 If SKIPCD is 1, repeat the process from step 3.1.

5.1 Perform the following computations:

OVERTM = HOURS−40

If OVERTM is less than zero set OVERTM to zero

GROSS = HOURS*RATE+OVERTM*.5*RATE

FICA = GROSS*FICAR

STAX = GROSS*STATER

FTAX = GROSS*FEDR

TTAX = FICA+STAX+FTAX

NET = GROSS−TTAX

6.1 Output the following items according to the formats specified in the problem: SSN, HOURS, OVERTM, GROSS, FICA, STAX, FTAX, TTAX, and NET.

6.2 Repeat the process from step 3.1.

In the second-level refinement of Procedure PAYROLL every step specifies clearly some action that can be translated directly into a ForTran statement. It is important to understand the use of STOPCD and SKIPCD, the two program switches. Recall that if any of the four major data items N, FICAR, STATER, and FEDR is in error, processing should stop. Thus, in steps 1.3 through 1.6, if at any point "bad" data are encountered, we set STOPCD to 1. At step 2.1, if STOPCD is found to be 1, that indicates that at least one of four major input data is incorrect, in which case we stop any further processing.

SKIPCD is used in a way similar to STOPCD in connection with the data for a given employee. If SKIPCD is 1, that means that at least one of the input data for a given employee is incorrect, so we skip the output associated with that employee. The only time processing goes beyond step 4.1 is when SKIPCD has not been set to 1.

Now we present the ForTran program PAYROL corresponding to the preceding procedure.

ForTran solution to "Procedure PAYROL"

```
          PROGRAM PAYROL
C*******************************************************************
C*          PROGRAM DEFINITION                                    *
C*              THIS PROGRAM VERIFIES AND PROCESSES TWO SETS OF    *
C*              DATA. FIRST SET CONSISTS OF ONE DATA CARD          *
C*              CONTAINING:                                        *
C*                 A. N=TOTAL EMPLOYEE CARDS TO BE READ, O<N<100.  *
C*                 B. FICAR= FICA RATE, O < FICAR < .10.           *
C*                 C. STATER= STATE TAX RATE, O < STATER < .15.    *
C*                 D. FEDR= FEDERAL TAX RATE, O < FEDR < .25       *
C*              THE SECOND SET CONSISTS OF N DATA CARDS            *
C*              EACH CONTAINING THE FOLLOWING THREE ITEMS:         *
C*                 A. SSN= SOCIAL SECURITY NUMBER.                 *
C*                 B. HOURS= TOTAL HOURS WORKED, O<= HOURS < 80.   *
C*                 C. RATE= PAY/REGULAR HOURS, 3.0 <= RATE < 50.0. *
C*              AFTER THA DATA ARE VERIFIED THE FOLLOWING          *
C*              RESULTS ARE COMPUTED:                              *
C*              GROSS= REGULAR HOURS * RATE + OVERTIME*            *
C*                       1.5*RATE                                  *
C*              STAX= GROSS * STATER                               *
C*              FTAX= GROSS * FEDR                                 *
C*              FICA= GROSS * FICAR                                *
C*              TTAX= FICA + STAX + FTAX                           *
C*              NET= GROSS - TTAX                                  *
C*              OVERTM= HOURS - 40                                 *
C*******************************************************************
C*          VARIABLE DEFINITION                                   *
C*              GROSS = GROSS PAY                                  *
C*              FICAR = FICA TAX RATE                              *
C*              STATER = STATE TAX RATE                            *
C*              FEDR = FEDERAL TAX RATE                            *
C*              FICA = FICA TAX                                    *
C*              FTAX = FEDERAL TAX                                 *
C*              STAX = STATE TAX                                   *
C*              HOURS = TOTAL HOURS WORKED                         *
C*              RATE = HOURLY PAY RATE                             *
C*              OVERTM = OVERTIME HOURS WORKED                     *
C*              N = TOTAL NUMBER OF EMPLOYEES                      *
C*              SSN = SOCIAL SECURITY NUMBER                       *
C*              STOPCD = A VARIABLE TO BE SET TO 1 AS SOON AS      *
C*                       ANY ERROR OCCURS IN THE FIRST SET OF      *
C*                       DATA.                                     *
C*              SKIPCD = A VARIABLE TO BE SET TO 1 AS SOON AS      *
C*                       ANY ERROR OCCURS IN EMPLOYEE DATA.        *
C*******************************************************************
```

```
C*
      REAL NET, GROSS, FICAR, STATER, FEDR, FICA
      REAL FTAX, HOURS, RATE, OVERTM, STAX
      INTEGER N, SSN, STOPCD, SKIPCD
C*
      READ*,N, FICAR, STATER,FEDR
      STOPCD=0
C*
C*              VERIFY THE INPUT
C*              (WHEN YOU LEARN ELSEIF STATEMENT
C*               REWRITE THE TWO VERIFICATION
C*               SEGMENTS USING ELSEIF STATEMENT)
C*
      IF(N .LE. 0  .OR. N .GE. 100) THEN
        PRINT*, ' ERROR IN INPUT ',N
        STOPCD=1
      ENDIF
C*
      IF(FICAR .LE. 0 .OR. FICAR .GE. .10) THEN
        PRINT*,' ERROR: FICA RATE INCORRECT ',FICAR
        STOPCD=1
      ENDIF
C*
      IF(STATER .LE. 0 .OR. STATER .GE. .15)THEN
        PRINT*,' ERROR: STATE RATE INCORRECT ',STATER
        STOPCD=1
      ENDIF
C*
      IF(FEDR .LE. 0 .OR. FEDR .GE. .25) THEN
        PRINT*, ' ERROR: FEDERAL RATE INCORRECT ',FEDR
        STOPCD=1
      ENDIF
C*
      IF(STOPCD.EQ.1)STOP
C*
C*******************************************************************
C*              AT THIS POINT EVERY ITEM IN THE FIRST DATA       *
C*              CARD IS CORRECT. THE NEXT N CARDS ARE READ       *
C*              ONE-BY-ONE. IN CASE OF ERROR THE DATA CARD       *
C*              IS IGNORED AND THE NEXT CARD IS PROCESSED.       *
C*******************************************************************
C*
  30     IF(N.LE.0)STOP
         N=N-1
         READ*,SSN,HOURS,RATE
         SKIPCD=0
```

```
C*
C*              VERIFY THE INPUT
C*
      IF(SSN .LE. 111111111 .OR. SSN .GE. 999999999) THEN
        PRINT*,' ERROR IN SOCIAL SECURITY ',SSN
        SKIPCD=1
      ENDIF
C*
      IF(HOURS .LE. 0 .OR. HOURS .GE. 80) THEN
        PRINT*,'ERROR IN HOURS ',HOURS
        SKIPCD=1
      ENDIF
C*
      IF(RATE .LE. 3 .OR. RATE .GE. 50) THEN
        PRINT*, ' ERROR IN RATE ',RATE
        SKIPCD=1
      ENDIF
C*
      IF(SKIPCD .EQ. 1)GOTO 30
C*
C*             COMPUTE THE RESULTS AND REPORT THEM.
C*
      OVERTM = HOURS-40
      IF(OVERTM.LT.0)OVERTM=0
      GROSS = HOURS * RATE + OVERTM * .5 * RATE
      FICA = GROSS * FICAR
      STAX = GROSS * STATER
      FTAX = GROSS * FEDR
      TTAX = FICA + STAX + FTAX
      NET = GROSS - TTAX
      PRINT*, ' SOCIAL SECURITY NUMBER ',SSN
      PRINT*, 'TOTAL HOURS ',HOURS,' OVERTIME ',OVERTM
      PRINT*, ' GROSS WAGE $',GROSS
      PRINT*, ' FICA TAX    $',FICA
      PRINT*, ' STATE TAX   $',STAX
      PRINT*, ' FEDERAL TAX $',FTAX
      PRINT*, ' TOTAL TAXES $',TTAX
      PRINT*, '-------------------------------------'
      PRINT*
      PRINT*, ' NET PAY      $',NET
      PRINT*
      PRINT*
   GOTO 30
   END
```

Sample output of Program PAYROL

Finally, we show a sample output of the preceding program where the input data for N, FICAR, STATER, and FEDR are, respectively,

2   .07   .1   .23

and where the individual employee data are

111224444   46   7.8
111225555   48   15

```
SOCIAL SECURITY NUMBER        111224444
TOTAL HOURS          46.0000  OVERTIME    6.00000
GROSS WAGE      $   382.200
FICA TAX        $    26.7540
STATE TAX       $    38.2200
FEDERAL TAX     $    87.9060
TOTAL TAXES     $   152.880
----------------------------------

NET PAY         $   229.320


SOCIAL SECURITY NUMBER        111225555
TOTAL HOURS          48.0000  OVERTIME    8.00000
GROSS WAGE      $   780.000
FICA TAX        $    54.6000
STATE TAX       $    78.0000
FEDERAL TAX     $   179.400
TOTAL TAXES     $   312.000
----------------------------------

NET PAY         $   468.000
```

This example has served to illustrate the very common job assigned to computers of verifying input data to help avoid errors in results owing to human error in entering such data. The program also illustrates the process of producing understandable output through the use of appropriate labels.

In the preceding sections we have been discussing the two-way selective structure and its two special cases. Now we turn our attention to the more general k-way selective structure, called the CASE structure by some authors because certain high-level languages actually use the CASE statement to define this structure. Since ForTran 77 does not provide for the CASE statement, we choose to refer to this structure as the k-way selective structure.

k-WAY SELECTIVE STRUCTURE

Before we discuss this structure in a general way we consider an example.

Example

Problem: Develop a procedure to compute N factorial ($N!$) where N is a positive integer less than or equal to 15. If the integer input for N is 0, report the result as 1. If the input for N is negative or is greater than 15, report an appropriate message and stop the process. Recall that $N! = (N)(N-1)(N-2)\dots 1$ and that the word capacity of most computers does not permit storage of integers much greater than 15!.

Diagrammed solution: Figure 3.2 gives a solution in the form of a diagram. This diagram summarizes the solution to this problem. In order to develop a ForTran program from this procedure we need to expand on the action "compute $N!$" to the level that it translates directly into For-Tran statements.

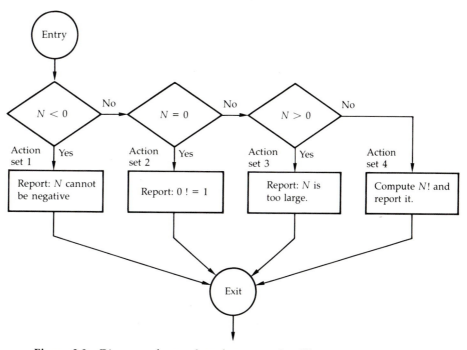

Figure 3.2   Diagram of procedure for computing $N!$

English solution to procedure FACTOR: The box containing action set 4 is the one for which we must develop further details. One possibility for those details follows:

1. Set FACT equal to *N*.
2. Set *M* equal to *N*.
3. Decrease *M* by 1.
4. Compute FACT = FACT * *M*.
5. If *M* is greater than 1 repeat the process from step 3.
6. Report FACT as *N*!

ForTran solution to procedure FACTOR (1):   Now, from the diagram of Figure 3.2 and the above details for action set 4, we can develop the following ForTran program.

```
      PROGRAM FACTOR
C*****************************************************************
C*         PROGRAM DEFINITION                                   *
C*             THE READER SHOULD PROVIDE PROBLEM AND            *
C*             VARIABLE DEFINITION FROM THE DESCRIPTION         *
C*             OF THE PROBLEM IN THE TEXT.                      *
C*****************************************************************
C*       THE READER SHOULD PROVIDE COMMENTS
C*
      INTEGER N,M,FACT
        READ*,N
C*
      IF(N .LT. 0)THEN
          PRINT*,N, ' IS NEGATIVE'
          STOP
        ENDIF
C*
      IF(N .EQ. 0)THEN
          PRINT*,'0 FACTORIAL = 1'
          STOP
        ENDIF
C*
      IF(N .GT. 15)THEN
          PRINT*,N, ' IS GREATER THAN 15'
          STOP
        ENDIF
C*
      FACT = N
        M = N
```

```
C*
 5       IF(M .LE. 1)GOTO 10
           M = M - 1
           FACT = FACT * M
         GOTO 5
C*
10       PRINT*, N, 'FACTORIAL IS ',FACT
         STOP
         END
```

If after studying the preceding program you wonder about the possibility of checking twenty or more conditions, we would have to tell you that such a situation is indeed possible. That would be a definitely more complex version of a *k*-way selective structure than the example just given. After we have discussed the general *k*-way selective structure and its ForTran implementation we shall present a second program for the factorial problem. You will agree that the second program is easier to follow and much shorter. Note that Figure 3.2 clearly shows *three* conditions being tested and *four* action sets, one of which is executed depending on the result of the condition test.

Figure 3.3 shows a *k*-way structure in the form of a diagram similar to Figure 3.2. At the entry point, condition 1 is tested. If it is true, action

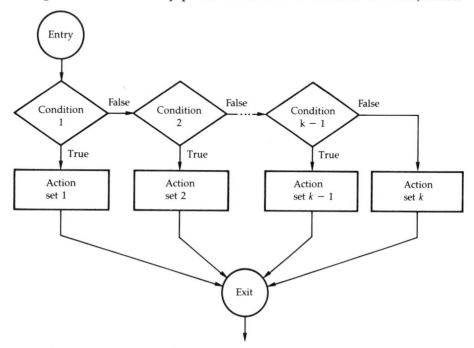

Figure 3.3   Diagram of *k*-way selective structure

set 1 is executed before the exit point is reached. If condition 1 is false, then the second condition is checked and if that is true, action set 2 is executed and control goes to the exit point. In general, if the ith condition is true, action set i is executed (while all other action sets are ignored) before control passes to the exit point. If the ith condition is false, condition $i+1$ is tested, and so on until as many as $k-1$ conditions may be tested and exactly one of k action sets is executed before control goes to the exit point.

General form of ForTran implementation of k-way selective structure

```
IF(Condition 1) THEN
   } Action set 1
ELSEIF(Condition 2) THEN
   } Action set 2
ELSEIF(Condition 3) THEN
   } Action set 3
      .
      .
      .
ELSEIF(Condition k−1) THEN
   } Action set k−1
ELSE
   } Action set k
ENDIF
```

As seen in the above general form, the k-way selective structure begins with an IF—THEN statement and ends with an ENDIF. Between these two statements appear sets of statements that we have called *action sets*, only one set of which is executed depending on the truth of a given condition. The first condition is tested in the first IF—THEN statement. If the condition is true, action set 1 is performed before transferring to the statement following the ENDIF statement. If the result of testing condition 1 is false, then condition 2 is tested. In general, condition i is tested only if the preceding $i-1$ conditions have all been false.

You have undoubtedly noticed the new ForTran statement shown in the general form of the k-way selective structure—namely, the ELSEIF statement. Its general form is

```
ELSEIF(Condition) THEN
```

where "condition" is any relational expression. If the relational expression is true, then the statements immediately following the ELSEIF are executed until another ELSEIF or ELSE statement is encountered, at which time transfer goes to the statement immediately following the first ENDIF that occurs next in the program.

Example

Now we present a second program for solving the factorial problem discussed previously. This new program uses the ELSEIF statement just described.

ForTran solution to procedure FACTOR (2: ELSEIF)

```
        PROGRAM FACTOR
C**********************************************************************
C*          PROGRAM DEFINITION                                       *
C*              THE READER SHOULD PROVIDE PROBLEM AND                *
C*              VARIABLE DEFINITION FROM THE DESCRIPTION             *
C*              OF THE PROBLEM IN THE TEXT.                          *
C**********************************************************************
C*
        INTEGER N,M,FACT
        READ*,N
C*
        IF(N .LT. 0)THEN
          PRINT*,N, ' IS NEGATIVE'
        ELSEIF(N .EQ. 0)THEN
          PRINT*,'0 FACTORIAL = 1'
        ELSEIF(N .GT. 15)THEN
          PRINT*,N, ' IS GREATER THAN 15'
        ELSE
          FACT = N
          M = N
5         IF(M .LE. 1)GOTO 10
            M = M - 1
            FACT = FACT * M
          GOTO 5
10        PRINT*, N, 'FACTORIAL IS ',FACT
        ENDIF
        STOP
        END
```

The form, readability, and understandability of this program is obviously an improvement over the first one presented for the factorial problem. Here is another program example using the *k*-way selective structure.

Example

Problem: Suppose the college class status of a student is given as 1, 2, 3, 4, or 5 to indicate freshman, sophomore, junior, senior, or graduate student, and suppose we call the class code CLASS. Suppose further that we input student identification number, IDNO, and CLASS for each student until a class code of 0 is encountered, at which time the program stops. For each student, output the identification number and the appropriate word corresponding to class code. If an invalid code is encountered, skip any further processing for that student.

Program

```
1       PROGRAM STUDNT
C***********************************************************
C*         PROGRAM DEFINITION                              *
C*             THE READER SHOULD PROVIDE PROBLEM AND        *
C*             VARIABLE DEFINITION FROM THE DESCRIPTION     *
C*             OF THE PROBLEM IN THE TEXT.                  *
C***********************************************************
C*
2       INTEGER CLASS, IDNO
3       READ*, CLASS,IDNO
C*
4    30 IF(CLASS .EQ. 0)STOP
5        IF(CLASS .EQ. 1)THEN
6           PRINT*, IDNO,' FRESHMAN'
7        ELSEIF(CLASS .EQ. 2)THEN
8           PRINT*, 'SOPHOMORE'
9        ELSEIF(CLASS .EQ. 3)THEN
10          PRINT*, IDNO,'JUNIOR'
11       ELSEIF(CLASS .EQ. 4)THEN
12          PRINT*, IDNO,'SENIOR'
         ELSEIF(CLASS .EQ. 5)THEN

            PRINT*, IDNO, 'GRAD'
13       ELSE
14          PRINT*,'INVALID CLASS CODE ',CLASS, IDNO
15       ENDIF
16       READ*, CLASS, IDNO
17       GO TO 30
18       END
```

Suppose that the data processed by this program are 2 and 111222 on the first card and 0 and 0 on the next card. Line 3 of the program would result in 2 being stored in variable CLASS and 111222 being stored in variable IDNO. Line 4 would determine the condition (CLASS .EQ. 0) to be false, so line 5 would be executed next. Then it would be determined that the condition (CLASS .EQ. 1) is also false, and line 7 would result in a true condition, so that line 8 would be next executed, causing the printing of the line

```
111222 SOPHOMORE
```

Following the execution of line 8, program control passes to line 15. When this statement is executed control is transferred to line 16. There new values are input to variables CLASS and IDNO—namely, 0 and 0. Next, line 17 transfers control to line 4, and since CLASS is now 0, the condition in line 4 is true, resulting in the execution of the STOP statement, which stops any further processing.

In this example, if class code is something other than 0, 1, 2, 3, 4, or 5, line 14 would be executed, indicating an invalid class code. In this case no student information would be printed—only the message regarding an invalid class code.

Although the various cases of the *k*-way selective structure are implemented in ForTran in somewhat different ways (IF-THEN-ELSE, IF-THEN, logical IF, and ELSEIF structures), we shall follow the practice of other authors and refer to any of these implementations as the block IF structure.

In the next section we discuss a more complicated block IF structure known as the nested block IF.

Nested block IF structure

Recall from our discussion of the two-way selective structure that we identified a set of instructions as the IF block and another set as the ELSE block. No restrictions were placed on the kinds of statements that could be included within these two blocks of statements. Therefore, it would be possible to include another IF-THEN, or an IF-THEN-ELSE statement. If either of these were to happen, the resulting structure is called a *nested block IF structure*. This structure is such that it results in one block IF structure with its ENDIF between another block IF and its ENDIF, hence the adjective *nested*. The inner structure is nested within the outer structure. It is essential that each block IF be paired with its own ENDIF. Appropriate indentation of statements is critical to the understandability of a program containing nested block IF statements. Consider the following example.

Example

Problem: Given the shift code identifying the work shift for an employee (1 = night shift, anything else = day shift) and the number of hours worked during a shift (HOURS), compute the wages earned by an employee during a shift if these rules apply: night-shift employees earn $10.00 per hour for all hours worked, whether overtime, holiday, or regular hours. Dayshift employees are of two groups. The first group earns $6.00 per hour for regular hours (REGH), $9.00 per hour for hours worked more than eight during a shift (OVERTM); and $10.00 per hour for Sunday and holiday hours. The second group earns $8.00 per hour for regular hours, $11.00 per hour for overtime hours, and $12.00 per hour for Sunday and holiday hours. The program could also be written by using an ELSEIF statement. As an exercise, the reader should rewrite the following program changing the statement "IF (SHIFT .NE. 1) THEN" to the statement "IF (SHIFT .EQ. 1) THEN."

Program

```
      PROGRAM  WAGES
C*******************************************************************
C*        PROGRAM DEFINITION                                       *
C*            SEE THE TEXT FOR COMPLETE DESCRIPTION OF             *
C*            THE PROBLEM.                                          *
C*******************************************************************
C*        VARIABLE DEFINITION                                      *
C*            SHIFT  IS TO INDICATE THE DAY OR NIGHT SHIFT*
C*            SHIFT = 1 INDICATES NIGHT SHIFT AND ANY              *
C*            OTHER VALUE INDICATES DAY SHIFT.                     *
C*            DAY  IS 7 FOR HOLIDAYS AND SUNDAYS AND IS            *
C*            ANY OTHER VALUE FOR OTHER DAYS.                      *
C*            GROUP  IS THE GROUP IDENTIFIER. GROUP               *
C*            = 1 INDICATES GROUP 1 AND ANY OTHER                  *
C*            VALUE INDICATES GROUP 2.                             *
C*            HOURS  IS TOTAL HOURS WORKED IN A SHIFT.            *
C*            REGH  IS REGULAR HOURS WORKED IN A SHIFT.           *
C*            OVERTM  IS OVERTIME WORKED IN A SHIFT.              *
C*            WAGE  IS TOTAL WAGE EARNED IN A SHIFT.              *
C*******************************************************************
C*
      INTEGER DAY, SHIFT, GROUP
      REAL  WAGE, HOURS, OVERTM
```

```
C*
      OVERTM = 0
      READ*, DAY, SHIFT, GROUP, HOURS
      IF(HOURS - 8 .GT. 0)OVERTM = HOURS - 8
      REGH = HOURS - OVERTM
C*
      IF(SHIFT .NE. 1)THEN
        IF(GROUP .EQ. 1)THEN
          IF(DAY .EQ. 7)THEN
            WAGE = HOURS * 10
          ELSE
            WAGE = REGH * 6 + OVERTM * 9
          ENDIF
        ELSE
          IF(DAY .EQ. 7)THEN
            WAGE = HOURS * 12
          ELSE
            WAGE = REGH * 8 + OVERTM * 11
          ENDIF
        ENDIF
      ELSE
        WAGE = HOURS * 10
      ENDIF
      PRINT*, WAGE
      STOP
      END
```

Example

We illustrate further the use of block IF structures in the following example.

Problem: Write a program to maintain the correct balance in a checking account where input consists of a transaction code and a transaction amount. The sentinel datum is 'S' for transaction code and '0' for transaction amount. Assume a transaction code of 'D' for a deposit and 'W' for a withdrawal. The output (including headings) should be as follows:

TRANSACTION	AMOUNT	BALANCE
DEPOSIT	XXXX.XX	XXXXX.XX
WITHDRAWAL	XXXX.XX	XXXXX.XX
.	.	.
.	.	.
.	.	.

Algorithm   (in complete English statements)

1. Produce headings and initialize variables.
2. Enter the transaction code and amount for the first transaction.
3. Process steps 4, 5, and 6 repeatedly until a transaction code of S is encountered, at which point stop all processing.
4. If the transaction code indicates a deposit has been made, then compute the new balance.
5. If the transaction code indicates a withdrawal and the withdrawal does not cause an overdraft, compute the new balance and report the amount withdrawn and the new balance. If an overdraft condition results, compute the amount of the overdraft and report it.
6. Enter the transaction code and amount for the next transaction.

Next we increase the level of detail in this algorithm and abbreviate the statements to form a sort of pseudocode closer to ForTran code.

Algorithm   (in pseudocode)

1.0 Output headings.
1.1 Initialize balance to 0.
2.0 Input transaction code (tr code) and amount of the first transaction.
3.0 If tr code = 'S' then stop all processing.
4.0 If tr code = 'D' then do steps 4.1 and 4.2.
   4.1 Compute balance = balance + tr amount.
   4.2 Output 'DEPOSIT', tr amount, and balance.
5.0 If tr code = 'W' then do steps 5.1 through 5.6.
   5.1 If balance $\geqslant$ tr amount then
      5.2 Compute balance = balance − tr amount.
      5.3 Output 'WITHDRAWAL', tr amount, balance.
   5.4 ELSE
      5.5 Compute overdraft = tr amount − balance.
      5.6 Output 'WITHDRAWAL', tr amount, overdraft, 'OVERDRAWN'.
6.0 If tr code is none of the above, report an error message.
7.0 Repeat the process from step 2.0.

From this latest version of the algorithm we can easily write the following ForTran program.

ForTran program

```
        PROGRAM TRANS
C*************************************************************
C*          PROGRAM DEFINITION                               *
C*              THIS PROGRAM READS A TRANSACTION CODE         *
C*              AND A TRANSACTION VALUE.                      *
C*              TRANSACTION VALUE IS A DEPOSIT IF             *
C*              THE CODE IS 'D'.                              *
C*              TRANSACTION VALUE IS A WITHDRAWAL IF          *
C*              THE CODE IS 'W'.                              *
C*              IF THE CODE IS 'S' THE PROGRAM STOPS.         *
C*                                                           *
C*              FOR EACH TRANSACTION THE TRANSACTION          *
C*              CODE AND VALUE AS WELL AS NEW BALANCE         *
C*              ARE REPORTED. IF A WITHDRAWAL CREATES         *
C*              OVERDRAW THE AMOUNT OF OVERDRAW IS            *
C*              REPORTED.                                     *
C*************************************************************
C*          VARIABLE DEFINITION                               *
C*              BAL  = ACCOUNT BALANCE                        *
C*              AMOUNT = TRANSACTION AMOUNT                   *
C*              CODE  = TRANSACTION CODE                      *
C*                      S   FOR STOP                          *
C*                      D   FOR DEPOSIT                       *
C*                      W   FOR WITHDRAWAL                    *
C*************************************************************
C*
        REAL BAL, AMOUNT
        CHARACTER CODE
C*
        BAL = 0
        READ*,CODE, AMOUNT
   10   IF(CODE .EQ.'S')STOP
          IF(CODE .EQ. 'D')THEN
            BAL = BAL + AMOUNT
            PRINT*,'DEPOSIT ',AMOUNT, BAL
          ELSEIF(CODE .EQ. 'W')THEN
            IF(BAL .GE. AMOUNT)THEN
              BAL = BAL - AMOUNT
              PRINT*, 'WITHDRAWAL ',AMOUNT, BAL
            ELSE
              PRINT*,' OVERDRAWN ',AMOUNT - BAL
            ENDIF
          ELSE
```

```
      PRINT*, 'ERROR IN CODE
      ENDIF
      READ*,CODE, AMOUNT
   GOTO 10
   END
```

The understandability of the preceding program can be improved when other repetitive structures become available to us.

PROGRAM ERRORS

In this section, we demonstrate the process of locating and removing errors in a program that might typically have been submitted for execution as an error-free program. Two types of errors are exemplified. The first kind consists of errors due to misspelling of special words or due to incorrect punctuation. Such errors are called *syntax* errors. Most syntax errors are discovered and identified by the compiler and result in error messages called *diagnostic messages*, or, simply, *diagnostics*. To correct them one reenters the lines of the program in which errors were detected; then tries executing the program again.

A second kind of error in computer programs is the result of improper conception of the problem or incorrect procedures for solving the problem. In either case some incorrect logical thinking has occurred; hence this kind of error is called a *logic error*. If a program has *logic* errors it may be syntactically correct and may correctly accept input data. It may even execute in what seems to be a correct run, yet the output is incorrect or it may come to an unexpected halt without processing all input data. The following example will demonstrate the process of removing errors from a simple program.

Example
Problem: Input a positive number and round it to the nearest integer. If the number is exactly halfway between two integers, round it to the next higher one. The program should check that the input datum is indeed positive and, if it isn't, should print an appropriate error message. If input datum is positive, output should give the rounded value.

Solution (English statements):   The solution to this problem may be put in the form of the following procedure.

1. Input a number A.
2. If A is negative, report an appropriate message and stop processing.
3. Add 0.5 to the value of A. The integer part of this sum is the rounded value of A.
4. Report the result and stop processing.

A flawed ForTran solution: Here is a program that corresponds approximately to this procedure except that some errors are deliberately included. Examine it and see how many errors in it seem obvious to you. Then compare your responses with the analysis that follows the program.

```
070        PROGRAM ROUND
080        REAL A
090        INTEGER I
120        A = A + .5
130        IF)A .LT. .5)GOTO 20
140        READ*, A
150        I = A
160        PRINT*,  THE ROUNDED VALUE OF,A, IS ,I
180 20     PRINT*,  BAD INPUT
190        STOP
200        END
```

```
*** ERROR LIST FOR THIS PROGRAM ***
120                     A=A+.5
   (120) - CAUTION A IS NOT SET ABOVE
130                     IF)A .LT. .5) GO TO 20
   (130) - FATAL----MISSING ) OR EXTRA (
   (130) - FATAL----MISSING ( OR EXTRA)
                        END
                     CAUTION STATEMENT NUMBER 20 WAS NEVER USED
THE FOLLOWING ERROR HAS OCCURRED AT LINE 120 OF THE SOURCE PROGRAM
MODE 4 ERROR -- ATTEMPT TO USE UNSET DATA AREA, OR ZERO DIVIDED BY ZERO.
ENSURE CORE IS INITIALIZED - USE MNF (Z) - AND CHECK DIVISIONS
```

Analysis: An error message calls attention to actual errors and situations that are likely to be errors. Following the printing of a copy of line 120 of the program is the message

```
CAUTION A IS NOT SET ABOVE
```

This means that previous to line 120, no value has been assigned to location A by a READ statement or by an assignment statement. This *may* not be an error because, if the program does not place a value in a memory location, the compiler uses the last value stored there. Usually, however, such a situation is an oversight by the programmer and should be corrected. To eliminate this error we need to enter a new line, 115, that inputs a number into location A. To do this we can move the READ from line 140 to line 115 and eliminate line 140 completely.

The next error messages refer to line 130. The compiler cannot exactly determine what the error is but it points out that there is something wrong with the parentheses in that there are not the same number of left- and right-facing parentheses. A look at line 130 reveals quickly that the character to the right of IF should have been a right-facing parenthesis instead of a left-facing parenthesis. The correction is made by reentering line 130 with the right-facing parenthesis following IF.

Finally, after printing a copy of the END statement, two diagnostic messages are given: one stating that the line whose reference number is 20 was never used although it was given a reference number, and the second that at line 120 an unset location was used. Actually, the line with reference number 20 *was* referred to in line 130 but because of the error with the left parenthesis, line 130 was never analyzed any further. Thus the correction of the left parenthesis will automatically eliminate the caution message about the line with reference number 20. The final error message, about the use of an "unset data area," will be eliminated when the position of the READ statement is changed as indicated previously. Following is a listing of the program corrected as just described and a run of the corrected program.

Partially-corrected ForTran solution

```
070        PROGRAM   ROUND
080        REAL A
090        INTEGER I
115        READ*, A
120        A = A + .5
130        IF(A .LT. .5)GOTO 20
150           I = A
160           PRINT*,'  THE ROUNDED VALUE OF',A,' IS ',I
180  20    PRINT*,'  BAD INPUT'
190        STOP
200        END
```

If we assume that the input datum was 23.79, the output produced by this program is as follows:

```
THE ROUNDED VALUE OF   24.2900    IS    24
BAD INPUT
```

The output as shown above indicates that although the program now is free of syntax errors, there must be at least one logical error somewhere in it because the number input is 23.79 but the output statement refers to 24.2900 as if that were the number input. Furthermore, the output message BAD INPUT appears even though the input data of 23.79 is valid in that it is indeed a positive number. A second look at the program reveals that if, in line 120, we change A on the left of the equals sign to something different, say X, and in line 150 change the A to X, then the correct input value is printed in the output line and the number is correctly rounded. In other words, a logical error was committed in letting A denote the number originally entered for rounding as well as the number used in the rounding process. Thus, the originally entered number is not available for printing at line 160. The other error regarding the incorrect appearance of the output line

    BAD INPUT

can be corrected by adding a STOP statement immediately following line 160.

Here is the program corrected as now suggested. Also shown are the results of two executions of the finally corrected program.

Correct but poorly structured ForTran solution

```
070        PROGRAM   ROUND
080        REAL A
090        INTEGER I
115        READ*, A
120        X = A + .5
130        IF(A .LT. .5)GOTO 20
150          I = X
160          PRINT*,'  THE ROUNDED VALUE OF ',A ,' IS ',I
170          STOP
180 20     PRINT*,'  BAD INPUT'
190        STOP
200        END
```

If the input datum is 23.79 the output is

    THE ROUNDED VALUE OF   23.7900   IS   24

If the input datum is 123.47 the output is

```
THE ROUNDED VALUE OF    123.470    IS    123
```

Although the preceding program is a correct program it is not very well structured in the sense that it is not very understandable. Compare the following program with the preceding one.

Correct, improved ForTran solution

```
PROGRAM  ROUND
REAL A
INTEGER I
READ*, A
IF(A .LT. .5)THEN
  I = A + 0.5
  PRINT*, '  THE ROUNDED VALUE OF',A,' IS ',I
ELSE
  PRINT*,'  BAD INPUT'
ENDIF
STOP
END
```

This last program eliminates a GO TO statement and a STOP statement as well as one variable, X. The program preceding this last one was developed not only to demonstrate errors in syntax and logic, but also to illustrate a program that has virtually no structure.

It should be understood that the process demonstrated in this section—that of first developing an incorrect and poorly structured program and finally ending up with a correct and structured program—is *not* the one to follow. Attention should be given to structure as well as to correctness from the time of developing the procedure for solving a given problem. Usually a well-written algorithm leads to a correct and well-structured program. In the preceding example the algorithm we first developed is correct and well structured. Had we followed it carefully it would naturally have yielded the very last program shown. All too often, however, inexperienced programmers will write poor programs simply because they are not careful in developing the pseudocode procedure or they are careless in producing the program from the pseudocode. We urge you to consider correctness and good structure at every stage of program development.

INDENTATION

The observant reader will have noticed that many of the program examples given in this chapter have some lines indented beyond others. In some cases there is even indentation followed by further indentations. Although this process of indenting program lines is not required by ForTran 77, it is done to make the programs more understandable. If you review the examples given you will notice that indentation is used to associate the indented program lines with each other. For example, in the last program example preceding this section, the two lines

```
I=A+0.5
PRINT*,'THE ROUNDED VALUE OF',A,'IS',I
```

are indented to make it clear that these two statements are executed if the relational expression in the preceding IF-THEN is true. Then when these two statements are complete, control passes to the END statement. That is, indented lines in a group provide visual evidence of the manner in which the logic of a program flows. This helps to make the program more readable and understandable.

For further examples of the use of indentation see the programs given in Exercise 9 at the end of this chapter.

SUMMARY

In this chapter we have considered ForTran statements needed to be able to build repetitive processes called loops into programs. These are the GO TO and IF statements, both of which may change the order of executing ForTran statements from the usual physical-position order. We have also discussed logical IF statements and the statements necessary to form the block IF structures. Here are general forms of the statements discussed in this chapter:

1. Logical IF statement. Form: "IF (Condition) statement," where "Condition" represents a simple or compound relational (logical) expression and "statement" represents any executable ForTran statement except for another IF statement.
2. Block-IF structure. Form:
   IF (Condition) THEN
       } IF block
   ELSE
       } ELSE block
   ENDIF

In the general form, IF-block and ELSE-block represent one or more executable ForTran statements. When the "condition" is true, the statement(s) in the IF block is (are) the next one(s) executed, and the ELSE-block statements are completely ignored, with the statement following the ENDIF being executed following the IF-block statements. If the "condition" is false, the IF block is ignored and the ELSE-block statements are executed.

It is legitimate to omit the ELSE statement and ELSE-block statements. In this case, if the "condition" is true, the IF-block statement(s) is (are) executed, after which the statement following ENDIF is executed. If "condition" is false, transfer of control goes directly to the statement following ENDIF.

3. ELSEIF statement. This statement is available to increase the alternative actions in the block-IF structure. Form:

IF (Condition 1) THEN
  } Block 1
ELSEIF (Condition 2) THEN
  } Block 2

     .

     .

     .

ELSEIF (Condition n) THEN
  } Block n
ELSE
  } ELSE block
ENDIF

If "condition 1" is true, the statement(s) in Block 1 is (are) executed, after which the statement immediately following ENDIF is executed. If "condition 1" is false, "condition 2" is tested for true or false. If it is true, Block 2 statements are executed, followed by the statement immediately after ENDIF. In general, when "condition i" is true, the statements in Block i are executed followed by the statement after ENDIF. If "condition i" is false, the next condition is tested and similar actions occur as described above. If none of the conditions is true, and an ELSE statement is present, the ELSE block statements are executed. If there is no ELSE statement nor ELSE block, program control goes to the statement following ENDIF.

EXERCISES

1. Answer the following questions:
   (a) What is meant by a mixed-mode expression? Give three examples.
   (b) What are the symbols used in ForTran for the various arithmetic operations?
   (c) What is the priority of arithmetic operations?
   (d) What is the effect of parentheses on this priority?
   (e) What are the relational operators in ForTran?
   (f) What are the logical operators in ForTran?
   (g) What action occurs when a GO TO statement is executed?
   (h) Identify the type of IF statement in each of the following:
      (1) `IF(A.GT.B)STOP`
      (2) `IF(A.LT.A+B)GO TO 20`
      (3) `IF(A.LT.B.OR.A.GT.C)A=B+C`
      (4) `IF(A*B.GT..NOT.A)A=-B`
   (i) Give a definition of a loop.
   (j) What is meant by an infinite loop?
   (k) What two methods are used to control the number of times a loop is executed?
2. For each of the following IF statements, indicate whether or not the statement is syntactically correct. If not correct, identify the error and propose a correction.
   (a) `IF(A+B)GO TO 10`
   (b) `IF(A.GT.B)10`
   (c) `IF(B)2,3`
3. For each of the following simple logical IF statements, indicate whether or not the statement is syntactically correct. If not correct, identify the error and propose a correction.
   (a) `IF(A+B.GT.A)A+B=C`
   (b) `IF(STOP.EQ.END)END=0`
   (c) `30 IF(A.GT.B)GO TO 30`
   (d) `IF(A+B.EL.A)GO TO 10`
   (e) `IF(A.GT.5)GO TO 5`
   (f) `IF('A'.LT.AB)AB='AB'`
   (g) `IF(A.NE.2HAB)GO TO 20`
   (h) `IF(A.E.AB)GO TO 20`
   (i) `IF(A+B.GE.A-B)A+B=X-B`
   (j) `IF(A+B.LE.A-B)PRINT*,A+B,A-B`

4. For each of the following compound logical IF statements, indicate whether or not the statement is syntactically correct. If not correct, identify the error and propose a correction.

  (a) `IF(A.GT.B.AND..LT.C)GO TO 10`

  (b) `IF(A+B.OR.2.LT.5)GO TO 20`

  (c) `IF(A*B.AND.A-B.GT.A+B)A=B`

  (d) `IF(A.GT.B.AND.(A.LT.B.OR.B.LT.C).AND.A.LT.C)A=B+C`

  (e) `IF(.NOT.A.EQ.B)A=-B`

  (f) `IF(A.OR.B.NOT.C.OR.D)GO TO 10`

  (g) `IF(A.GT.(B*A)**2.OR.(A*B/2).AND.B.LT.5.6)GO TO 40`

  (h) `IF(((A.GT.B).AND.(B.LE.C)).OR.((A.NE.2*C).AND.(A.EQ.C)STOP`

  (i) `IF((A.GT.B).OR.(B.GT.C).OR..NOT.(A.LT.B))GO TO 50`

  (j) `IF(A+B+C.LE.A-B+C.AND.THIS.EQ.THAT)SET X TO 5`

  (k) `IF(A+B.GT.A-B)LET B=-A`

  (l) `IF(A+B.EQ.2.5)THEN A=5.5`

 (m) `IF(A*B**2.NE.B*A**2)ELSE B=A**2`

  (n) `IF(2*3..GT.A.AND.2.EQ.B)GO TO 20`

  (o) `IF(A.EQ.B)GO TO=20`

5. Determine whether or not there are errors in any of these uses of IF statements.

  (a)
```
10   IF (A.LT.B)THEN
         X=A+B
         GO TO 10
     ELSE
         Y=A+B
     ENDIF
```

  (b)
```
10   IF(A-B.EQ.A+B)THEN
         IF(B.EQ.0)THEN
             B=B+1
             GO TO 10
         ELSE
             B=0
             GO TO 10
         ENDIF
     ENDIF
```

(c)
```
IF(A+B.GT.A.AND.A-B.LT.B)THEN
    A=A+B
    IF(A.GT.100.)A=100
ELSE
    A=200
ENDIF
```

(d)
```
IF(A+B .LT. 5.5 .AND. A-B .GT. 3)THEN
    IF(A*B .EQ. 6 .OR. A*B .GT. 8)THEN
        IF(X.EQ.2.1)THEN
            A=X*B
        ELSE IF(X.LT.2.0)
            A=X/B
        ELSE
            A=X*A
        ENDIF
    ENDIF
```

(e)
```
IF(A.AND.B.EQ.A.AND.C)THEN
    A=A*C
ELSE
    IF(A+B.GT.X)THEN
        IF(X.GT.5)THEN
            A=X+1
            PRINT*,A
        ELSEIF(X.LT.5.0)
            A=X-1
            PRINT*,A
        ENDIF
    ELSE
        PRINT,'INCORRECT INPUT'
        STOP
    ENDIF
ENDIF
```

(f)
```
IF(IFLAG.EQ.1)THEN
      PRINT *, 'JANUARY'
ELSEIF(IFLAG.EQ.2.AND.ICOLD.LE.-10)
     PRINT*,'FEBRUARY NORTHEAST'
ELSEIF(IFLAG.EQ.3)PRINT*,'SPRING IS COMING.'
ELSEIF(IFLAG.GT.3.AND.IFLAG.LT.7)
     PRINT*,'SPRING IS HERE.'
      IF(IFLAG.GE.7.AND.IFLAG.LT.10)PRINT*, 'HOT SUMMER'
ELSEIF(IFLAG.GT.9)PRINT*,'BACK TO FALL.'
ENDIF
```

(g)
```
      IF(FLAG.NE.PASS)THEN
20    KPASS=KMIN
      KMIN=TEMP
      TEMP=KPASS
      ELSEIF(FLAG.EQ.PASS)
      PRINT*,'ALL SORTED'
      PASS=MIN=TEMP=0
      GO TO 20
      ENDIF
```

(h)
```
10    IF(SPACE.GE.100)THEN
      PRINT*,'SPACE IS AVAILABLE.'
      IF(CAR.GE.5)THEN
          PRINT*,'SHIPMENT SHOULD START.'
      ELSEIF(CAR.LT.5)
          PRINT*,'WAIT FOR CARS TO ARRIVE.'
      ENDIF
      SPACE=200
      GO TO 10
      ENDIF
```

```
(i)     IF(A.LT.145)THEN
          PRINT*,A, 'IS NOT ENOUGH.'
        ELSE
          IF(A.GT.350)THEN
          PRINT*,A, 'TOO MANY'
          ELSE
              LOW=1
              HIGH=12
              IF(LOW.LT.HIGH)THEN
                MID=(LOW+HIGH)/2
              IF(A.LT.AREA)THEN
                  HIGH=MID-1
              ELSE
                  IF(A.GT.AREA)THEN
                    HIGH=MID+1
                  ELSE
                    NUM=NUM+1
                    LOW=HIGH+1
                  ENDIF
              ENDIF
              GO TO 10
              ENDIF
          ENDIF
        ENDIF
```

6. Study each of the following programs and determine the output.

```
(a)     PROGRAM SCORE
        INTEGER ISCORE
        PRINT*,'INPUT YOUR TEST SCORE'
        READ*,ISCORE
        IF(ISCORE.GE.90)PRINT*,'YOUR GRADE IS A.'
        IF(ISCORE.LT.90.AND.ISCORE.GE.80)PRINT*,'YOUR GRADE IS B.'
        IF(ISCORE.LT.80.AND.ISCORE.GE.70)PRINT*,'YOUR GRADE IS C.'
        IF(ISCORE.LT.70.AND.ISCORE.GE.60)PRINT*,'YOUR GRADE IS D.'
        IF(ISCORE.LT.60)PRINT*,'YOUR GRADE IS F.'
        END
```

Assume an input value of 62 in determining output for the above program.

(b) Modify the program at (a) so that the output is

```
YOUR SCORE IS ___ YOUR GRADE IS ___
```

(c)
```
     PROGRAM GRADE
     REAL TOTAL
     INTEGER N,SCORE,NSTUDT
     CHARACTER GR
     READ*,N,TOTAL
     PRINT*,'STUDENT NO.','SCORE','GRADE'
100  READ*,NSTUDT,SCORE
     SCORE=(SCORE/TOTAL)*100
     GR='A'
     IF(SCORE .LT. 90) GR='B'
     IF(SCORE .LT. 80) GR='C'
     IF(SCORE .LT. 70) GR='D'
     IF(SCORE .LT. 60) GR='F'
     PRINT *,NSTUDT,SCORE,GR
     N=N-1
     IF(N .GT. 0) GO TO 100
     END
```

Specific input for the above program will not be given but your task this time is to *describe* what the output would be.

(d) After you have determined the output of the program in (c), rewrite it, inserting appropriate comment lines to make the program as understandable as possible.

(e)
```
     PROGRAM ROUND
     READ*,X
     N=X+0.5
     PRINT*,N
     END
```

Describe the output of this program as related to the input, whatever the input is.

(f)
```
        PROGRAM EVEN
        READ*,N
        I=(N/2)*2-N
        IF(I.EQ.0)PRINT*,N,'IS AN EVEN NUMBER'
        IF(I.NE.0)PRINT*,N, 'IS AN ODD NUMBER'
        END
```

Describe what this program does as related to the input regardless of the specific number provided as input.

(g)
```
        PROGRAM SEVEN
        INTEGER N,KOUNT
        READ*,N
        IF(N.LT.7)GO TO 20
        KOUNT=0
        I=0
     15 I=I+7
        N=N-1
        KOUNT=KOUNT+1
        IF(N.GE.7)GO TO 15
        PRINT*,'THERE ARE',KOUNT,'SEVENS IN',N
        STOP
     20 PRINT*,'THERE ARE NO SEVENS IN',N
        END
```

Describe the output of the above program with respect to the number provided as input for N.

(h)
```
        PROGRAM MULTP
        INTEGER N,M,MULT,FACT,RESULT
        IF(M .GT. N) FACT=M
        IF(M .GT. N) MULT=N
        RESULT=0
     10 MULT=MULT-1
        RESULT=RESULT+FACT
        IF(MULT .GT. 0)GO TO 10
        PRINT*,N,'X',M,'=',RESULT
        END
```

Describe the output of this program with respect to the input.

7. The following program is supposed to compute the sum of the even integers from 2 to 1000. However, the program contains some logical errors, so it does not perform the desired task. Find and correct these logical errors.

```
    ISUM=2
    I=2
30  ISUM=ISUM+I
    I=I+2
    IF(I.LT.1000)GO TO 30
    PRINT*,ISUM
    END
```

8. The following program is supposed to input an integer greater than 10 and output the digits of the integer vertically with the units digit first. For example, if the integer input were 9368 the output should be

```
8
6
3
9
```

Check to see if there are any logical errors in this program.

```
    PROGRAM FIVE
    INTEGER NUMB,NEWNUM,DIGIT
    READ*,NUMB
15  IF(NUMB.LT.10) GO TO 35
    NEWNUM=NUMB/10
    DIGIT=NUMB-NEWNUM*10
    NUMB=NEWNUM
    PRINT*,DIGIT
    GO TO 15
35  PRINT*,NUMB
    END
```

9. Review the following programs for correctness. For those that are incorrect identify the error(s), and suggest corrections so that the program will run with no syntactical or logic errors. If the program as given is free of errors, specify its output with respect to the input data.

(a)
```
20 READ,N
      IF(N.LT.0)THEN
        M=-N
        PRINT*,M,'IS THE ABSOLUTE VALUE OF',N
      ELSEIF(N.EQ.0)
        PRINT*,TRY AGAIN'
        GO TO 20
      ELSEIF(N.GT.0)
        PRINT*,'THE ABSOLUTE VALUE OF A POSITIVE'
        PRINT*,'NUMBER IS THE NUMBER ITSELF.'
        PRINT*,'TRY AGAIN'
        GO TO 20
      ENDIF
      END
```

(b)
```
      READ*,N,M
      L=N+M
      LL=N-M
      LLL=M-N
      IF(L.EQ.LL)THEN
        PRINT*,'M MUST BE ZERO.'
      ELSEIF(L.EQ.LLL)
        PRINT*,'N MUST BE ZERO.'
      ENDIF
      PRINT*,'NEITHER M NOR N IS ZERO.'
      END
```

(c)
```
10 READ*,N,M
   IF(N.LE.0)THEN
      PRINT*,'INCORRECT INPUT FOR N.'
      PRINT*,'N MUST BE A POSITIVE INTEGER.'
      PRINT*,'ENTER N AND M AGAIN.'
      GO TO 10
   ELSE
      IF(M.LE.0)THEN
         PRINT*,'INCORRECT INPUT FOR M.'
         PRINT*,'M MUST BE A POSITIVE INTEGER.'
         PRINT*,'ENTER N AND M AGAIN.'
         GO TO 10
      ENDIF
   ENDIF
   END
```

(d)
```
   READ*,A,B,C
   IF(A.EQ.0)THEN
      X= -C/B
      PRINT*,B,'*X+', C, '=0'
      PRINT*,'IMPLIES X=', X
   ELSE
      D=B*B-4*A*C
      IF(D.EQ.0)THEN
         X= -B/2.
         PRINT*,A,'*X*X+',B, '*X+', C, '=0'
         PRINT*,'IMPLIES X=', X
      ELSEIF(D.GT.0.)
         X1=(-B+D)/(2.*A)
         X2=(-B-D)/(2.*A)
         PRINT*,A, '*X*X+', B, '*X+', C, '=0'
         PRINT*,'IMPLIES X=', X1, 'AND X=', X2
      ENDIF
   ENDIF
   END
```

```
(e)     READ*,X,Y,Z
        IF(X.GE.Y)THEN
           IF(X.GT.Z)THEN
              IF(Y.GT.Z)THEN
                 PRINT*,X,Y,Z
              ELSE
                 PRINT*,X,Z,Y
           ELSE
              PRINT*,Z,X,Y
        ELSE
           IF(X.GE.Z)THEN
              IF(Y.LE.Z)THEN
                 PRINT *,Z,Y,X
              ELSE
                 PRINT*,Y,Z,X
              ENDIF
           ELSE
              PRINT*,Y,X,Z
           ENDIF
        ENDIF
        END
```

10. In Exercise 9(b) there is a logic error resulting in the printing every time the program is run of the output

```
NEITHER M NOR N IS ZERO.
```

Correct the error so that the output produced corresponds to the input data.

11. In Exercise 9(c) there are two logical errors. As given, the program checks separately for the validity of data input for N and M but requests both values to be reentered although only one may be invalid. A second error pertains to the fact that only one of M or N are checked for validity at a given time but since both M and N must be reentered, an error could occur in the one not checked. Rewrite the program to correct these errors in logic.

12. You will recognize that Exercise 9(d) deals with solutions to a quadratic equation where D is the discriminant—namely, B^2-4AC. The program as given handles the situations where D is zero or greater than 0 but does nothing with the case in which D is less than 0. Modify the program to include this case as well. Also, change the program so that the PRINT statement causing the printing of the equation A*X*X+B*X+C=0 occurs only once in the program.

13. In Exercise 9(e), the program was intended to input three numbers and output them in descending order. As given it will work correctly for certain sets of three numbers but not for *any* three numbers. Modify the program so that it is correct for any three numbers given as input.

14. The following program is intended to find the first half of string ST and output it. Review the program to see if it does the job correctly.

```
PROGRAM HALF
INTEGER M
CHARACTER ST*10
READ*,ST
M=LEN(ST)
ST=ST*1:M/2)
PRINT*,ST
END
```

15. The following program is supposed to reverse the order of the characters in string ST. Find the errors in it.

```
PROGRAM REVERSE
CHARACTER ST*10, STR*10
INTEGER M
READ*,ST
STR=' '
M=10
10 STR=STR//ST(M:M)
   M=M-1
   PRINT*,STR
END
```

The rest of these exercises define a situation for which you are to write a ForTran program. The word in parentheses immediately to the right of the exercise number indicates the field of application of that exercise. The exercises are arranged approximately in order of increasing difficulty.

16. (Computer Science)
    (a) Input two integers, M and N, and compute the sum of the integers from M to N inclusive. Output M, N, and the sum, including some identifying information.
    (b) Same as for part (a) except only the odd integers.
    (c) Same as for part (b) except only the even integers.

17. (Computer Science)
Input any number, determine its integer part and its fractional part, then produce the output

```
INTEGER PART XXXX FRACTIONAL PART .XXXX
```

where the x's represent appropriate digits. For example, if the number input were 396.8210, the output should be

```
INTEGER PART 396 FRACTIONAL PART .8210
```

18. (Mathematics)
From mathematics, given any positive integer N, the number called N factorial (denoted $N!$) is defined as follows: $N! = N \cdot (N-1) \cdot (N-2) \cdots (2) \cdot (1)$. For example, $5! = 5 \cdot 4 \cdot 3 \cdot 2 \cdot 1 = 120$. Write a ForTran program that inputs any positive integer N (N less than or equal to 15), computes $N!$, and produces the output XX FACTORIAL = XXXXX where the X's represent appropriate digits. Thus, if the number input for N were 6, the output should be

```
6 FACTORIAL = 720
```

19. (Mathematics)
Input the coefficients A, B, and C for a second-degree polynomial equation $AX^2 + BX + C = 0$. Produce the following output:
   (a) The equation in the form (A)*X**2+(B)*X+(C). Thus, if A, B, and C were input as 2, -5, and 6, the equation should be printed as (2)*X**2+(−5)*X+(6)=0.
   (b) The real roots of the equation if it has any. *Hint:* Use the discriminant of the equation to determine if there are any real roots and use the quadratic formula to compute them.
   (c) If there are no real roots, an appropriate message should be printed.

20. (Science and Computer Science)
Input a single-digit code and a number. Assume that if the code is positive, the number input will be in English units and should be converted to metric units. If the code is negative, the reverse situation is true. The output is to be of the form

```
_____ POUNDS = _____ KILOGRAMS
```

Use the following codes and units:

Code	English unit	Metric unit
1	Pound	Kilogram
2	Gallon	Liter
3	Degrees Fahrenheit	Degrees centigrade
4	Yard	Meter
5	Square yards	Square meter
6	Cubic yards	Cubic meter
7	Mile	Kilometer
8	Stop any further program action	

Thus, if the two numbers input were 1 and 3.3, the output would be

```
3.3 POUNDS = 1.5 KILOGRAMS
```

21. (General)

This program is to input data from which can be printed a monthly report of activity in a checking account. The first line (card) of input data is to provide the beginning balance. All other lines (cards) of input data each contain a transaction code (1 = deposit, 2 = withdrawal, 0 = no more transactions), a six-digit date (in the form DDMMYY), and the amount of the transaction. Assume that transactions are input in chronological order.

Output is to be a report as follows (including headings):

```
TRANSACTION     DATE      AMOUNT     BALANCE
                                     1357.62
DEP             051282      17.50    1375.12
WDR             051482     100.00    1275.12
     .            .           .          .
     .            .           .          .
     .            .           .          .
```

Note that the beginning balance is alone on the first line following the headings. Note also that the numeric codes 1 and 2 will need to be translated into "DEP" and "WDR," respectively, for the output report.

22. (General)

This program provides a simple analysis of gasoline usage. The input consists of the following: beginning mileage, ending mileage, gallons gasoline used, cost per gallon. The output is the following including headings:

TOTAL MILES	GALLONS USED	MILES/GAL	TRIP COST
XXXX.X	XXX.X	XX.X	XXX.XX

23. (General)

This program computes current value, A, of an amount invested at compound interest. The input is principal (PRIN), annual interest rate (RATE), compounding period (P, a fraction of a year as 1, ½, ¼, $\frac{1}{12}$, and $\frac{1}{365}$), and the number of years (N) for which the money is invested. The formula to use is $A = PRIN(1.0+RATE*P)^{(N/P)}$. Output is to be:

____DOLLARS INVESTED AT____PERCENT FOR____YEARS WILL HAVE THE VALUE____

24. (Mathematics and Computer Science)

This program computes the Nth power of a number X, where N is a positive integer, by doing repeated multiplications.

(a) Input N and X, and compute X^N by performing an appropriate number of multiplications. Output should be

____ RAISED TO THE ____ POWER = ____

(b) Same as part (a), except this time use the minimum number of multiplications. For example, compute X^5 as $Y=X*X$, then $X^5=Y*Y*X$. This requires only three multiplications rather than four as in $X*X*X*X*X$.

25. (Business, Mathematics, Computer Science)

This program computes a schedule of monthly payments for a mortgage of P dollars borrowed at an annual rate of RATE and for N years. The input values are the principal borrowed, P; the annual interest rate, RATE; and the number of years, N. Use this formula for monthly payment:

$$\text{Monthly payment} = \frac{P*R*(R+1)^M}{(R+1)^M - 1}$$

where $R = RATE/12$ and $M = N*12$. The output should be a payment schedule as follows:

MONTH	INTEREST PD	PRINCIPAL PD	BALANCE
.	.	.	.
.	.	.	.
.	.	.	.

26. (General)

Write a program that outputs a multiple choice test item, then accepts as input a response to the item. If the answer selected is correct, output the word "CORRECT," then stop. If the answer selected is wrong, output the word "INCORRECT" and the message "TRY AGAIN.", then output the test item again. Repeat until the correct choice is input. An example of the type of test item to print is the following:

```
THE CURRENT U.S. PRESIDENT IS
1) RONALD REAGAN
2) JAMES CARTER
3) TED KENNEDY
4) NONE OF THE ABOVE
```

27. (Business, Computer Science)

COMPS, Inc. uses five different salary schedules for its employees as follows:

Schedule	Average salary	No. of employees
1	$12,000	10
2	15,000	17
3	18,000	20
4	21,000	18
5	26,000	8

The corporation plans to give raises such that the overall increase is 8%.

Write a program to compute and output the following:
1. Total money needed for the raise.
2. Overall average pay for all employees before the raise.
3. Overall average pay for all employees after the raise.
4. New average salary for each schedule and the dollar amount of the raise for each group of employees.

28. (Business, Computer Science)

The real estate tax on residential homes is computed as follows:

Assessed valuation	Computed tax
$30,000 or less	$800
$30,000 < Assessed valuation ≤ $50,000	$800 + 1% of assessed valuation over $30,000
$50,000 < Assessed valuation ≤ $80,000	$800 + 1.2% of assessed valuation over $30,000
$80,000 < Assessed valuation ≤ $120,000	$800 + 1.4% of assessed valuation over $30,000
Assessed valuation > $120,000	$800 + 1.5% of assessed valuation over $30,000

Write a program to input the assessed valuation of a home and compute the real estate tax. Output the assessed valuation and the amount of the tax.

29. (Mathematics, Computer Science)

Write a program to input any positive real number of at most twelve digits. Output, one at a time on a separate line, each of the digits in the number beginning with the leftmost one.

30. (Probability, Computer Science)

A container has ten red balls numbered $00, 01, \ldots, 09$ and ten black balls numbered $10, 11, \ldots, 19$. Assume there are available outside of the container as many additional similarly numbered red and black balls as needed for the process about to be described.

Write a program using the RAND function described in Chapter 2 to simulate the drawing of balls from the container. (You might use the occurrence of 0 to represent red and 1 to represent black.) The procedure for drawing and adding balls to the container is as follows:

1. Determine the color of the ball to be drawn.
2. Determine the number on the ball to be drawn.
3. Add to the container a ball of opposite color but identical number to that on the ball drawn.

The output should be as follows:

	BEFORE DRAW		AFTER DRAW	
DRAW	NO. RED	NO. BLACK	NO. RED	NO. BLACK
1	10	10	. . .	. . .
2	. . .	. . .	. . .	. . .

Continue this process until 100 balls have been drawn.

CHAPTER 4
Problem Solving with ForTran

Earlier chapters in this book have introduced you to a number of structures useful in developing procedures for solving problems. We have also discussed several ForTran statements associated with these structures. In this chapter we direct our attention toward developing a correct program, one that has good structure and understandability.

A CORRECT PROGRAM

How do we decide when a program is *correct*? One response might be, "By looking at the output it produces." Let's consider the validity of that answer.

Suppose we want to write a ForTran program to compute the average of three numbers. Assume, further, that the three numbers are 5, 7, and 9. Clearly the average of these three numbers is 7.

Now look at this program:

```
PROGRAM SIMPLE
A=7
WRITE(6,*)A
END
```

Its output is the number 7, so if program *output* were the criterion for correctness, we would conclude that the above program is correct for solving the problem of finding the average of the three numbers 5, 7, and 9. In fact, the program produces correct output for any problem where we want to find the average of a set of numbers as long as the average is 7. However, it is not very likely that anyone would have had in mind a program like program SIMPLE when thinking about a program for finding the average of three numbers.

More likely would be a program that accepts as input any three numbers, then uses output statements to print the result of computing their average. But in that case, how could we be sure that the output is correct? It would make little sense to talk about testing the program for every possible set of three numbers since there are infinitely many sets of three numbers. Even testing the program for a large collection of sets doesn't seem to make much sense because that alone would not assure that the program is correct for sets of numbers not tested.

Therefore, there must be a more reasonable way to determine the correctness of a program other than testing it for all possible cases. An important part of such a reasonable way is to try to understand the computation that takes place when the program is executed.

Looking at the short program previously given as one that produces the average of three specific numbers, it is quite obvious that there are two executable statements, neither of which cause any computation whatsoever. Hence, the program cannot be a correct program for computing the average of three numbers even though it produces the correct result for the average of the three numbers 5, 7, and 9. Really, all the program does is produce the number 7 as output. We arrive at this conclusion by reading and understanding the program.

Not only must a good (or correct) program produce correct results, therefore, but it must also be understandable by someone examining it for the first time. It must be possible to look at it or study it and tell what is accomplished by it. An understanding of a program is dependent on knowing about each statement in the program, but it is also necessary to comprehend the relationships among the statements making up the program. That is, the organization of the separate statements into a complete program must be understood if we are to judge the goodness of a program. This organization is known as the *structure* of a program and it is knowledge of its structure that will help us understand a program.

As an aid in developing the concept of program structure, let's consider a problem somewhat more complex than computing the average of three numbers. Then we will generate a ForTran program that solves the problem.

A sample problem and its solution

Suppose we want to use a computer to find the variance of a set of numbers. Variance is a term from the field of statistics and may be defined as follows: the number obtained by finding the average of the sum of the squares of the differences between the mean (average) and each number in the set.

So the problem can then be restated by giving a few more details obtained from the definition of variance.

Problem: Computing variance.   Given a set of numbers, compute the sum of the squares of the differences obtained by subtracting the mean of the set from each number in the set, and dividing the sum by the number of numbers in the set. Then output the result.

Let's consider what all is involved in solving the problem. Of course, the numbers making up the set must be identified or provided as input. We need to determine how many numbers there are in the set. The sum of the numbers must be computed and the mean obtained by dividing that sum by the number of members in the set. Then the computed mean must be subtracted from each member of the set, the resulting differences must be squared, and the sum of those squared differences must be accumulated. Finally, the total sum of the squared differences must be divided by the number of members and the resulting quotient must be output as the desired solution, which is the variance of the given set of numbers.

What we have just done is to spell out a plan, or procedure, for solving the problem. Now let's restate the procedure again, this time in a form that shows more clearly the steps to be done in obtaining the solution.

1. Input the set of numbers and compute their mean.
2. Compute the sum of the squares of the differences of each number in the set and the mean.
3. Output the quotient obtained by dividing the sum of squares by the number of members in the set.

Now, let's recapitulate our progress from the start. (See Figure 4.1.)

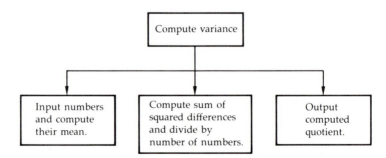

Figure 4.1   First level of subdivision

The preceding diagram shows what is happening as we solve the problem. From a statement of the problem we develop somewhat independent subproblems each of which must be completed in the order specified if the solution is to be obtained. Clearly, the three steps in this first level of subdivision are far from being directly translatable into statements in a ForTran program. Therefore we proceed to subdivide the three steps into other steps.

As you review Figure 4.2 it should become apparent that we have moved from a simple statement of the main problem to progressively smaller steps in the process of solving the problem. Even at the present level of subdividing steps, however, we are not to the point where each step can be translated into a ForTran statement. That's our goal, since we began by saying we wanted to use a computer to solve the problem.

Before we develop another level of subdivision, consider some questions that need to be answered in order to further specify steps in arriving at the solution. If you look at Figure 4.2, the second step in the second level of subdivided steps states

Determine number of numbers (members).

Nothing is given there or anywhere in the original statement of the problem to answer the question, "How do we determine the number of members in the set for which we are to compute the variance?" There are several answers to that question, among which might be the following: (1) Input the number of members in the set. (2) Count the numbers entered until you encounter a number unlike anything expected in the set, say, 99999. The first answer implies that before we input the numbers for which we want the variance, we input a number specifying the num-number of numbers in the set for which we want the variance. In other words, we know ahead of time how many numbers are in the set. There are many practical situations where this will be the case.

The second answer implies that we do *not* know ahead of time just how many numbers are in the set for which we want the variance but that a final dummy number, a number unlike anything expected in the set of which we are to compute the variance, is provided. This implies, further, that we must count the numbers input until the dummy number is encountered, at which time we know we have input all numbers in the set. Such a count, or tally, would give us the number needed in the computation of the mean. As explained in Chapter 3, dummy numbers like the one we have just described are sometimes called *trailer* or *sentinel data*. Either name indicates that such a number is not to be used in the problem computation but has the specific purpose of signalling the end

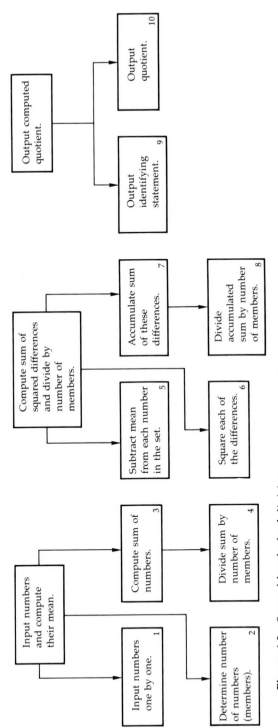

Figure 4.2   Second level of subdivision

of a set of data. Great care must be used in selecting sentinel data so that they really are different from any potential member of the set of numbers being processed.

Look again at Figure 4.2, particularly at the third step in the second level of subdivided steps. It states

Compute sum of numbers.

When human beings compute the sum of a set of numbers with pencil and paper, they will usually write the numbers in a vertical alignment and compute the sum first with the rightmost column of digits, then move progressively to the left, and finally have the sum written at the bottom of the column of numbers. If an electronic calculator is used, the person will usually clear the calculator, then enter the numbers one by one pressing the "+" button after keying each number. When the last number has been keyed, the sum of the numbers is displayed by pressing an appropriate key to display the result. In this latter method it is conceivable that as each number is entered on the calculator, it is counted in some manner so that when the last number is entered, the total count will reflect the number of numbers entered in computing the sum. Such a total count, or tally, would be needed in calculating the mean.

Look once more at Figure 4.2, this time at the fifth step. This step reads

Subtract mean from each number in the set.

A question here might be, "How do we have available for this second use the numbers in the original set?" That is, the set of numbers was input once as specified in the first step of Figure 4.2. How do we make those same numbers available at the fifth step for the computation of the differences between each of the numbers and the mean of the set? In Chapter 5 you will learn a better method for handling a situation like this, but for now let's assume that when the set of numbers originally input at step 1 is needed again at step 5, the numbers should be input one by one a second time. Later in this section we use another method for computing variance that requires the input of the numbers only once.

Having considered some questions related to further subdividing the steps in Figure 4.2, we turn now to a continuation toward our goal of developing a collection of steps that can be easily translated into ForTran statements. Figure 4.3 shows a third level of subdividing steps to the point where each step is directly translatable into a ForTran statement. Before making such a translation, however, notice some other things about the procedure detailed in Figure 4.3.

Procedure VARIANCE

1.0 Initialize COUNT and SUM to zero.

1.1 Input NUMBER.

1.2 If NUMBER is not equal to 99999 then increase COUNT by 1,
add NUMBER to SUM,
input a new NUMBER,
and repeat this step from the beginning until a NUMBER equal to 99999 is encountered.

1.3 Else
compute MEAN as : MEAN=SUM/COUNT.

2.0 Then initialize SUM2 to zero.

2.1 Input NUMBER.

2.2 If NUMBER is not equal to 99999 then
compute SQUARE=(NUMBER−MEAN)*(NUMBER−MEAN),
add SQUARE to SUM2,
input a new NUMBER,
repeat this step from the beginning until a number equal to 99999 is encountered.

2.3 Else
compute VARIANCE=SUM2/COUNT.

3.0 Then output the results.

Figure 4.3   Third level of subdividing steps

Step 1.0 in Figure 4.3 has no counterpart in Figure 4.2. Its purpose is to ensure that memory locations for accumulating sums are set to zero before any accumulating occurs. Step 1.2 in Figure 4.3 is where we determine whether or not there are more numbers to input. Whenever the condition stated in the first line of step 1.2 is true, the actions indicated following the condition, one per line, are all executed. When the condition is false, the action at step 1.3 is taken. That is, step 1.3 is performed only when the trailer datum 99999 is encountered, since the trailer indicates that the input of numbers is complete and the mean is computed.

Now we write a ForTran program from the steps in Procedure VARIANCE.

```
      PROGRAM VARIAN
C********************************************************************
C*         PROGRAM DEFINITION                                      *
C*              SEE THE TEXT FOR COMPLETE DESCRIPTION OF THE        *
C*              PROBLEM.                                            *
C********************************************************************
```

```
C*          VARIABLE DEFINITIONS.                                      *
C*              NUMBER = A SINGLE MEMBER OF THE SET OF NUMBERS.  *
C*              SUM = THE SUM OF THE NUMBERS.                          *
C*              SUM2 = THE SUM OF THE SQUARES OF THE DIFFERENCES*
C*              OF THE NUMBERS IN THE SET AND THEIR MEAN.          *
C*              SQUARE = THE SQUARE OF THE DIFFERENCE BETWEEN   *
C*              A GIVEN NUMBER IN THE SET AND THEIR MEAN.           *
C*              VAR = VARIANCE.                                        *
C*              MEAN = MEAN OF THE SET OF NUMBERS.                 *
C**********************************************************************
C*          IMPORTANT NOTE                                             *
C*              YOU MUST HAVE AT LEAST ONE NUMBER OTHER THAN    *
C*              99999 TO ENTER OR THIS PROGRAM WILL NOT WORK    *
C*              CORRECTLY.                                             *
C**********************************************************************
        REAL NUMBER, MEAN, SUM, SUM2,  VAR, SQUARE
        INTEGER COUNT
C*
C*                  INITIALIZATION
C*
        COUNT=0
        SUM=0
        SUM2=0
        READ(5,*)NUMBER
C***********************************
C*              COMPUTE MEAN         *
C***********************************
  3     IF(NUMBER .EQ. 99999)GOTO 7
            COUNT=COUNT + 1
            SUM=SUM + NUMBER
            READ(5,*)NUMBER
        GO TO 3
  7     MEAN=SUM/COUNT
C*************************************
C*              COMPUTE VARIANCE.      *
C*************************************
        READ(5,*)NUMBER
 10     IF(NUMBER .EQ. 99999) GOTO 15
            SQUARE=(NUMBER-MEAN)*(NUMBER-MEAN)
            SUM2=SUM2 + SQUARE
            READ(5,*)NUMBER
        GO TO 10
 15     VAR=SUM2/COUNT
        WRITE(6,*) COUNT,VAR
        STOP
        END
```

There are likely those readers who question the method used in this procedure and the resultant program. Such readers will be aware of a so-called computational formula that identifies variance as follows:

$$\text{Variance} = \frac{1}{n}\left(\sum_{i=1}^{n} X_i^2 - n\,\overline{X}^2\right)$$

Using this formula would have made it possible to input the set of numbers only once rather than twice as our procedure required. The reason for choosing the other method was that it fit better into the philosophy of this chapter. We are learning about problem solving, and using a method of solution that is based on definitions rather than on mysterious formulas seems more in keeping with the goal.

Let's review what was done in the variance problem. Figure 4.4 shows a progression from problem statement to computer generation of the solution. It is important to have a precise statement of the problem, one that provides an adequate explanation of what is to be accomplished. Arriving at such a statement of the problem may require rewriting the problem until it is clear in its implications.

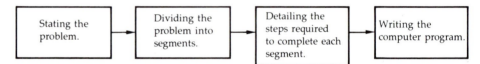

Figure 4.4   Diagram of problem-solving technique

Once the problem is understood, it is usually necessary to compartmentalize various aspects of it into segments, each of which can be readily seen as a computational task. Then each computational task is broken down into steps each of which can usually be accomplished by a single ForTran statement but may sometimes require more. Finally, the ForTran statements corresponding to each detailed step are written, with indentation used as necessary. The result is a good, understandable program. It will have structure or organization and can be called a structured program. You would be wise to develop the skills needed to write such programs.

A second procedure for computing variance.   Before we leave the variance problem, let's consider a different procedure for solving the problem, one that makes use of the alternative formula given previously. Figure 4.5 shows such a procedure at a level of subdivision comparable to that shown in Figure 4.2. After analyzing the procedure in this way, we can subdivide further the steps from Figure 4.5 as shown in Figure 4.6. Figure 4.6 is therefore comparable to Figure 4.3.

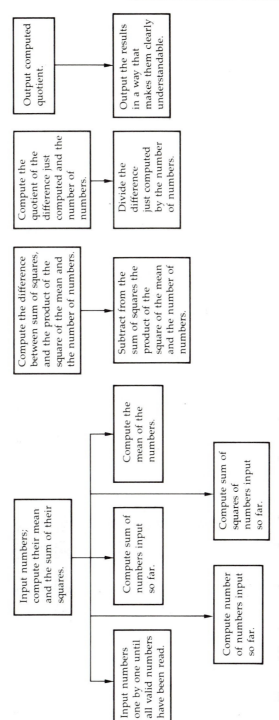

Figure 4.5   A second procedure for computing variance

The level of detail in Figure 4.6 brings the procedure very close to that required in order to write a computer program easily for the procedure.

Procedure VAR2

1.0 Initialize to zero the variables COUNT (the number of numbers in the set),
SUM (the sum of the numbers) and SUMSQ (the sum of the squares of the numbers).
2.0 Input NUMBER.
3.0 If NUMBER is equal to 99999, then proceed to step 4.0
   3.1 Add NUMBER to SUM.
   3.2 Add square of NUMBER to SUMSQ.
   3.3 Increase COUNT by 1.
   3.4 Input the next NUMBER.
   3.5 And repeat step 3 from the beginning.
4.0 Compute MEAN=SUM/COUNT.
5.0 Compute VAR=(SUMSQ−COUNT*MEAN*MEAN)/COUNT.
6.0 Output the results.
7.0 End.

Figure 4.6   Detailed steps for second procedure for computing variance

The program that follows begins with a PROGRAM statement and REAL and INTEGER statements that specify the types for the variables used. Following these are comment statements defining variables and describing actions occurring in the first part of the program. Remaining statements correspond almost one-to-one to steps shown in Figure 4.6.

Skill in programming has nothing to do with being able to develop the second procedure for computing variance, and its program; rather, what is required is a more complete knowledge of facts related to the problem to be solved. The process of developing a procedure is often referred to as *analysis* and it is the job of a good *analyst* to develop the most efficient procedures as well as ones that are well-structured. In this book we will be more concerned with developing good programming skills, including generation of understandable procedures, than we will be with developing the most efficient procedures. Understand that the latter objective is a very important one for a programmer/analyst to achieve but it is not a primary goal of this book.

```
      PROGRAM VAR2
**********************************************************************
*          PROGRAM DEFINITION                                       *
*                SEE THE TEXT FOR COMPLETE DESCRIPTION OF THE        *
*                PROBLEM.                                            *
**********************************************************************
*          VARIABLE DEFINITIONS.                                    *
*                NUMBER = A SINGLE MEMBER OF THE SET OF NUMBERS.     *
*                SUM = THE SUM OF THE NUMBERS.                       *
*                SUMSQ = THE SUM OF THE SQUARES OF THE NUMBERS       *
*                VAR = VARIANCE OF THE SET OF NUMBERS                *
*                MEAN = MEAN OF THE SET OF NUMBERS                   *
**********************************************************************
*          IMPORTANT NOTE                                           *
*                YOU MUST HAVE AT LEAST ONE NUMBER OTHER THAN        *
*                99999 TO ENTER OR THIS PROGRAM WILL NOT WORK        *
*                CORRECTLY.                                          *
**********************************************************************
      REAL NUMBER, MEAN, SUM, SUMSQ,  VAR
      INTEGER COUNT
*
*                INITIALIZATION
*
      COUNT=0
      SUM=0
      SUMSQ=0
      READ(5,*)NUMBER
***********************************************************
*                COMPUTE MEAN AND VARIANCE               *
***********************************************************
3     IF(NUMBER .EQ. 99999)GOTO 5
         SUM = SUM + NUMBER
         SUMSQ = SUMSQ + NUMBER * NUMBER
         COUNT = COUNT + 1
         READ(5,*) NUMBER
      GOTO 3
*
5     MEAN = SUM / COUNT
      VAR = (SUMSQ - COUNT * MEAN * MEAN)/COUNT
      WRITE(6,*) ' THE VARIANCE IS ',VAR
      STOP
      END
```

A PROBLEM FROM REAL LIFE

Now we consider solving a larger, more complex problem, with an eye toward demonstrating techniques that are likely to help the beginning programmer learn how to start solving a problem and how to write programs that are easier to understand and, therefore, easier to verify for their correctness.

The baker's problem

An owner of a small bakery knows that he currently has the following resources limiting any production schedule he may determine: 120 pounds of cookie mix, 32 pounds of icing mix, and 15 hours of bakery labor per day. There is enough oven capacity to bake as many cookies as the other resources will allow.

He knows also that it takes 0.6 pound of cookie mix and 0.1 hour of bakery labor to make one dozen sugar cookies; and 1.0 pound of cookie mix, 0.4 pound of icing mix, and 0.15 hour of bakery labor to make one dozen iced cookies.

Finally, he knows that he makes a profit of 31 cents per dozen on the iced cookies and 20 cents per dozen on the sugar cookies. How many dozens of each kind of cookie should he make so as to realize the most profit?

Solution. From the last paragraph of the problem statement it is clear that the equation for the total profit (in dollars) is

$$P = .20 \, S + .31 \, I \qquad\qquad (Eq.\ 4.1)$$

where P is the profit, S is the number of dozens of sugar cookies, and I the number of dozens of iced cookies. The problem, of course, is to find the values for S and I that make P as large as possible. For example, if $S = 100$ and I is 100, the profit would be $51.00. However, are there resources available to produce 100 dozens of each kind of cookie? Furthermore, within the resource limitations, how do we determine the *best* combination?

In order to understand the restrictions involved in this problem, we consider some tables made from the preceding information. Consider Table 4.1, which shows five possible combinations of ways to use the supply of cookie mix for the two kinds of cookies. Of course, there is a very large number of such combinations, so simply tabulating the information cannot provide enough of a basis for making the final choice about use of the cookie-mix resource to yield the greatest profit.

TABLE 4.1

Combination	Pounds of cookie mix used for sugar cookies	Pounds of cookie mix used for iced cookies	Dozens of sugar cookies produced	Dozens of iced cookies produced
1	120	0	200	0
2	90	30	150	30
3	60	60	100	60
4	30	90	50	90
5	0	120	0	120

Nevertheless, Table 4.1 gives information useful in constructing a graph showing the relationship between production of sugar cookies and iced cookies. (See Figure 4.7.)

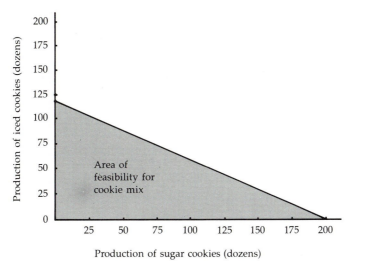

Figure 4.7   Combinations of cookie production from 120 pounds cookie mix

The two rightmost columns in Table 4.1 are reflected in Figure 4.7, as are a large number of other combinations of cookie production. The area under the slanted line in Figure 4.7 represents feasible combinations of cookies that can be produced from the 120 pounds of cookie mix.

Using a similar analysis with respect to the available bakery labor yields Table 4.2 and Figure 4.8.

TABLE 4.2

Combination	Hours of labor used for sugar cookies	Hours of labor used for iced cookies	Dozens of sugar cookies produced	Dozens of iced cookies produced
1	15	0	150	0
2	12	3	120	20
3	6	9	50	50
4	3	12	30	90
5	0	15	0	100

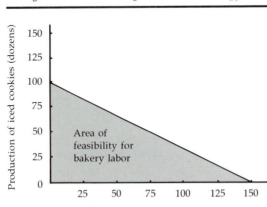

Figure 4.8   Combinations of cookie production from fifteen hours bakery labor

Since the amount of available icing mix has no effect on the number of sugar cookies made but limits only the number of iced cookies that can be produced, a simple computation, 32 divided by .4, reveals that 80 dozens iced cookies are the most that could be made (32 pounds of icing mix are available and the mix is used at the rate of .4 pound per dozen). Figure 4.9 shows a graph of that relationship.

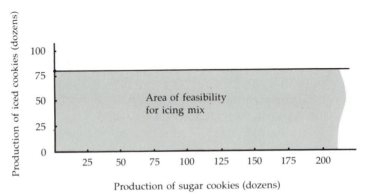

Figure 4.9   Combinations of cookie production from thirty-two pounds of icing mix

Now we construct the line graphs shown in Figures 4.7 through 4.9 on the same pair of axes. This composite graph appears in Figure 4.10.

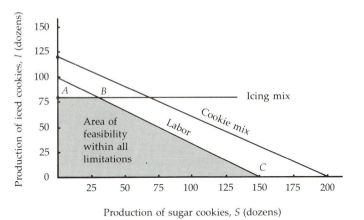

Figure 4.10   Composite graph of the three limitations on cookie production

To this point we have produced tables and graphs so as to understand the *resource* limitations placed on cookie production in this bakery. We return now to a consideration of Equation (4.1), the profit equation. The task is to maximize P within the resource limitations graphically displayed in Figure 4.10. There are several mathematical techniques that accomplish this task but it is beyond the purpose of this book to consider them. Nevertheless, it should be stated that perhaps the most commonly used method is called the simplex method, one of several linear programming techniques.

There is, however, a procedure that can be easily understood and that will finally bring us to the use of a computer program. That method is simply trial and error. We try combinations of S and I that lie within the shaded area of Figure 4.10. A plan for arriving at various combinations might be to start at point A in Figure 4.10, then move to the right to point B, then down and to the right to point C. (Note that the "cookie mix" line lies above the shaded area of feasibility and so has no effect on our choices of S and I.) Or, we could begin at point C, move to the left and upward to point B, and finally to the left to point A. When considering this latter path, it is obvious that the maximum point for which we are searching is *not* on the path from B to A because on that path I (dozens of iced cookies) stays constant at 80 while S (dozens of sugar cookies) decreases, hence P (profit) would decrease. Therefore, the maximum profit must be obtained when S and I correspond to points somewhere on segment BC.

Segment BC is a portion of the "labor" line on which lie the points $(0, 100)$ and $(150, 0)$. Therefore, applying the slope-intercept formula for the equation of a line gives

$$I = \left(\frac{100-0}{0-150} \right) S + 100 \qquad \text{(Eq. 4.2)}$$

Equation (4.2) reduces to

$$I = -\frac{2}{3} S + 100 \qquad \text{(Eq. 4.3)}$$

Thus, Equation (4.3) shows the relationship between I and S on segment BC.

The coordinates of point B can be obtained by solving simultaneously the equation for segment AB and that for BC. Since line AB represents a constant value of 80 for I and allows for *any* value of S between 0 and 30, its equation is

$$I = 80 \qquad \text{(Eq. 4.4)}$$

Solving simultaneously,

$$\left. \begin{array}{l} I = 80 \\[2mm] I = -\dfrac{2}{3} S + 100 \end{array} \right\} \qquad \text{(Eq. 4.5)}$$

gives the solution $S = 30$ and $I = 80$. Therefore, point B is $(30, 80)$. From Figure 4.10, it can be seen that point C is $(0, 150)$.

Here are the steps we have taken so far, together with the remaining steps needed to solve the baker's problem. The overall process coincides with that diagrammed in Figure 4.4.

1. Formulate a concise, accurate representation of the problem to be solved.

   (This was done when we produced Equation (4.1), together with the recognition that Equation (4.1) must be used to find the largest possible value for P for various values of S and I.)

2. Divide the problem into segments.

   (This was done when we produced Tables 4.1 and 4.2, and Figures 4.7 through 4.10. Then, these were used to determine Equations (4.3) and (4.4) so as to find the S and I coordinates of point B in Figure 4.10. Coordinates for point C were obtained directly from Figure 4.10.)

3. Detail the steps required to complete each segment.

   (This remains to be done.)

4. Write the computer program to determine the solution.
   (This remains to be done.)

Suppose we choose as the path in Figure 4.10 to use for computing the trial values of P the path from C to B. Point C has 150 as its S value while at point B it is 30. All along the path, Equation (4.3) tells us that

$$I = -\frac{2}{3}S + 100$$

Figure 4.11 shows a diagram of the solution process as completed thus far.

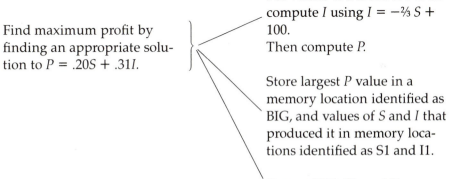

Find maximum profit by finding an appropriate solution to $P = .20S + .31I$.

For values of S from 150 to 30 compute I using $I = -\frac{2}{3}S + 100$.
Then compute P.

Store largest P value in a memory location identified as BIG, and values of S and I that produced it in memory locations identified as S1 and I1.

Output BIG, S1, and I1.

Figure 4.11   First- and second-level steps in solving the baker's problem

Next, we refer to Figure 4.11 to specify further detailed steps in the solution. Figure 4.12 shows these further details as third-level steps.

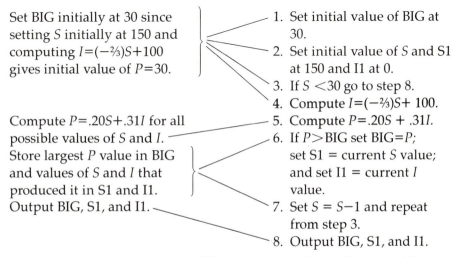

Set BIG initially at 30 since setting S initially at 150 and computing $I=(-\frac{2}{3})S+100$ gives initial value of $P=30$.

Compute $P=.20S+.31I$ for all possible values of S and I. Store largest P value in BIG and values of S and I that produced it in S1 and I1.
Output BIG, S1, and I1.

1. Set initial value of BIG at 30.
2. Set initial value of S and S1 at 150 and I1 at 0.
3. If $S < 30$ go to step 8.
4. Compute $I=(-\frac{2}{3})S+ 100$.
5. Compute $P=.20S + .31I$.
6. If $P>$BIG set BIG$=P$; set S1 = current S value; and set I1 = current I value.
7. Set $S = S-1$ and repeat from step 3.
8. Output BIG, S1, and I1.

Figure 4.12   Second- and third-level steps in solving the baker's problem

As in previous examples, we next develop ForTran statements from the third-level steps. The following program shows this transition.

Third-level steps and ForTran program

```
                          PROGRAM BAKER
      C*******************************************************
      C*           PROGRAM DEFINITION                        *
      C*                  SEE THE TEXT FOR A COMPLETE         *
      C*                  DESCRIPTION OF THE PROBLEM.         *
      C*******************************************************
      C*           VARIABLE DEFINITION                       *
      C*                  S   IS TOTAL DOZENS SUGAR COOKIES   *
      C*                  I   IS TOTAL DOZENS ICE COOKIES.    *
      C*                  P   IS COMPUTED PROFIT.             *
      C*                  BIG  IS MAXIMUM PROFIT.             *
      C*                  S1 & I1 ARE THE S AND I VALUES      *
      C*                         USED IN COMPUTING BIG.       *
      C*******************************************************
                          REAL S, I,P,BIG, S1,I1
      C*
      C*                  INITIALIZATION
      C*
                          BIG = 30
                          S = 150
                          S1 = 150
                          I1 = 0
      C*
      C*******************************************************
      C*                  THE MINIMUM VALUE FOR S IS 30, SEE  *
      C*                  PREVIOUS DISCUSSIONS. HERE WE COMPUTE *
      C*                  I AND BIG CORRESPONDING TO CURRENT  *
      C*                  VALUE OF S AND WE REMEMBER THE I & S *
      C*                  VALUES FOR THE MAXIMUM PROFIT.      *
      C*******************************************************
      C*
         25               IF(S .LT. 30)GOTO 30
                          I = (-2.0/3) * S +100
                          P = 0.2 * S + 0.31 * I
                          IF(P .GT. BIG)THEN
                             BIG = P
                             I1 = I
                             S1 = S
                          ENDIF
                          S = S - 1
                          GOTO 25
      C*
         30               WRITE(6,*)' SOLUTION FOLLOWS'
                          WRITE(6,*)' MAXIMUM PROFIT ',BIG
                          WRITE(6,*)' DOZENS SUGAR COOKIES ',S1
                          WRITE(6,*)' DOZENS ICE COOKIES ',I1
                          STOP
                          END
```

1.0   Set initial value of BIG at 30.

2.0   Set initial value of S and S1 at 150 and initial value of I1 at zero.

3.0   If S has not reached its final value of 30 then perform steps 3.1 through 3.5. Else proceed to step 4.

3.1   Compute I= -2/3S+100

3.2   Compute P= .20S+.31I

3.3   If P>BIG then
        set BIG = P
        set S1 = S
        set I1 = I.

3.4   Decrement S by 1(or any other appropriate value needed)

3.5   Repeat the process from step 3.

4.0   Report the results.

5.0   Terminate the process.

As you can tell by studying the preceding program, the solution method is a trial-and-error procedure in which values of P, the total profit from both kinds of cookies, are repeatedly computed as the number of dozens of sugar cookies decreases from 150 to 30 in steps of .5, and the number of dozens of iced cookies increases from 0 to 80 in accordance with the equation $I = -\frac{2}{3}S + 100$. These ranges were obtained from an analysis of the resource limitations defined in Figure 4.10. For each different computed value of P, the largest value is saved.

Note in the preceding program, the program statement $S = S-1$, where values for S are decreased repeatedly by 1. Actually, the decrement value of 1 was arbitrarily selected and might well have been smaller or larger. It was arbitrarily selected so as to start computations with S at 150 and to stop when S becomes 30. Recall from our previous discussion that S ranges from 150 to 30 as we move from point C to point B (Figure 4.10).

In the first part of this chapter we have discussed the process of producing a set of steps, which we have called a procedure, for solving a given problem. We continue now to provide further examples and introduce new concepts with a view toward developing skill in problem solving.

ALGORITHMS

Another word for *procedure* is *algorithm.* The word algorithm comes from the name of a Persian mathematician, Abu Ja'fur Monammedibu Musa *al-Khowarizmi*, who lived around the time 825 A.D. This man is remembered for having written the first known book on algebra.

An algorithm is a recipe, method, technique, or routine for accomplishing a task. In computer science, algorithms are so important that a very careful definition should be given.

Definition

An algorithm is a set of rules for solving a specific problem. It has five notable features, as follow: (1) it terminates after a finite number of steps; (2) the actions specified in each step must be precisely and unambiguously defined; (3) it has a finite number of quantities given to it initially before the algorithm begins (these are called "inputs"); (4) it has one or more quantities derived from the inputs by applying the algorithm (these are called "outputs"); and (5) all operations specified in the algorithm must be effective, meaning that they can be done exactly and in a finite length of time by a person using pencil and paper.

We now present seven examples to clarify the concept of an algorithm.

Example 1

Problem: Given N numbers—call them $X_1, X_2,...,X_N$—find the largest number in the set.

Initially, let's call the first number in the set LARGE. Our plan will be to compare each subsequent number in the set with LARGE and whenever a comparison indicates LARGE to be a smaller number, we replace LARGE by the larger number. When all numbers in the set have been compared to LARGE and replacements made as just described, LARGE may be reported out as the largest number in the set. Here is the algorithm.

Algorithm:

1. Input $X_1, X_2,...,X_N$.
2. Set index, I, to 2.
3. Set LARGE equal to X_1.
4. If X_I is larger than LARGE, then set LARGE equal to X_I.
5. Increase index I by 1.
6. If there are more X values to compare to LARGE, then repeat the process from step 4.
7. Else report the value of LARGE.
8. End.

Let's examine this algorithm for the properties included in the definition of an algorithm given previously: (1) it terminates at step 8 after a finite number of repetitions; (2) each step is precise and unambiguous as to what action should be taken; (3) there are a finite number (N) of inputs; (4) there is one output; (5) each action specified in the algorithm can be done exactly and in a finite length of time with pencil and paper (in other words, a solution exists). Therefore, all five properties of an algorithm are present in this example.

Example 2

Problem: Given two integers, find their greatest common divisor (GCD). (The GCD of two integers is the largest integer that divides both of the original integers yielding an integer quotient.)

Algorithm:

1. Input two integers, calling the locations where they are stored M and N.
2. Find the larger of M and N and store a copy of it in LARGE. Store a copy of the other integer in SMALL.

3. Divide LARGE by SMALL. Store the integer part of the quotient in Q and the remainder in R.
4. If $R=0$, output SMALL as the GCD, then stop.
5. Set LARGE=SMALL and SMALL=R.
6. Repeat the process from step 3.

As with Example 1, the algorithm in Example 2 has the five properties of a true algorithm. It terminates after a finite number of steps. The action specified in each step is clear and precise. There are two inputs and one output. Each action specified can be done exactly and in a finite amount of time.

Example 3
Problem: Find the first 100 positive integers divisible by 7 and print them.

Algorithm:

1. Set COUNT = 1 and DIVISOR = 7.
2. Output DIVISOR as a number divisible by 7.
3. Increase DIVISOR by 7.
4. Increase COUNT by 1.
5. If COUNT is less than 100, then repeat the process from step 2.
6. End.

The reader should check for the presence of the five necessary properties in the preceding algorithm as well as in the algorithms of the next three examples. Be careful, for not all of the algorithms have all five properties.

Example 4
Problem: Find all positive integers divisible by 7.

Algorithm:

1. Set $N=7$.
2. Output N.
3. Set $N=N+7$.
4. Repeat the process from step 2.

Note that this algorithm is faulty because it is impossible to solve the problem. There are an *infinite* number of positive integers divisible by 7 so one could never find them all. Similarly, the algorithm does not have the property of finiteness.

Example 5

Problem: A positive integer greater than 1 is called a *prime number* if its only integer divisors are 1 and the given number itself. For example, 2, 3, 5, 7, and 11 are the first five prime numbers. The problem is to develop an algorithm to find the first 100 prime numbers.

The following algorithm is based on the above definition of prime number. We follow the procedure that if we are given an odd integer, N, as a candidate for a prime number, we search for a divisor for N. The search begins at 3 and checks all *odd* numbers up to and including the square root of N. Why *odd* numbers and why the square root of N? From the definition of prime number, the only *even* prime number is the number 2. Furthermore, no odd number can have an even divisor, thus only odd numbers can be candidates for divisors.

Regarding the upper limit for divisor candidates, the largest value possible is $N/2$, for the following reason. If p is a divisor of N, then the following equation is true:

$$N = p \times q$$

where q is the quotient upon dividing N by p. The largest that p can be and yet not exceed q is the case where p and q are equal, and they are equal if both are the square root of N.

Our procedure will be as follows: given N, we start from 3 and check for a divisor of N. If such a divisor is found, then N is not a prime number and we proceed to the next odd integer, never exceeding the square root of N. If the candidate for divisor reaches this maximum value without being a divisor, then N is a prime number and should be so reported.

We keep track of the number of prime numbers we find so that when 100 of them have been reported, we stop.

Algorithm:

1.0   Output 2, 3. (These are the first two prime numbers.)
2.0   Set COUNT=2. (Count is the number of prime numbers output thus far.)
3.0   Set NUM=5. (NUM is the candidate for the next prime number.)
4.0   Set LIMIT=Integer part of ($\sqrt{\text{NUM}}$).
    (LIMIT is the value beyond which we need not try for more divisors of NUM.)
5.0   Set I=1.
6.0   Set I=I+2. (I is a potential divisor of NUM.)
7.0   If I > LIMIT, then do steps 7.1 through 7.4. Else, do steps 8.1 through 8.2.

7.1 Output NUM.

7.2 Set NUM=NUM+2. (Generate a new candidate for a prime number.)

7.3 Set COUNT=COUNT+1. (Increase by 1 the count of prime numbers output thus far.)

7.4 If COUNT ⩾ 100, then stop the process. Else repeat the process from step 4.

8.0 If NUM is divisible by I, then set NUM=NUM+2 and repeat the process from step 4.0.

8.1 Else set $I=I+2$.

8.2 Repeat the process from step 7.0.

For Example 5, besides checking for the presence of the five characteristics of an algorithm, run through the steps of the algorithm (using pencil and paper) so that you understand how it works.

Example 6

Problem: Given the positions on a chess board occupied by each of your pieces and the pieces of your opponent, determine what the best move is for you to make next.

Algorithm:

1. Input positions of all chessmen for both players.
2. Check all possible moves for each of your chessmen; keep a record of opponent's men that could be captured in each case and a record of all possible subsequent moves that could be made by each of your men in the next four turns.
3. Evaluate each possible current move.
4. Output the move that is most valuable.

Although you may intuitively feel that the preceding algorithm does not satisfactorily meet the condition of precise action stated in each step, it does meet the condition of finiteness since there are at most sixteen chessmen for each player and there are at most sixty-four squares on which chessmen can be placed, so the total number of all possible moves must be finite. That fact is correct, but being *theoretically* finite is quite different from being *feasibly* finite. The total number of possible moves is theoretically finite but it is somewhere around 16 times 2^{64}, a very, very large number. It is so large that even if we were using a computer that could perform ten million operations per second (which some of today's computers can do) it would take that computer approximately fifteen years of steady calculating to check out all possible moves. Clearly that is not feasibly finite. You must, therefore, always be alert to this matter of feasible finiteness as you develop the skills needed to produce useful algorithms from which computer programs are to be written. You must acquire the ability to think critically about each problem as well as about the algorithm you develop to solve it. Don't accept the first algorithm you produce for a given problem without analyzing it for efficiency. Often careful thought can convert an unusable algorithm into one that efficiently provides the needed solution.

Example 7

Problem: Find the first 100 numbers that are approximately divisible by 3.

Algorithm:

1. Set N to approximately 3.
2. Output N.
3. Increase COUNT by 1.
4. If COUNT is greater than or equal to 100, stop the process.
5. Else increase N by 3.
6. Repeat the process from step 2.

This algorithm is faulty because it has one step that is not well defined. In step 1, N is to be set *approximately* to 3. What does it mean to be "approximately" 3? Of course, the trouble has its roots in the statement of the problem, where it was written "approximately divisible by 3." This example is included to alert you to the importance of well-defined processes. It is impossible to translate into a computer program statement any operation that is not precisely defined.

CONSIDERATIONS WHEN DEVELOPING ALGORITHMS

Now that you have a better concept as to what an algorithm is and some possible pitfalls of algorithm development, we alert you to some techniques and considerations that will help make you an effective developer of algorithms and, thus, an effective computer programmer.

How to start algorithm development

As previously stated, the acquisition of skills in generating algorithms is so important to a good programmer that some universities offer entire courses devoted to the topic of algorithm development. We will not appeal to any sophisticated techniques in this book, but we do at this time remind you of the fact that earlier in this chapter we attempted to identify some steps useful when beginning to produce a procedure (or algorithm) that will lead to a computer program. We repeat those steps here:

1. State precisely the problem to be solved, including the identification of given information and the specification of results desired.
2. Identify and list major substeps, each of which must be accomplished to obtain the desired results.
3. For each major substep, identify and list second-level steps each of which must be accomplished to complete the major substep.
4. For each second-level step, produce a program statement.

Uniqueness in algorithms

A question that may come to your mind is this: Is there a *unique* algorithm for each problem to be solved? The answer is no. Theoretically, there are as many different algorithms for a given problem as there are people who produce them. Of course, it is quite likely that many, if not most, of the algorithms would differ in very insignificant ways, leading one to believe that although there is no unique algorithm for solving a given problem, there probably is a most common one.

Our concern in this book is to open your thinking to the fact that there are usually several procedures for accomplishing a given task with a computer. Your goal, beyond producing an algorithm that solves the problem, should be to consider some aspects of efficiency. Which algorithm takes least computer time? Which algorithm is easiest to understand? The answers to these questions are important, although not as important as "Which algorithm correctly solves the problem?" In this book we will help you learn skills that will help you to consider all three questions.

Formulating an algorithm

Of most concern to beginning programmers is the matter of starting to write an algorithm. "How do I start? How do I write down the ideas I have for solving the problem?"

Earlier in this chapter we presented some techniques for developing an algorithm and, in fact, a computer program. The underlying concept is to proceed from stating the problem, to identifying individual minor tasks. This method is usually called the *top-down* approach. Some people refer to the step just preceding the writing of program instructions as the writing of semicode or pseudocode. The tasks at that stage are detailed enough that they can readily be coded into program statements. We prefer not to formalize those delineated tasks by assigning to them a name; rather, think of them as the last step preceding the actual writing of program instructions for solving a given problem.

We remind you also that understandability is a most important characteristic of good programs. If you cannot understand what is being done by a program or section of it, you cannot be sure that it solves the problem.

Validating an algorithm

We consider next the *validity* of algorithms and thus the validity of computer programs. Recall that early in this chapter we discussed an example of a program that provided the correct output for a given problem but that could not in any manner be construed as a correct solution to the problem. Such a program does not pass the test of validity. A valid algorithm (program) produces not only the correct output for one given set of input data but does so for *all* sets of correct input data. Since it is generally impossible to test an algorithm or program for *all* sets of input data, the only way that program validity can be determined is by understanding the logic of the program (or algorithm) so that one can decide if it really would correctly process all sets of valid input data and thus produce correct output.

We consider another example to illustrate further what we mean.

Example 8

Problem: Given any positive integer, determine whether or not it is a prime number. (See Example 5 in this chapter for a definition of prime number.)

Algorithm:

1. Set $I=1$.
2. Input a positive integer as NUM.
3. Compute LIMIT=integer part of the square root of NUM.
   (LIMIT is the maximum value for any potential divisor of NUM.)
4. Set $I=I+1$. (I is a potential divisor of NUM.)
5. If NUM is divisible by I, then the output is NUM "IS NOT PRIME"; stop the process.
6. If $I >$ LIMIT then output NUM, and indicate that it is prime; then stop the process.
7. Repeat the process from step 4.

When this algorithm is applied to the integer 1, the output is "1 IS

PRIME," yet according to the definition of a prime number, 1 is *not* prime. Similarly, when the algorithm is applied to the integers 2 and 3, the output is "2 IS NOT PRIME" and "3 IS NOT PRIME," though both 2 and 3 really are prime numbers. If you test the algorithm for 4, 5, and, in fact, for any other positive integer, it provides the correct output. Since this algorithm is understandable, it is possible to determine its validity.

The changes required to make it completely valid are minor ones and the reader should try to determine what those changes should be.

Techniques for testing an algorithm

We turn our attention now to some methods whereby algorithms can be tested for correctness. This testing process consists of two major areas: *debugging* and *profiling*.

Debugging techniques.   Debugging is the process of finding and eliminating errors in an algorithm or computer program. Here are some debugging techniques.

1. There is supporting theory for proving whether or not an algorithm is correct. The process is similar to that of proving the truth of mathematical theorems. However, this theory is beyond the level of computer science knowledge required for this course, so we simply mention its existence.

2. An algorithm can be tested by pencil and paper for correctness in processing at least a few sets of input data. This is often a very useful technique for reviewing the logic of an algorithm. Depending on the algorithm, of course, this method can be a very time-consuming process. It is especially important to run tests with sets of all types of input data. For example, if positive and negative numbers are permissible as input data, it is essential to run tests with enough appropriate mixes of positive and negative numbers to make certain that the algorithm properly handles all special cases.

3. An algorithm can be translated into program instructions by using a specific programming language and then executed on a computer system with selected sets of input data for which known results exist. As with pencil-and-paper testing (method 2), it is important to run tests for all types of input data to assure that errors will not occur with special cases. Recall from Chapter 3 that program syntax errors are an additional source of frustration when debugging algorithms by this method. The process of writing program statements from algorithm steps is the source of such syntax errors.

Methods 2 and 3 are useful in detecting the *presence* of errors but are of no help in guaranteeing their *absence*. Only method 1 does this. However, since we are not examining that method in detail, we must be content with methods that detect the presence of errors and with elimination of errors.

Profiling.   Once an algorithm has been debugged and a computer program has been written for it, and the program has been debugged, the program is executed by using a variety of acceptable input data. For each execution, various statistics are collected and analyzed so as to determine if the program can be improved by making changes indicated by the statistics. The statistical analysis as well as the collection of data require skills beyond the scope of this book.

The quality of an algorithm

We have already alluded to the fact that some algorithms are better than other ones in solving a given problem. What properties are important to look for in a good algorithm? The topic of algorithm analysis is broad and deep, and there are entire courses devoted to the topic in computer science curricula. We shall not consider the topic in detail but will simply mention some factors that the beginner can consider.

1. *An algorithm should be relatively fast.*

By this we mean that the number of operations required should be minimal. It is often very difficult, if not impossible, to compute the exact amount of time required to execute a given algorithm. For this reason it is equally difficult to determine which of two or more algorithms for a given problem can be executed in the least amount of time. However, this dilemma is not sufficient reason to ignore the matter of speed completely. Being aware of the need to *consider* speed is enough to cause most algorithm developers to recognize an obviously poor algorithm. As a case in point study the following example:

Example 9

Problem: Given two positive integers N and M, determine if N is divisible by M.

Algorithm A:

1. Input N and M.
2. Compute $K=$integer part of (N/M).
3. Compute $I=(K*M)-N$.

4. If $I=0$, output N "IS DIVISIBLE BY" M; then stop.
5. Else output N "IS NOT DIVISIBLE BY" M; then stop.

Notice that the time required to execute the preceding algorithm is independent of the input data. Now consider algorithm B.

Algorithm B:

1. Input N and M.
2. Set $K=N$.
3. Compute $K=K-M$.
4. If $K=0$, then report N "IS DIVISIBLE BY" M.
5. If $K>0$, then repeat the process from step 3.
6. If $K \leq 0$, then report N "IS NOT DIVISIBLE BY" M.
7. End.

It will take you only a little examination of algorithm B to realize that it also solves the problem originally given. However, it is definitely dependent on the input data. If N is large with respect to M, the loop consisting of steps 3, 4, and 5 must be executed many times, thus requiring a longer period of time. It would definitely be advantageous to use algorithm A to solve this problem.

From Example 9 it is clear that when you are developing algorithms it is important to consider the amount of time required to execute them. Even more consideration could have been given to that matter in Example 8. In order to increase the efficiency of the algorithm still more, we could have eliminated from consideration as prime numbers all even numbers. Analysis like that is what increases algorithm efficiency.

2. *An algorithm should require as little computer memory as possible and still produce the required output in an acceptable amount of time.*

When an algorithm has this property, consideration has been given to the trade-off between computer memory and speed of executing the algorithm. There are many situations such that if there were unlimited computer memory for the implementation of the algorithm it could be designed to take very little time for execution. But computer memory is a resource that costs money just as is the computer time required to execute an algorithm. The programmer must weigh these two costs as well as other considerations associated with computer implementation of an algorithm, then choose the algorithm design that optimally meets all considerations. This means that some applications (as, for example, controlling the navigation system of a space vehicle carrying human beings) are so important that speed of execution is the major consideration. There are others for which speed of execution is not as critical as saving money by using less computer memory.

3. *An algorithm should be structured.*

The structure an algorithm has bears directly on its understandability. We have already discussed the importance of understandability especially as it relates to validating an algorithm.

As to what we mean by a structured algorithm, let us at this point say that it has to do with the organization and format in which the algorithm is written. Earlier in this chapter you saw the importance of a simple thing like indentation as related to the understandability of a program. The same sort of thing is true for algorithms. Because algorithm and program structure are important concepts, we shall now devote two sections to the topic.

Writing a structured algorithm (program)

In a previous subsection we stated that a good algorithm (and thus a good program) should be structured. By saying that an algorithm has *structure* we mean that it has these properties:

1. The physical layout (format) of the steps is such that close relationships among various groups of steps are visually apparent. A common technique for doing this is appropriate use of indention (see earlier section) and modules, which are self-contained segments within the algorithm as a whole.

2. The flow of operations in the algorithm is such that by starting at the beginning it is readily possible to proceed toward the end by making decisions and performing steps that generally yield forward progress. That is, steps in the algorithm should not result in jumping forward and backward until the comprehension of any cohesive logical unit is made completely impossible.

Throughout the rest of this book our examples reflect the above two major considerations.

Some helps for achieving structured programs

Since writing algorithms and programs that have structure is so important to their understandability we enumerate here some suggestions for producing structured programs.

1. Develop a clear and explicit statement of the problem to be solved.

This is comparable to knowing the destination of a trip before starting on the way: if you know where you're going you'll be more apt to get there. In the realm of problem solving, unless you can clearly identify the information you have on hand and the new information that must be developed by processing the on-hand material, you are not likely to develop a satisfactory algorithm for solving the problem.

2. Write easily understood statements of subtasks that must be done in order to solve the given problem. If necessary, write equally understandable statements of second-level tasks within the subtasks.

This is called the top-down approach. First you follow step 1 and develop a clear statement of the problem. But progress toward solution is likely to be snaillike unless you can see *segments* of the total problem, thus separating the main problem into subtasks. If the first attempt at developing subtasks results in situations too complex to resist solution, restate each one into sub-subtasks (which we have called second-level tasks) and repeat the process if necessary, until a statement of tasks to be completed is such that each task can be accomplished by a single program statement.

3. Write program instructions in the programming language of your choice, using the subtask statements and sub-subtask statements as you do.

This step in solving the problem is easy if you know an appropriate programming language. As discussed in step 2 above, at this point the problem solution is in the form of small tasks. Select a program statement that accomplishes each stated task, then get the resulting program to run correctly.

4. As you write program instructions, use the logical power of the IF-THEN-ELSE structure to avoid unnecessary jumping back and forth within the program.

This step is intended to encourage you to think about the logical flow of the program you develop. Make every attempt to have it flow smoothly from beginning to end with no branching back and forth. Where branching is necessary, it should be obvious *why* it is essential; then it will be readily understood by anyone reading the program.

5. As you write the program instructions, avoid as much as possible the use of GO TO statements. Use them only where program understandability is diminished by using alternate types of program statements, or when they are essential to the writing of loops.

This step is closely related to step 4 and justification for it is included in the comments accompanying step 4.

6. Insert comments throughout your program but don't try to clarify illogical program organization by excessive use of comments. Rather, rewrite the program instructions to conform to the concept of good logic flow as identified in property 2 of structured algorithms (preceding section).

    Comments should be used to further clarify an already comprehensible situation but *not* to try to elucidate a poorly structured program. Do, however, use comments freely. They help significantly when modifying a program previously developed.

7. Use the programming concepts of subprograms and functions (to be discussed in a later chapter) to divide the program into modules.

    The use of subprograms and functions is simply taking advantage of the program-language facilities designed to carry out the process described in step 2.

8. Use indentation of certain program statements to make your program more readable.

    The practice of indenting program statements provides visual evidence that certain statements are to be considered as a group of related statements. This is especially helpful in the case of iterative structures and block-IF structures.

We conclude this chapter by presenting three problems with complete solutions. The first two assume that input data for the problems were previously stored on secondary computer storage devices such as magnetic disks, which are then used in the solution. This situation has not yet occurred in examples discussed in this book. As you recall, input has been entered on terminal or card reader while output has been on terminal or printer.

Since the concept of data files on devices such as magnetic disks is a new one, we shall take the space here to discuss data files in ForTran.

DATA FILES IN FORTRAN

A working definition of a data file might be as follows: a collection of information stored in a specified format on either a magnetic tape or magnetic disk.

The functional unit of information in a data file is called a *logical record,* or more often just a *record.* Although the record is the functional unit for processing information in a data file, the record itself is made up of segments of information called *data elements,* which, in turn, are made up of individual *characters.* A record may consist of only blanks, in which case it is called a *blank record.*

Now let's consider an example of the kind of information that could be included in a data file in the light of the terms just described. Suppose we ask every one of the thirty-two students in Computer Science 140 to punch a card containing name, student identification number, and total credits registered for. If we collect these punched cards we would have a file of data containing thirty-two logical records, each record consisting of the three data elements: name, student identification number, and credits. This collection of information is not yet a data file according to our working definition of data file. Nevertheless, if we were to input these cards in some specific format and output the data in a specified format onto a magnetic tape or magnetic disk, we would have a data file.

To further clarify the concept of a data file, we present another example. Suppose we ask the twenty-eight students in Computer Science 150 to fill out forms on which have been designated 20 spaces for a name, 30 spaces for a local address, 30 spaces for a home address, 7 spaces for a telephone number, 15 spaces for the name of a subject-major, 15 spaces for the name of a subject-minor, 3 spaces for total credits completed, and 4 spaces for a cumulative grade-point-average (GPA). You will find that a total of 124 spaces are thus reserved for information to be filled in by each student. If these data are punched into two cards for each student, there would be fifty-six physical records (two cards for each of twenty eight students) but only twenty-eight logical records (one set of data for each student). Each logical record consists of eight data elements (name, local address, home address, telephone number, major, minor, total credits, and GPA). Note that any reference made in this section to a record will always mean logical record. If these fifty-six cards were input with appropriate formats, and then output (again in appropriate format) to a magnetic tape or magnetic disk, we would have a data file. The device (tape unit or disk unit) would have an identifying number associated with it called the *logical unit number.* The logical unit number is needed by the programmer so that the program using a given data file can contain specific information about where processing should proceed in order to have access to those data. What we have just described is the original use of logical unit number. On present-day computers, however, logical unit numbers are used to refer to portions of whatever secondary storage is available to the computer system.

There are three common methods of organizing data files on magnetic media: *sequential, indexed-sequential,* and *random.* We shall discuss only sequential files here. Data files consist of collections of records and when organized as a sequential file records are read or written one after the other—that is, sequentially. When accessing (reading or writing) the nth record in a sequential data file it is necessary to process the $n-1$ records that sequentially precede it in the file. This means that in some way, the program must process these $n-1$ records so that they are by-passed in accessing the nth record, the one that we assume is the object datum for the given program.

Data files are usually assigned names, again for the purpose of making positive identification of the data file. For example, a data file containing the information from the previous example might be assigned the name STUDNT because the data are associated with students. The allowable length of a name is dependent on the particular compiler or computer system on which the processing takes place.

We proceed now to a discussion of some ForTran statements necessary for processing data files. A more complete discussion of data-file organization and ForTran facilities for doing file processing is given in Chapter 8.

The OPEN statement

This statement is used to prepare a file for either input (to be read from) or output (to be written into). Its general form is discussed in Chapter 8 but we present here a somewhat restricted form adequate for our current needs:

$$\text{OPEN}\left(u, \text{FILE}='\text{filename}', \text{STATUS}=\begin{Bmatrix}'\text{OLD}'\\'\text{NEW}'\end{Bmatrix}\right)$$

where u is the unit number assigned to the file, 'filename' is the name of the file, 'OLD' is the STATUS if the file already exists in the system and 'NEW' is the STATUS if the file is just being created.

A file whose status is designated as 'NEW' should be used only with a WRITE statement—that is, for output. A file whose status is designated as 'OLD' is usually intended to be used with a READ statement—that is, for input. An 'OLD' file *may* be used for output but you must be aware that you will lose existing information on the file since the output process erases existing information as the new information is written.

Examples of the OPEN statement follow:

(a) `OPEN(1,FILE='STUDNT',STATUS='OLD')`

(b) `OPEN(3,FILE='TEMP',STATUS='NEW')`

In example (a), the existing file 'STUDNT' is associated with unit 1 and is prepared for input or output. In example (b), file 'TEMP' is made ready for output, hence it is created for the first time and is associated with unit 3.

If an existing data file is opened as 'NEW,' or a nonexistent file is opened as 'OLD,' the result will be a fatal error message indicating that the program will not run until the error in file status is corrected.

The REWIND statement

This statement is used to position a data file at its starting point so that the first record is ready for input or output. Its form is

REWIND u or REWIND (u)

where u is the unit number associated with the file. Specific examples are
(a) REWIND 3
(b) REWIND (4)
In example (a), unit 3 is positioned so that the first record of the data file is ready for reading or writing, while example (b) does the same for the data file associated with unit 4.

Reading from a data file

To read data from a data file, use the following form of the READ statement:

READ $(u,*,END=n)$ var-list

where u is the unit number associated with the file, *refers to unformatted input, and n is the reference number of the statement to be executed next when the end of the file is reached. Var-list refers to the list of variables into which data are to be read. Here are three specific examples:
(a) READ(2,*,END=20) A,B
(b) READ(7,*,END=30) NAME,ADDRES
(c) READ(1,*) X,Y,Z
In example (a), one record from the file associated with unit 2 is accessed and two data elements are read, the first into variable A and the second into variable B. If reading should be attempted when the last record of the file has been passed, the statement executed next in the program is the one with reference number 20 (END=20).

In example (b), one record from the file associated with unit 7 is accessed and two data elements are read, the first into variable NAME and the second into variable ADDRES. If reading should be attempted past the last record in the file, program control transfers to statement 30.

Example (c) causes one record from the file associated with unit 1 to be read; three data elements are stored, one in variable X, one in variable Y, and the third in variable Z. In this example there is no "END=." Its inclusion is optional. However, if it is omitted, as in this example, and reading is attempted beyond the last record of the file, an error results that causes the program to be terminated.

Unit number 5 on most systems is reserved for the card reader or terminal, depending on the type of computer system. Therefore the statements

```
READ (5,*) A
```

and

```
READ *,A
```

both read one record (card or line) of data from a card reader or terminal; one data element is read from the record and stored in variable A.

Writing to a data file

The general form of a statement to write one record into a data file is the following:

WRITE (u,*) output-list

where u and * are as described in the previous section and "output-list" refers to constants or variables to be output to the data file associated with unit u. Here are two specific examples:

(a) `WRITE (3,*) NAME,STREET,CITY,STATE,ZIP`

(b) `WRITE(1,*) HOURS`

In example (a), one record consisting of five data elements is output to the data file on unit 3 in unformatted form. In example (b), one record consisting of one data element is output to the data file on unit 1, also in unformatted form.

The CLOSE statement

This statement is used to close a file that is no longer needed by the program. Every file used by a program should be closed before the end of the program so as to properly disconnect the file. The general form of the statement is

CLOSE (*ul,u2,* . . . ,*uk*)

where *ul,u2,* . . . ,*uk* are *k* different unit numbers. There may be only one unit number specified. Here are examples:

(a) CLOSE (1)

(b) CLOSE (3,7,8)

In example (a), the file on unit 1 is properly disconnected. In example (b), three files, 3, 7, and 8, are all properly disconnected.

Program that utilizes data files

An example of a program using data files follows. Study it and then read the analysis that follows it.

```
1            PROGRAM  FILES
2            CHARACTER  NAME*20
3            INTEGER  STID, COUNT
4            REAL  GPA
5            OPEN(3,FILE='GPA3',STATUS='NEW')
6            OPEN(2,FILE='STUDNT',STATUS='OLD')
7            REWIND 2
8            COUNT = 0
9   10       READ(2,*,END=100)NAME,STID,GPA
10             IF(GPA .GT. 3.0)THEN
11                WRITE(3,*)NAME,STID,GPA
12                COUNT = COUNT + 1
13             ENDIF
14           GOTO 10
15  100      WRITE(6,*)COUNT,'STUDENTS HAVE GPA OVER 3.0'
16           CLOSE (2,3)
17           STOP
18           END
```

Analysis

The preceding program reads an existing file called STUDNT and writes data about those students with GPA greater than 3.0 into a new file called GPA3. The program also counts the number of student records written into file GPA3. Let's review some of the program details.

At line 5 the output file, GPA3, is opened and associated with unit 3. Similarly, line 6 opens the input file STUDNT and associates it with unit 2. At line 7 the file at unit 2, the input file, is positioned to the first record in the file in preparation for reading at line 9. Note that line 9 has reference number 10 to which control returns every time line 14 is executed. Each time line 9 is executed a new record consisting of 3 data elements is read. This process continues until the end of file STUDNT is reached, at which time program control transfers to line 15, which has reference number 100.

At line 10 the value of GPA from the latest record read from file STUDNT is tested to determine whether or not it is greater than 3. If it is, line 11 is executed; this writes the entire record of which the tested GPA is a part onto unit 3, with which the file GPA3 was associated because of line 5. Whenever this writing onto unit 3 occurs, it is immediately followed by the execution of line 12, causing the value of COUNT to be increased by 1 (it has an initial value of zero because of line 8). Thus when program control ultimately passes to statement 100 (at line 15) variable COUNT will contain a count of all student records with GPA greater than 3.

Line 16 causes the proper closing of both files, the one on unit 2 and the one on unit 3.

The preceding program results in the creation of a permanent data file called GPA3. Thus if the program is executed a second time on the same computer system, line 5 will result in a fatal error message because GPA3 is opened as a 'NEW' file. Once this program has been run, GPA3 cannot be opened as a 'NEW' file, only as an 'OLD' file.

Now we are ready to present the three problems and their solutions that were mentioned earlier.

PROBLEMS WITH COMPLETE SOLUTIONS

Problem 1

Assume there exists a data file named BOOKS that contains two pieces of information about every book in the public library: (1) a seven-digit unique book number, and (2) a code indicating the current status of the book, where

0 means book is ready for use
1 means book is checked out
2 means book is lost
3 means book is out for repair

Assume that file BOOKS is sorted in ascending order of book number. The task is to develop an algorithm and a program that accepts as input a book number and produces as output an appropriate message indicating the status of the book. If the book is not in the library, a message should so indicate. The program is to be developed for an interactive computer system and should, therefore, communicate with the user.

First-level steps in a solution algorithm might be as follows:

1. Remind user to enter the number of the book desired, then input that number.
2. By checking the permanent file called BOOKS for the specified number and associated status code, determine appropriate message to be output.
3. End.

Second-level steps in the algorithm might be the following:

1.0. Output message reminding user to input book number.
1.1. Input book number of book wanted, NUM.
2.0. Read the file BOOKS until you find the number of the desired book or until all data in the file have been read.
2.1. If the desired number is found, prepare appropriate output based on the code of the book.
2.2. If the desired number is not found, report that the book is not in the file BOOKS.
3.0. End.

Here is a ForTran program developed from the preceding algorithm:

```
        PROGRAM BOOK
C*******************************************************
C*          PROGRAM DEFINITION                         *
C*              THIS PROGRAM READS A BOOK NUMBER        *
C*              THEN IT READS THE FILE "BOOKS" UNTIL    *
C*              IT FINDS THE BOOK NUMBER JUST READ.     *
C*              BASED ON THE CODE NUMBER ASSOCIATED     *
C*              WITH THIS BOOK, APPROPRIATE MESSAGE     *
C*              IS PRINTED.                             *
C*******************************************************
C*          VARIABLE DEFINITION                         *
C*              CODE HAS ONE OF THE FOLLOWING VALUES:   *
C*              0 = THE BOOK IS READY                   *
C*              1 = THE BOOK IS CHECKED OUT             *
C*              2 = THE BOOK HAS BEEN LOST              *
C*              3 = THE BOOK IS OUT FOR REPAIR          *
C*                                                      *
C*              NUM = NUMBER OF THE BOOK REQUESTED      *
C*              INUM = NUMBER OF THE BOOK READ FROM     *
C*              FILE "BOOKS"                            *
C*******************************************************
```

```
C*
      INTEGER CODE, NUM, INUM
        OPEN(1,FILE='BOOKS',STATUS='OLD')
        REWIND(1)
        PRINT*,' ENTER YOUR BOOK NUMBER.'
C*
      READ*,NUM
 10     READ(1,*,END=100)INUM,CODE
        IF(INUM .LT. NUM) GO TO 10
        IF(INUM .EQ. NUM) THEN
           IF(CODE .EQ. 0) THEN
              PRINT*,' THE BOOK IS READY.'
           ELSEIF(CODE .EQ. 1) THEN
              PRINT*,' THE BOOK IS CHECKED OUT.'
           ELSEIF(CODE .EQ. 2) THEN
              PRINT*,' THE BOOK HAS BEEN LOST.'
           ELSEIF(CODE .EQ. 3)THEN
              PRINT*,' THE BOOK IS NOT AVAILABLE FOR CHECK OUT.
           ELSE
              PRINT*,' ERROR IN CODE ',CODE
           ENDIF
        ELSE
           PRINT*, ' BOOK NUMBER ', NUM, ' NOT IN LIBRARY.'
        ENDIF
        CLOSE(1)
        STOP
100     PRINT*, ' BOOK NUMBER ', NUM, ' NOT IN LIBRARY.'
        CLOSE(1)
        STOP
        END
```

In order to demonstrate some sample runs of this program we assume the contents of file BOOKS to be as follows:

Book number	Status code
1234567	0
1234568	1
1234569	2
1234570	0
1234571	0
1234572	1
1234573	3

Following is a sample run of the program.

```
ENTER YOUR BOOK NUMBER.
1234567
THE BOOK IS READY.

RUN COMPLETE.
```

The first line of output is the result of the first PRINT statement in the program. The second line is not part of the output but is rather the input needed by the first READ statement in the program, READ*,NUM. This is the book number of the book requested by the user, and is entered as input. Based on checks made by the program, the third line is printed by the program. The last line, RUN COMPLETE, is *not* produced by the user's program but by the software of the computer system being used, and is, therefore, likely to vary somewhat from system to system.

Three more executions of the program appear below:

```
ENTER YOUR BOOK NUMBER.
1234569
THE BOOK IS NOT AVAILABLE FOR CHECK OUT.

RUN COMPLETE.
```

```
ENTER YOUR BOOK NUMBER.
1234566
BOOK NUMBER              1234566 NOT IN LIBRARY.

RUN COMPLETE.
```

```
ENTER YOUR BOOK NUMBER.
1234568
THE BOOK HAS BEEN CHECKED OUT.

RUN COMPLETE.
```

Problem 2

As a part of the bank-statement reconciliation process, company management decides to have its data processing staff automate certain aspects of the process. Here are the steps management wants done:

Phase 1:

1. For every check issued by the company, prepare a punched card containing the check number and the amount of the check.
2. Sort data in ascending order of check number.
3. Produce a printed list of the cards prepared at step 2.
4. Produce a data file to be stored electronically in the computer's auxiliary storage from the sorted data in step 2.

Phase 2:

1. From the bank statement received at the end of the month prepare a punched card for each check that contains the check number and the amount paid on that check by the bank.
2. Sort data in ascending order of check number.
3. Determine which paid checks showed different amounts paid than amounts issued. For any checks where these amounts differ, output the check number, the amount for which the check was written, and the amount paid for the check by the bank.
4. Produce a list of those checks that are outstanding.
5. Report the total amount of difference between issued amounts and paid amounts.
6. Report the total amount of outstanding checks.

Solution. We must know whether the cards being processed are for phase 1 or phase 2, so we decide to precede all check data cards with a card that has punched in it either 1 or 2 in column 1. Such a card is often called a header card. In order to stop the repetitive process of reading data cards, we agree that a sentinel card containing 0 0 will follow the last check-data card whether it be for phase 1 or phase 2.

Because the problem consists of two distinct phases we divide the solution into two parts which we call Procedure PHASE1 and Procedure PHASE2.

Procedure PHASE1:

1.0 Input check number and amount.
2.0 If the input is the sentinel data, then stop processing.
2.1 Output the check number and amount on unit 1 as well as on the printer terminal.
2.2 Repeat the process from step 1.0.
3.0 End.

This procedure is a very simple one that inputs all data and outputs them onto a file and on paper. Following is a ForTran program segment that carries out the steps in the above procedure.

```
      READ *,NUMWR,AMTWR
100 IF(NUMWR .GT. 0) THEN
          WRITE(1,*) NUMWR,AMTWR
          WRITE(6,*) NUMWR,AMTWR
          READ*,NUMWR,AMTWR
      GO TO 100
      ENDIF
      CLOSE(1)
      STOP
```

In the preceding program segment the first WRITE statement places a copy of each check number and amount on a storage device identified by "1" in the parentheses following WRITE. We shall refer to this newly created set of data as file 1 in a subsequent discussion.

The second phase of the problem requires that we input data including check numbers and corresponding amounts for checks that have cleared the bank. We shall call these "paid checks." We have assumed in this problem that both the data in file 1 and the second-phase data called paid checks are sorted in ascending order of check number. The procedure we are about to present does not check for a violation of this assumption. Consequently, with errors in input data, the performance of the program is unpredictable. As an exercise, you might revise the procedure to include some checking for faulty input data and provide for an appropriate message. Such a feature of a procedure or a program is called an *error-recovery feature*.

Procedure PHASE2:

1.0 Input paid-check data (NUMPD and AMTPD) from card.
2.0 If no more paid-check data exist, go to step 4.0.
3.0 Input written-check data (NUMWR and AMTWR) from unit 1 until ...
    (a) the number of the check paid (NUMPD) is the same as the number of the check written (NUMWR). When a match occurs, if the amount paid (AMTPD) is not the same as the amount written (AMTWR), then report NUMPD, AMTPD, and AMTWR, update the total amount of difference between written and paid amounts, read the next paid-check, and proceed to step 2.0.

or . . .

(b) the check is determined to be outstanding, at which time report the check and update the total amount outstanding and proceed to step 2.0.

4.0 Report all remaining written-check-data from unit 1 as outstanding and add the amounts of these checks to TOTOUT.

4.1 Report TOTOUT.

5.0 End.

```
      PROGRAM CHECKS
C*******************************************************************
C*          PROGRAM DEFINITION                                    *
C*              SEE THE TEXT FOR COMPLETE DESCRIPTION OF THE       *
C*              PROBLEM.                                           *
C*******************************************************************
C*          VARIABLE DEFINITION                                   *
C*              PHASE=1 OR 2 TO INDICATE PHASE 1 OR PHASE 2.       *
C*              NUMWR = THE CHECK NUMBER FOR A WRITTEN CHECK.      *
C*              NUMPD = THE CHECK NUMBER FOR A PAID CHECK.         *
C*              AMTWR = AMOUNT OF A WRITTEN CHECK.                 *
C*              AMTPD = AMOUNT OF A PAID CHECK.                    *
C*              TOTOUT = TOTAL OF AMOUNTS OF CHECKS OUTSTANDING.*
C*              TDIFF  = TOTAL DIFFERENCES BETWEEN WRITTEN         *
C*                       AMOUNT AND PAID AMOUNT OF ALL THE         *
C*                       CHECKS.                                   *
C*******************************************************************
      REAL AMTPD,AMTWR,TOTOUT, TDIFF
      INTEGER NUMWR,NUMPD,PHASE
C*
C*          INITIALIZATION
C*
      TOTOUT = 0
      TDIFF = 0
      READ*, PHASE
C*
      IF(PHASE .EQ. 2) GOTO 200
C*
C*******************************************************************
C*          ****PHASE 1****                                       *
C*              THIS PHASE IS TO MAKE FILE1 FROM THE LIST OF       *
C*              THE WRITTEN CHECKS.                                *
C*******************************************************************
```

```
C*
          OPEN(1,FILE='FILE1',STATUS='NEW')
          REWIND 1
          READ*,NUMWR,AMTWR
100       IF(NUMWR .EQ.0) GOTO 105
              PRINT*,NUMWR,AMTWR
              WRITE(1,*)NUMWR,AMTWR
              READ*,NUMWR,AMTWR
          GO TO 100
 105      CLOSE(1)
          STOP
C***********************************************************************
C*        ****PHASE 2****                                             *
C*              THIS PHASE USES THE DATA FILE 'FILE1'                 *
C*              CREATED IN PHASE ONE.  IT ALSO READS DATA             *
C*              CARDS ASSOCIATED WITH CHECKS PAID BY THE BANDK.       *
C*              THE DATA ON THE CARDS ARE COMPARED WITH THOSE         *
C*              ON DATA FILE 'FILE1' AND APPROPRIATE                  *
C*              PRINTED OUTPUT IS PRODUCED.                           *
C***********************************************************************
 200      OPEN(1,FILE='FILE1',STATUS='OLD')
          REWIND 1
          READ*,NUMPD,AMTPD
 300      IF(NUMPD .EQ. 0) GO TO 1000
          READ(1,*,END=400)NUMWR,AMTWR
C***********************************************************************
C*              IF THE END-OF-FILE MARK IS ENCOUNTERED               *
C*              ON FILE 'FILE1' BEFORE A MATCH IS FOUND              *
C*              BETWEEN NUMPD AND NUMWR, THERE IS AN                 *
C*              ERROR IN THE DATA FOR NUMPD, AND THE                 *
C*              PROGRAM TERMINATES.                                  *
C***********************************************************************
          IF (NUMPD .EQ. NUMWR) THEN
             IF (AMTPD .NE. AMTWR) THEN
               TDIFF = TDIFF+AMTWR-AMTPD
               PRINT*,NUMPD,AMTWR,AMTPD
             ENDIF
             READ*,NUMPD,AMTPD
          ELSEIF(NUMPD .GT. NUMWR)THEN
             TOTOUT = TOTOUT + AMTWR
             PRINT*,NUMWR,AMTWR,'OUTSTANDING'
          ELSE
             PRINT*, NUMPD,AMTPD, ' ERROR IN INPUT'
             CLOSE(1)
             STOP
          ENDIF
          GOTO 300
```

```
C*
 400      PRINT*, ' ERROR IN FILE1 OR TRANSACTION.'
          PRINT*, ' FILE1 ENDED BEFORE TRANSACTIONS DID.'
          CLOSE(1)
          STOP
C**************************************************************
C*                AT THIS POINT ALL PAID CHECKS ARE PROCESSED.      *
C*                THE REMAINING OUTSTANDING CHECKS, IF ANY,         *
C*                ARE NOW PRINTED OUT.                              *
C**************************************************************
 1000     READ(1,*,END=2000)NUMWR,AMTWR
            TOTOUT = TOTOUT + AMTWR
            PRINT*,NUMWR,AMTWR,'OUTSTANDING'
          GO TO 1000
 2000     PRINT*,' TOTAL OUTSTANDING CHECKS ', TOTOUT
          PRINT*, ' TOTAL DIFFERENCE PAID    ',TDIFF
          CLOSE(1)
          STOP
          END
```

Sample run. Now we show a run of the program just described where the input data cards were

1		*PHASE is 1, to indicate creation of FILE1.*
12	23.45	
15	34.56	
18	123.45	
20	134.56	
25	234.78	
27	345.67	
39	2345.00	
40	345.70	
45	23.45	
46	567.00	
47	564.00	
0	0	

The printed output produced is as follows:

12	23.45
15	34.56
18	123.45
20	134.56
25	234.78
27	345.67

```
39    2345.00
40     345.70
45      23.45
46     567.00
47     564.00
```

A second run of the program was done with these input data:

```
 2                              PHASE is 2, to indicate a Phase 2 run only.
12      23.45
15      34.56
27   34567.00
47    1564.00
 0       0
```

The output produced is as follows:

```
18      123.450    OUTSTANDING
20      134.560    OUTSTANDING
25      234.780    OUTSTANDING
27      345.670      34567.0
39      2345.00    OUTSTANDING
40      345.700    OUTSTANDING
45      23.4500    OUTSTANDING
46      567.800    OUTSTANDING
47      564.000      1564.00
TOTAL OUTSTANDING CHECKS       3773.74
TOTAL DIFFERENCE PAID         -35221.3
```

Problem 3

The City Utilities Department wants to computerize its billing process. To begin with, only its residential customers will be included and only the electricity and water billing will be computerized. The Utilities Department provides its data processing staff the following information: (1) the number of customers it has, (2) the base electricity rate per kilowatt hour (KWH), and (3) the base water rate per 100 cubic feet. The staff uses the following rate schedule for the two utilities:

Electricity	Water
First 500 KWH = base rate	First 50 100-cu.-ft. = base rate
Next 1500 KWH = 90% of base rate	Next 150 100-cu.-ft. = 80% of base rate
Over 2000 KWH = 85% of base rate	Over 200 100-cu.-ft. = 70% of base rate
Minimum billing of $5.00	Minimum billing of $2.00

The Utilities Department produces for each customer a computer card on which are punched: (1) the customer number, (2) the number of KWH's used this billing period, and (3) the total number of 100-cu.-ft. of water used this period, appearing on each card in that order from left to right.

For each customer the Utilities Department wants a bill as follows:

```
CUSTOMER NUMBER    XXXXXX
ELECTRICITY USED   XXX.XX    CHARGE   XXX.XX
WATER USED         XXX.XX    CHARGE   XXX.XX
TOTAL MONTHLY CHARGE                  XXX.XX
```

It also wants a Summary Report after printing all customer bills to appear as follows:

```
          SUMMARY REPORT
TOTAL ELECTRICITY USED (KWH)        XXXXXX.XX
TOTAL WATER USED (100-cu.-ft.)      XXXXXX.XX
TOTAL ELECTRICITY CHARGE            XXXXXX.XX
TOTAL WATER CHARGE                  XXXXXX.XX
TOTAL MONTHLY CHARGE                XXXXXX.XX
```

Algorithm: Here is an algorithm for solving this problem:

1. Input the number of customers, N; the base electricity rate per KWH, ELCTR; and the base water rate per 100-cu.-ft., WATR.
2. Compute the two reduced electricity rates (ELCTR2 and ELCTR3) and the two reduced water rates (WATR2 and WATR3).
3. While the number, N, of customers remaining is greater than zero, execute repeatedly steps 4 through 12. When all the customers are processed, then output the final report and stop all processing.
4. Input customer number, electricity used, and water used.
5. Decrease by 1 the number of customers remaining to be processed.
6. Separate the number of KWH of electricity used into three rate divisions (one or more may have a value of 0).
7. Separate the number of 100-cu.-ft. of water used into three rate divisions (one or more may have a value of 0).
8. Compute the electricity charge.
9. Compute the water charge.
10. Check that minimum charges for both are met.
11. Output the customer billing.
12. Update the total amounts of utilities used and total charges.

Program

```
      PROGRAM UTILIT
C*************************************************************
C*            PROGRAM DEFINITION                            *
C*                GIVEN APPROPRIATE INPUT DATA, THIS PROGRAM *
C*                COMPUTES AND PRINTS CUSTOMER UTILITIES     *
C*                BILLINGS FOR ELECTRICITY AND WATER ONLY.   *
C*                IT ALSO PRODUCES A SUMMARY REPORT.         *
C*************************************************************
C*            INPUT DATA DESCRIPTION                         *
C*                CARD 1:                                    *
C*                    TOTAL NUMBER OF CUSTOMERS, N           *
C*                    ELECTRICITY RATE PER KWH, ELCTR        *
C*                    WATER RATE PER 100 CU. FT., WATR       *
C*                CARD 2 AND FOLLOWING N-1 CARDS:            *
C*                    CUSTOMER NUMBER, NUM                   *
C*                    KWH'S OF ELECTRICITY USED, ELCT        *
C*                    100-CU-FT OF WATER USED, WATER         *
C*************************************************************
C*            VARIABLE DEFINITIONS                           *
C*                ELCTC   ELECTRICITY CHARGE FOR 1 CUSTOMER  *
C*                ELCTR   BASE ELECTRICITY RATE PER KWH      *
C*                ELCTR2  SECOND REDUCED RATE                *
C*                ELCTR3  THIRD REDUCED RATE                 *
C*                ELCT    KWH'S OF ELECTRICITY USED BY 1 CUSTOMER*
C*                ELCT1   FIRST SEGMENT OF ELCT              *
C*                ELCT2   SECOND SEGMENT OF ELCT             *
C*                ELCT3   THIRD SEGMENT OF ELCT              *
C*                TELCT   TOTAL KWH'S OF ELECTRICITY USED    *
C*                TELCTC  TOTAL CHARGED FOR ALL ELECTRICITY  *
C*                TWATER  TOTAL NUMBER OF 100-CU-FT WATER USED *
C*                TWATRC  TOTAL WATER CHARGE                 *
C*                WATERC  WATER CHARGE FOR 1 CUSTOMER        *
C*                WATR    BASE WATER RATE PER 100-CU-FT      *
C*                WATR2   SECOND REDUCED RATE                *
C*                WATR3   THIRD REDUCED RATE                 *
C*                WATER   TOTAL WATER USED BY 1 CUSTOMER     *
C*                WATER1  FIRST SEGMENT OF WATER             *
C*                WATER2  SECOND SEGMENT OF WATER            *
C*                WATER3  THIRD SEGMENT OF WATER             *
C*                N       TOTAL NUMBER OF CUSTOMERS          *
C*                NUM     CUSTOMER NUMBER                    *
C*************************************************************
```

```
C*
        REAL ELCTC,ELCTR,ELCT,TELCT,TELCTC,TWATER,TWATRC,WATR
        REAL ELCTR2, ELCTR3, ELCT1, ELCT2,ELCT3, WATER, WATERC
        REAL WATER1, WATER2, WATER3, WATR2, WATR3
        INTEGER N,NUM
C*
        TELCT = 0
        TWATER = 0
        TELCTC = 0
        TWATRC = 0
C*****************************************************
C*              READ THE RATES AND BREAK THEM        *
C*              INTO REDUCED RATES.                  *
C*****************************************************
C*
        READ*,N,ELCTR,WATR
        ELCTR2=.90*ELCTR
        ELCTR3=.85*ELCTR
        WATR2=.80*WATR
        WATR3=.70*WATR
C*
C*****************************************************
C*              READ CUSTOMER DATA AND COMPUTE       *
C*              THE CHARGE FOR EACH CUSTOMER.        *
C*****************************************************
C*
  10    IF(N .LE. 0)GOTO 100
        READ*,NUM,ELCT,WATER
        N=N-1
C*
        IF(ELCT.LE.500)THEN
           ELCT1=ELCT
           ELCT2=0
           ELCT3=0
        ELSEIF(ELCT.LE.2000)THEN
           ELCT1=500
           ELCT2=ELCT-500
           ELCT3=0
        ELSE
           ELCT1=500
           ELCT2=1500
           ELCT3=ELCT-2000
        ENDIF
```

```
C*
        IF(WATER.LE.50)THEN
            WATER1=WATER
            WATER2=0
            WATER3=0
        ELSEIF(WATER.LE.200)THEN
            WATER1=50
            WATER2=WATER-50
            WATER3=0
        ELSE
            WATER1=50
            WATER2=150
            WATER3=WATER-200
        ENDIF
C*
        ELCTC = ELCT1*ELCTR+ELCT2*ELCTR2+ELCT3*ELCTR3
        WATERC = WATER1*WATR+WATER2*WATR2+WATER3*WATR3
C*
C********************************************************
C*              ENFORCE THE MINIMUM CHARGE.          *
C********************************************************
        IF(WATERC .LT. 2)WATERC = 2
        IF(ELCTC .LT. 5)ELCTC = 5
C********************************************************
C*              REPORT CUSTOMER DATA                 *
C********************************************************
        PRINT*,' CUSTOMER NUMBER...',NUM
        PRINT*
        PRINT*, ' ELECTRICITY USED ',ELCT, ' CHARGE ',ELCTC
        PRINT*,' WATER USED      ',WATER,'CHARGE',WATERC
        PRINT*,' TOTAL MONTHLY CHARGE','     ',ELCTC+WATERC
        PRINT*
C********************************************************
C*              UPDATE TOTALS FOR SUMMARY REPORT.*
C********************************************************
C*
        TELCT=TELCT+ELCT
        TELCTC=TELCTC+ELCTC
        TWATER=TWATER+WATER
        TWATRC=TWATRC+WATERC
      GOTO 10
C********************************************************
C*              PRINT THE SUMMARY REPORT             *
C********************************************************
```

```
C*
  100     PRINT*
          PRINT*,'          SUMMARY REPORT'
          PRINT*,'          ------- ------'
          PRINT*,' TOTAL ELECTRICITY USED    ',TELCT
          PRINT*,' TOTAL WATER USED          ',TWATER
          PRINT*,' TOTAL ELECTRICITY CHARGE ',TELCTC
          PRINT*,' TOTAL WATER CHARGE        ',TWATRC
          PRINT*
          PRINT*
          PRINT*,' TOTAL MONTHLY CHARGE       ',TWATRC+TELCTC
          STOP
          END
```

Sample run.   Finally, we show a sample run of the previous program using the following input data.

```
card 1: 3 .04 .50
card 2: 125 20  0
card 3: 126 5678 324
card 4: 127 6789 350

****PRGRAM OUTPUT FOLLOWS****

    CUSTOMER NUMBER...  125

    ELECTRICITY USED    20.0000    CHARGE     5.00000
    WATER USED            0.   CHARGE    2.00000
    TOTAL MONTHLY CHARGE          7.00000


    CUSTOMER NUMBER...  126

    ELECTRICITY USED    5678.00    CHARGE     199.052
    WATER USED          324.000  CHARGE   128.400
    TOTAL MONTHLY CHARGE         327.452


    CUSTOMER NUMBER...  127

    ELECTRICITY USED    6789.00    CHARGE     236.826
    WATER USED          350.000  CHARGE   137.500
    TOTAL MONTHLY CHARGE         374.326
```

```
          SUMMARY  REPORT
          -------  ------
    TOTAL  ELECTRICITY  USED        12487.0
    TOTAL  WATER  USED              674.000
    TOTAL  ELECTRICITY  CHARGE      440.878
    TOTAL  WATER  CHARGE            267.900


    TOTAL  MONTHLY  CHARGE          708.778
```

SUMMARY

In this chapter we have discussed what it means to have a program that is correct and good. Not only does this mean that the program must produce a correct solution to the problem but it must also be readable and understandable. Only if the reader of a computer program can understand what the program is accomplishing can that program be considered a good one. Such understandability is often enhanced by structuring the program into indented sections with each group of like-indented statements accomplishing a subtask of the whole. Examples were discussed in detail so as to provide the reader with concrete evidence of the techniques proposed.

Also in this chapter we elaborated on the technique of solving problems by developing increasingly detailed levels of action steps. The first level is simply a precise statement of the problem. The third level is a sequence of steps detailed enough that almost each step corresponds to a statement in the ForTran program that ultimately produces the desired solution. Of course, the third-level steps are somewhat determined by the knowledge that the resulting computer program will be written in ForTran. If a different programming language were used, third-level steps might be different from what have been given in this chapter.

Note further that the problem-solving process involves a considerable amount of analysis even before the solution steps are developed. When approaching a problem-solving situation, be certain that all information related to the problem is displayed in a form as useful as possible. In our case we constructed tables, graphs, and equations to help us determine the approach we would take in producing the desired solution.

As the computer program is being written, comments should be liberally interspersed with the action statements. This tactic adds to the understandability of the finished program especially when the program is being read at some time after it was completed.

As an aid to developing the solution to a problem we discussed these steps:

1. State the problem precisely.
2. Divide the problem into segments.
3. Develop each segment into detailed steps.
4. Write computer program statements corresponding to the detailed steps.

An algorithm was defined as a set of rules for solving a specific problem, and its properties were given as follows:

1. It terminates after a finite number of steps.
2. The actions specified in each step must be precisely and unambiguously defined.
3. It has a finite number of inputs.
4. It has one or more outputs.
5. All steps must be effective.
6. It is not necessarily unique.

We discussed a *good* algorithm as being one that

1. Requires relatively little time to be executed.
2. Requires as little computer memory as possible.
3. Is structured.

A *structured* algorithm has these properties:

1. Its physical layout makes it visually apparent that certain steps belong in a given group.
2. The flow of action proceeds generally from beginning to end.
3. It is divided into cohesive modules.

We presented three statements needed to process data files on magnetic media such as tape and disks. These statements were

```
OPEN(u,FILE='filename',STATUS='NEW')
CLOSE(u1,u2, . . . ,uk)
REWIND(u)
```

The OPEN statement prepares a data file for access by the program while the REWIND statement positions the file so that its first record is ready for input or output. The CLOSE statement properly disconnects the file.

EXERCISES

1. Answer the following questions related to algorithms and programs:
    (a) What other word is used synonymously with "algorithm"?
    (b) Give the computer science definition of "algorithm."
    (c) Distinguish between "theoretically finite" and "feasibly finite."
    (d) What is meant by a well-defined process?
    (e) What is meant by the "top-down" approach to algorithm development?
    (f) What is the role of uniqueness as related to algorithms?
    (g) What is meant by pseudocode? Where in the development of algorithms and computer programs does it fall?
    (h) What is the most important characteristic of a good program?
    (i) How can the validity of an algorithm or computer program be assured?
    (j) List and briefly explain three characteristics that a program should have if its quality is good.
    (k) What is meant by "debugging"? Identify three techniques for debugging a program.
    (l) What is meant by "profiling" a program?
    (m) What does it mean to say that an algorithm or program has "structure"?
    (n) List seven techniques for achieving a structured program.
2. In each of the following subexercises, a problem is stated and an algorithm is given. It is your task to determine which algorithms are correct and which are not. Where you find an incorrect algorithm, modify or rewrite it so as to correctly solve the problem if possible.
    (a) *Problem:* Given any two numbers, identify the larger one.
        Algorithm:
        1. Input the two numbers, X and Y.
        2. Compute their difference, D=X−Y.
        3. If D> 1 output "THE FIRST NUMBER IS LARGER." Otherwise output "THE SECOND NUMBER IS LARGER."
    (b) *Problem:* Given seven objects that appear identical although one of the objects weighs slightly less than the others, using a balance scale no more than twice, identify the object that weighs less.
        Algorithm:
        1. Put three of the objects on each side of the balance scale.
        2. If the scale balances, the object not on the scale is the lighter one. If not, repeat step 1 using the objects from the lighter side of the scale, placing one object on each side of the scale.

(c) *Problem:* Given any two positive integers, find their modular quotient.

(*Note:* The modular quotient of two integers is the remainder upon dividing the larger by the smaller. For example, the modular quotient of 3 and 7 is 1.)

Algorithm:

1. Input the two integers I and J.
2. Find the positive difference of I and J and call it K.
3. If K is less than either I or J, then output K and stop.
4. Replace the larger of I and J with K.
5. Repeat the process from step 2.

(d) *Problem:* We know that if a year designation is divisible by 4 it is a leap year and that February will have 29 days. For example, 1984 is divisible by 4 so it is a leap year. Given any year and the number of any day in that year (starting with 1 for January 1), identify the month and the day of the month for that day.

Algorithm:

1. Input YEAR and DAY.
2. Set $(NAME)_2$=FEB.
3. If YEAR is divisible by 4 set $(MONTH)_2$=29.
   Otherwise set $(MONTH)_2$=28.
4. Set the following values:

$(MONTH)_1$=31	$(NAME)_1$=JAN
$(MONTH)_3$=31	$(NAME)_3$=MAR
$(MONTH)_4$=30	$(NAME)_4$=APR
$(MONTH)_5$=31	$(NAME)_5$=MAY
$(MONTH)_6$=30	$(NAME)_6$=JUN
$(MONTH)_7$=31	$(NAME)_7$=JUL
$(MONTH)_8$=31	$(NAME)_8$=AUG
$(MONTH)_9$=30	$(NAME)_9$=SEP
$(MONTH)_{10}$=31	$(NAME)_{10}$=OCT
$(MONTH)_{11}$=30	$(NAME)_{11}$=NOV
$(MONTH)_{12}$=31	$(NAME)_{12}$=DEC

5. Do step 6 for i=1, 2, 3, . . . ,until DIFF is negative.
6. DIFF=DAY$-((MONTH)_1+(MONTH)_2+ \cdots +(MONTH)_i)$.
7. Compute D=DIFF+$(MONTH)_i$.
8. Output "THE GIVEN DAY IS DAY", D, "OF", $(NAME)_i$.

(e) *Problem:* Given a positive integer N, compute N factorial.
   Algorithm:
   1. Input N.
   2. Set 1 factorial=1.
   3. Compute K factorial = $K*(K-1)$ factorial for $K=1, 2, 3, \ldots, N$.
   4. Output N factorial.

(f) *Problem:* Given the number N, of different objects, and the integer K, compute the number of possible combinations of N objects taken K at the time.
   Algorithm:
   1. Input N and K.
   2. If N is less than K, output an error message, then stop.
   3. Set COMB(I,1)=I and COMB(I,I)=1 for any integer I.
   4. For $I=1, 2, 3, \ldots, N$ and $J=1, 2, 3, \ldots, K$,compute
      COMB(I,J)=COMB($I-1$,J)+COMB($I-1$,$J-1$).
   5. Output COMB(N,K).

(g) *Problem:* Find the smallest positive integer, N, for which N^2 is less then $2^{(N-2)}$.
   Algorithm:
   1. Set $N=1$.
   2. Compute $X=N*N$.
   3. Compute $Y=2^{(N-2)}$.
   4. If $X < Y$ then output N and stop the process.
   5. Increase N by 1.
   6. Repeat the process from step 2.

3. Write an algorithm for each of the following problems:
   (a) Given three numbers A, B, and C, find the solution of the quadratic equation $AX^2+BX+C=0$.
   (b) Given three numbers A, B, and C, find the fourth number X such that $A/B = C/X$.
   (c) A Pythagorean triple is a set of three positive integers A, B, and C such that $A^2=B^2+C^2$ (for example, $5^2=4^2+3^2$). Find the first twenty-five Pythagorean triples.
   (d) Two prime numbers are called twin primes if their difference is 2 (for example, 3 and 5). Find the first ten twin primes.
   (e) Given a number in its standard decimal notation (for example, 3765.23), determine the exponential notation for that number. (For example, 3765.23=3.76523E8.)
   (f) Simulate the game of tic-tac-toe.

4. Study each of the following programs and indicate the output.

(a)
```
READ*,X,Y
IF (X .GT. Y) THEN
  A=X
  X=Y
  Y=A
ENDIF
PRINT*,X,Y
END
```

Assume input values are 5. and 10., respectively.

(b)
```
READ*,X,Y
IF(X.LE.Y)PRINT*,X,Y
IF(X.GT.Y)PRINT*,Y,X
END
```

Assume input values are 5. and 10., respectively.

(c)
```
READ*,X,Y
IF(X.LE.Y) THEN
PRINT*,X,Y
ELSE
PRINT*,Y,X
ENDIF
END
```

Assume input values are 5. and 10., respectively.

Note that all three preceding programs produce identical output. In each case, two numbers are input and the output consists of those same two numbers printed in ascending order. Which of the three programs is easiest to understand? Which one is likely to run in the least amount of time? Which one do you like the best? Why?

The following exercises are intended to provide further practice in developing the skill of writing an algorithm (a procedure for accomplishing a task). A field of application is indicated for each exercise. We suggest that you do each exercise by (a) listing the main steps that must be accomplished to perform the specified task, (b) develop substeps as needed for each of the main steps, and (c) write ForTran statements corresponding to the substeps so that results can be obtained by computer.

5. (Computer Science)
You are given a list of the first forty United States presidents and

their ages at inauguration. Input the ages and determine the following:

(a) The age of the youngest president.
(b) The age of the oldest president.
(c) The number of presidents older than 50.
(d) The number of presidents older than 65.
(e) The number of presidents younger than 50.
(f) The average age of United States presidents at inauguration.

Number	President	Age at inauguration	Number	President	Age at inauguration
1	G. Washington	57	21	C. A. Arthur	50
2	J. Adams	61	22	G. Cleveland	47
3	T. Jefferson	57	23	B. Harrison	55
4	J. Madison	57	24	G. Cleveland	55
5	J. Monroe	58	25	W. McKinley	54
6	J. Q. Adams	57	26	T. Roosevelt	42
7	A. Jackson	61	27	W. H. Taft	51
8	M. VanBuren	54	28	W. Wilson	56
9	W. H. Harrison	68	29	W. G. Harding	55
10	J. Tyler	51	30	C. Coolidge	51
11	J. K. Polk	49	31	H. C. Hoover	54
12	Z. Taylor	64	32	F. D. Roosevelt	51
13	M. Fillmore	50	33	H. S. Truman	60
14	F. Pierce	48	34	D. D. Eisenhower	62
15	J. Buchanan	65	35	J. F. Kennedy	43
16	A. Lincoln	52	36	L. B. Johnson	55
17	A. Johnson	56	37	R. M. Nixon	56
18	U. S. Grant	46	38	G. R. Ford	61
19	R. B. Hayes	54	39	J. E. Carter	52
20	J. A. Garfield	49	40	R. Reagan	69

6. (Computer Science and Statistics)
   You are given a set of numbers ending with the sentinel number 999.999. Input the numbers, determine the sum of the absolute values of the numbers, and find their range (the difference between the largest number and the smallest one).

7. (Computer Science and Statistics)
   You have a set of numbers ending with the sentinel number −999.99. Input the set only once and, excluding the sentinel data, find
   (a) The number, N, of numbers in the set.
   (b) The number of negative numbers in the set.

(c) The sum of the negative numbers.

(d) The number of positive numbers in the set.

(e) The sum of the positive numbers.

(f) The arithmetic mean $\overline{X}$ (average) of all the numbers.

(g) The variance of the numbers where

$$\text{Variance} = \left(\sum_{i=1}^{N} X_i^2 - N*\overline{X}^2 \right) / N$$

8. (Mathematics)

Input the coefficients, A, B, and C of the quadratic equation
    $AX^2+BX+C=0$
and determine the roots as follows:

(a) If $A=0$, $X=-C/B$.

(b) If $A \neq 0$, compute $D=B^2-4AC$. Then compute X as follows: if $D>0$, then
    $$X1=(-B+\sqrt{D}\,)/(2*A)$$
    and
    $$X2=(-B-\sqrt{D}\,)/(2*A).$$
    If $D=0$, $X1=X2=(-B)/(2A)$.
    If $D<0$, there are no real roots.

9. (Computer Science and Statistics)

For each of an unknown number of students the input data are student ID number and six integers where each integer is the grade for a course, with 4=A, 3=B, 2=C, 1=D, and 0=F. The sentinel data are a student ID number of 999999 and 6 grades of 0. An example of one set of input data is

        821345   4 3 3 4 4 3

Note that the first datum is the student's ID number. The six digits that follow represent grades for six courses. In this example the student's grades are A, B, B, A, A, and B (4, 3, 3, 4, 4, 3). For each student, output the student ID number and the average of the six grades for that student.

10. (Statistics and Computer Science)

Input a single positive integer, N, and

(a) For $N<15$, compute N factorial = $N!$ = $N*(N-1)*(N-2) *\ldots*2*1$.

(b) For $20 \leq N < 100$, compute the sum of the squares of the integers $1, 2, 3, \ldots, N$.

(c) For $100 \leq N < 1000$, compute the sum of the odd integers 1, 3, 5, $\ldots, N$.

(d) For $N \geq 1000$, compute the square root of N; then stop.

11. (Mathematics)

For positive integers, X and Y, find the maximum value of the function $2X^2+3XY$ when X and Y are such that their sum is at least 2 but does not exceed 10. The output is the maximum value for the function and the integer values of X and Y that produce that maximum.

12. (Economics and Business)

When a new automobile is purchased at a cost of P dollars and retained for some number, k, of years, the cumulative utility, U, obtained from the automobile may be expressed as the following equation

$$U = \left(\frac{3P}{5}\right) \left(1+\frac{1}{3}+\frac{1}{5}+\cdots+\frac{1}{(2k-1)}\right)$$

as long as k does not exceed 10.

Experience also provides evidence that the current value of an automobile is given by the equation

$$V = P - \left(\frac{P}{3}\right) \left(1+\frac{1}{2}+\frac{1}{3}+\cdots+\frac{1}{k}\right)$$

where again P is the original price of the automobile. Write a program that will determine the value for k such that the annual decrease that is a part of the equation for V begins to exceed the annual utility that is a part of the equation for U. Please note that the equations for U and V are greatly simplified versions of similar equations developed by economists.

13. (Economics and Business)

In a free-enterprise society like that in the United States, the price of goods is affected by the availability of the goods (called supply) and the desire by people to purchase the goods (called demand). Assume that the price, P, of an item of which there is a supply, S, and for which there is demand, D, is given by the equation

$$P = 5 + \frac{250}{S} - 1.8(S-D)$$

where S is greater than D and $S-D$ does not exceed 20. The program has D as input and computes the integer value for S that makes P a maximum. Output consists of the maximum price and the values for S and D that yield that price.

14. (Economics and Computer Science)

Suppose that the utility value, U received by a person who consumes food, clothing, and recreation of amounts X_1, X_2, and X_3, respectively, is given by the equation

$$U = .5X_1 + .3X_2 + .2X_3$$

Assume also that P_1, P_2, and P_3 are the costs per unit, respectively, for food, clothing, and recreation. If $P_1 = \$10$, $P_2 = \$5$, and $P_3 = \$8$, and the total budget for these three items is $100, find the values for X_1, X_2, and X_3 that will make the value of U a maximum.

Hint: your program will probably have nested loops in which each of three loops controls the values for X_1, X_2, and X_3, respectively. For example, if we were to assume *only* food were purchased, we could buy ten units of food ($100/$10) and the value of U would be

$$U = (.5)(10) + (.3)(0) + (.2)(0) = 5$$

If only clothing were purchased for the $100, we could buy twenty ($100/$5) units of clothing and the value of U would be

$$U = (.5)(0) + (.3)(20) + (.2)(0) = 6$$

If only recreation were purchased for the $100, we could buy 12.5 ($100/$8) units and the value of U would be

$$U = (.5)(0) + (.3)(0) + (.2)(12.5) = 2.5$$

The maximum value for U in these instances is 6, which occurs when $X_1 = 0$, $X_2 = 20$, and $X_3 = 0$. It probably is true that the very largest value for U is obtained when a mix of food, clothing, and recreation are purchased. Your task is to write a program to find some approximation to that mix.

15. (Mathematics and Computer Science)

Given two integers, the greatest common divisor (GCD) of the two integers is the largest integer that divides both of them. For example, the GCD of 12 and 16 is 4 since it divides both numbers and it is the largest such divisor. Write a program that reads any two integers, finds their GCD, and reports the two numbers and their GCD with appropriate identifying information.

16. (Mathematics)

    An integer is called a *perfect number* if its value is equal to the sum of all its divisors including 1. For example, 6 is a perfect number because the sum of its divisors (1, 2, and 3) equals 6. Write a program to find and output the five smallest perfect numbers.

17. (Mathematics)

    In a right triangle, if c is the length of the hypotenuse and a and b are the lengths of the two legs, then $c^2 = a^2 + b^2$. Write a program to find the ten sets of smallest integer values for a, b, and c that satisfy this equation then output the ten sets of three numbers together with appropriate identifying information.

18. (Mathematics, Engineering, Computer Science)

    In mathematics, a very important number is Euler's number e defined as the limiting value of a certain expression. This number is also represented correctly by the equation

    $$e = 1 + \frac{1}{1!} + \frac{1}{2!} + \frac{1}{3!} + \cdots + \frac{1}{N!} + \cdots$$

    where $N!$ is as defined in Exercise 10(a). Write a program to compute e correct to twelve decimal places; then report that value.

    Hint: You will need to compute $N!$ so that $1/N!$ is less than .0000000000001.

19. (Business and Computer Science)

    This program will provide sales and profit reports for a portion of the business done by a university bookstore. The input consists of lines (cards) of data where each line (card) corresponds to a different book. For each book in stock the data are: (1) a 10-digit identifying number; (2) a 1-digit code indicating type of book; (3) the number of copies of this book ordered from publisher; (4) the number of copies of this book returned to publisher; and (5) the wholesale price of this book. The code for type of book is: 1 = required text, 2 = recommended text, 3 = reference book. The last line (card) of data has 9999999999 as the identifying book number.

    The program computes and outputs total sales for each of the three types of books, the total of all sales, the profit on each type of book, and the total of all profits. Profits are computed for each book title as follows: 30% of wholesale price if it is a textbook and, the price is less than $15; 25% of wholesale price if that price is $15 or more; and 20% of wholesale price if the book is a reference book regardless of price.

20. (Mathematics, Computer Science)

An integer is divisible by 9 if the sum of its digits is divisible by 9. For example, the number 729 is divisible by 9 (9 × 81) and the sum of its digits is 7+2+9=18. Of course, we know 18 is divisible by 9, but applying the same sum-of-digits rule to 18 gives 1+8=9, which shows that 18 is divisible by 9.

Write a program that accepts any integer as input and then applies the sum-of-digits rule to the original integer and to successive sums of digits until the last sum is exactly 9 or less than 9. If the last sum is exactly 9, the original integer is divisible by 9 and a message should be output saying

```
XXXXX is DIVISIBLE BY 9.
```

If the last sum is less than 9, output the message

```
XXXXX IS NOT DIVISIBLE BY 9.
```

In both of these output messages, XXXXX represents the original integer input to the program.

21. (General)

Write an interactive program (for an interactive system) to play the following game:

(a) Request the player to input age, month of birth, and year of birth, each as a two-digit number. For example, input of 16,05,64 would mean a person is sixteen years old and was born in May of 1964.

(b) Use the RAND function to determine randomly the positions each of the 6 input digits would hold in a six-digit number. Using the example just given, one might determine that 1 is in position 2, 6 in position 4, 0 in position 3, 5 in position 6, 6 in position 1, and 4 in position 5, resulting in the six-digit number 610645.

(c) Tell the player the computer has produced a six-digit number and the player will have at most fifteen chances to guess the number by keying the guess on the terminal.

(d) After each guess by the player, determine which digits, if any, in the guessed number are correct and output appropriate messages. For example, if the player guesses 512647 for the number generated in section (b) above, the output might be as follows:

```
CORRECT DIGITS ARE:
THE 1 IN POSITION 2
THE 6 IN POSITION 4
THE 4 IN POSITION 5
```

(e) Write the program so that play continues until the correct six-digit number is guessed or until the player has had fifteen attempts.

(f) At the end of each game, ask the player to enter 0 if no more playing is desired and 1 if another game is to be played.

22. (Business, Mathematics)

Write a program to perform certain computations related to the amortization of a loan, as follows. *Input*: Amount of principal, annual interest in percent, number of years for paying off the loan. *Output*: Monthly payment, total interest paid by end of designated time.

Also output an amortization table like this:

```
MONTHLY PAYMT    INTEREST PAID    PRINCIPAL PD    BALANCE
   . . .             . . .            . . .         . . .
   . . .             . . .            . . .         . . .
```

23. (General)

Write a program that accepts as input a real number that represents the amount of a semimonthly paycheck. Generate the following output:

```
YOUR PAYCHECK OF XXXX.XX YIELDS THIS MONEY:
$100 BILL      X
$50  BILL      X
$20  BILL      X
$10  BILL      X
$5   BILL      X
$2   BILL      X
$1   BILL      X
HALF DOLLAR    X
QUARTER        X
DIME           X
NICKEL         X
PENNY          X
```

24. (Statistics, Computer Science)

The citizens of Moorhead have completed surveys from which the following data were obtained:

(a) Respondent's sex (0 = male, 1 = female)

(b) Marital status (0 = single, 1 = married, 2 = other)

(c) Age

(d) Education (0 = less than high school, 1 = completed high school, 2 = some college but no degree, 3 = earned 4 yr. degree, 4 = earned graduate degree)

Write a program to input these data until an age of 0 is encountered and compute the following statistics:

(a) Percent of all respondents over 65.

(b) Percent of all respondents under 21.

(c) Percent of female singles.

(d) Percent of male singles.

(e) Percent of single females having a 4-year degree.

(f) Percent of single males having a 4-year degree.

(g) Average age of single males.

(h) Average age of single females.

(i) Assuming that an education code of 0 means an average of six years of education, a code of 1 means twelve years, a code of 2 means fourteen years, a code of 3 means sixteen years, and a code of 4 means nineteen years, determine the average years of education for males.

(j) Determine the average years of education for females. Your program should generate appropriate output for all of the above statistics.

CHAPTER 5
Subscripted Variables and Repeated Processes

In previous chapters we discussed the process of repeating a set of statements in a ForTran program. We learned that such sets of repeated instructions are called loops.

Since loops are such an effective method for utilizing the power of a computer system it is important that we become as adept as possible in using them. The ForTran language includes some capabilities specifically designed for writing program loops. We shall consider these structures in this chapter.

SUBSCRIPTED VARIABLES

We learned early in this book that values stored in computer memory must be associated with a name if they are to be used in a ForTran program. There are times when many different values may need to be processed and yet these values are all related to each other in some way. For example, a college professor may be dealing with the test scores of 200 students, or a university's registrar may be dealing with the records of 10,000 students. Certainly we don't want to insist that in order for the professor to use ForTran he must select 200 different variable names to correspond to the 200 test scores, or that the registrar must, similarly, invent 10,000 variable names in order to use ForTran to process the data on those 10,000 students.

Mathematicians long ago developed a notation for handling situations like these. It is called *subscript* notation. For example, if the professor wished to refer to those 200 scores by using the subscript notation of mathematics, she would write

$$T_1, T_2, T_3, \ldots, T_{200}$$

where the single group name T is used for all students' scores but the subscript $(1,2,3,\ldots,200)$ denotes the test score for a specific student. Similarly, the data for those 10,000 students could be referred to as

$$S_1, S_2, S_3, \ldots, S_{10,000}$$

where S is chosen as the group name for student data and the subscript identifies the data for a specific student.

In mathematics a set of numbers like the set T of student test scores is sometimes called a *vector* and the number of elements in the vector (as 200 in this case) is called the *dimension* of the vector.

Sometimes it is useful to consider a set of data as being arranged in rows and columns. For example, consider some data associated with the ten employees of a firm. In particular, suppose we have arranged these data in some order (maybe from the data for the employee with most seniority to the one with least) and have the hours worked by each for each day of the week. The resulting data might appear as follows:

8	10	8	7	8	4	0
8	9	8	8	8	2	0
8	8	8	8	8	2	1
8	7	8	8	8	1	1
8	8	8	8	8	0	0
8	8	8	8	7	0	1
6	8	8	6	7	0	0
8	5	8	8	6	0	0
7	8	8	8	7	0	0
8	6	6	8	6	0	0

Notice that there are ten rows and seven columns in the above set of data. Each row corresponds to an employee and each column corresponds to a day of the week beginning with—say, Monday. Thus, the number of hours worked on Tuesday by the employee with most seniority is 10. It is convenient to think of the set of data above as one collection and assign the name HOURS to it. Then each member of the set HOURS can be uniquely identified by its position as to the row and column in which it appears. Therefore, the number of hours worked on Tuesday by the first employee can be denoted as HOURS $_{1,2}$ where the first subscript indicates the row (the employee) and the second subscript the column (the day). In the above example HOURS $_{1,2}=10$.

Matrix

In mathematics an arrangement of data like that just discussed is called a *matrix*. Therefore, a matrix is a collection of objects (usually numbers) arranged in rows and columns and sometimes in layers of rows and columns. The notation used is that already mentioned—that is, the name of the whole set (as HOURS in our previous example) with subscripts specifying the row and column. If there is a situation where *sets* of rows and columns should be used (for example, if data for several companies like that in HOURS were to be processed), a third subscript could be added to denote the specific company. Although we will not use more than three subscripts in this book it is possible to imagine situations where it might be useful to have more than three subscripts. When a third subscript is used it is often referred to as the *layer* subscript to distinguish it from the row or column subscripts.

The matrix of ten rows and seven columns discussed earlier is called a 10-by-7 matrix, also denoted 10×7. In general, then, a matrix having m rows and n columns is called an $m \times n$ matrix. If there are more than one layer—say p layers—then the matrix may be called an $m \times n \times p$ matrix and any particular element in the matrix—say the element in the ith row, jth column, and kth layer—is denoted by $M_{i,j,k}$ where M is the matrix name.

Subscripts in ForTran

You know enough about the use of computer terminals and punched cards by this time that you realize that these machines make no provision for entering a number or letter spaced half a line below another as we have indicated with subscripts in mathematics. Yet with even the discussion so far in this chapter, you can easily imagine that it would be very useful in ForTran to have a capability equivalent to vectors and matrices in mathematics. That is, it would be worthwhile if something like subscripts were possible in ForTran. Fortunately, an equivalent does exist in ForTran.

The DIMENSION statement. Recall that our rules for naming variables allow the name to have a maximum length of six characters, only letters and digits are allowed, and the first character must be a letter. Furthermore, any such variable identifies a single storage location.

A subscripted variable like T_5 in mathematics is denoted T(5) in ForTran, and one like $HOURS_{1,2,3}$ is denoted HOURS (1,2,3) in ForTran. Notice that the vector or matrix names (T and HOURS) can be kept as used in mathematics but the subscripts are enclosed in parentheses and separated by commas rather than appearing below the name. Furthermore, in ForTran we refer to vectors and matrices (plural of matrix) as *arrays* and talk about them as single-subscripted, double-subscripted, or triple-subscripted arrays, or, as many authors say, one-dimensional, two-dimensional, and three-dimensional arrays.

In order to be able to use a subscripted variable in ForTran one must declare it as such in type-declaration statements such as INTEGER, REAL, and CHARACTER, or by the special type-declaration statement called the DIMENSION statement. In this section we discuss the DIMENSION statement and give examples of its use.

Here is the general form of this ForTran statement:

```
DIMENSION VAR1(n),VAR2(m,n),VAR3(m,n,p)
```

where VAR1, VAR2, and VAR3 are any acceptable ForTran variable names as previously defined, and where n, m, and p represent any positive integer constants within the limitations of the given computer memory. Here are some examples of specific applications of the DIMENSION statement.

Example 1

Problem: Suppose we want to use a subscripted variable to store the populations of each of the fifty states in the United States.

Solution: If we choose the array name STATE, then the DIMENSION statement would be as follows:

```
DIMENSION STATE(50)
```

This statement sets aside fifty adjacent computer words all having the name STATE but each one uniquely identified by an appropriate subscript. Thus, if array STATE contains the populations of states in the same order as an alphabetic ordering of the states, then

```
STATE(1) stores the population of Alabama
STATE(2) stores the population of Alaska
       .

       .

STATE(50) stores the population of Wyoming
```

Note that the DIMENSION statement causes nothing to be stored. It only sets aside adjacent memory locations. Other statements are required in order to actually store any numbers in those memory locations.

Other type-declaration statements may be used to declare a variable as a subscripted variable. Suppose, for example, that you want STATE to be an integer variable as well as an array of fifty locations. The statement to accomplish this is the following:

INTEGER STATE (50)

Thus there is no need to use a DIMENSION statement to declare the array. The INTEGER statement does both functions. Similarly, the REAL, CHARACTER, and other type-declaration statements may be used to perform two functions.

Example 2

Problem: Suppose this time we want to store the populations of each of the fifty states and of their capitals.

Solution: The following DIMENSION statement makes this possible:

```
DIMENSION STATE(50),CAPTAL(50)
```

This time fifty adjacent memory locations are set aside under the array name STATE and fifty under the array name CAPTAL. Note again that the DIMENSION statement does nothing with the *contents* of computer words. It simply sets aside the number of computer words specified inside the parentheses of each array name and allows for subsequent use of that variable name, including parentheses in the name. The order of storing information in these arrays is also not determined by the DIMENSION statement but by the order in which data are entered in conjunction with the READ statement that causes input to these two arrays.

Another method of storing data about state populations and capitals is to use a double-subscripted variable—say, STCAP. An appropriate DIMENSION statement would then be:

```
DIMENSION STCAP(50,2)
```

This statement sets aside 100 adjacent memory locations under the name STCAP as one array of fifty rows and two columns for a total of 100 words. This time STCAP (1,1) would store the population of Alabama and STCAP (1,2) would store the population of the capital of Alabama. Similarly, STCAP(2,1) would store the population of Alaska and STCAP(2,2) the population of the capital of Alaska, assuming that the order of data is in alphabetical order of state names.

If we also wanted the array STCAP to be an integer array, the following statements could be used:

```
DIMENSION STCAP(50,2)
INTEGER STCAP
```

However, the single statement

```
INTEGER STCAP(50,2)
```

accomplishes exactly the same result. In general, then, although the DIMENSION statement is available for the single purpose of declaring variables to be subscripted variables, one may as well accomplish this *along with* declaring a variable as real, integer, character, and so forth, by specifying the size and number of subscripts in the REAL, INTEGER, CHARACTER, or whatever statement is needed to declare the variable type.

The next example will demonstrate alternate methods for storing a given set of data. There is no intention of emphasizing one method over the others but as you gain programming experience you should be alert to choosing the method that makes the programming as easy as possible.

Example 3

Problem: Information about state populations is to be stored in a form such that reports can be extracted according to race (four to be considered) and sex, by state.

Solution: One way to accomplish this is by establishing subscripted variables as in these statements:

```
INTEGER MALE1(50),MALE2(50),MALE3(50),MALE4(50)
INTEGER FEMAL1(50),FEMAL2(50),FEMAL3(50),FEMAL4(50)
```

In this arrangement the four subscripted variables whose names begin with M would be used to store the number of *males* belonging to each of four races in each of the fifty states. Similarly, the four variables whose names begin with F would be used to store the number of *females* belonging to each of the four races in each of the fifty states. Clearly, it takes 400 computer words to get the appropriate categories and subcategories all included.

A second way to organize the population data so as to be able to produce reports according to state, race, and sex is by using the following INTEGER statement:

```
INTEGER RACE1(50,2),RACE2(50,2),RACE3(50,2),RACE4(50,2)
```

This time four double-subscripted variables are used where the variable names refer to the four races and the subscripts are related to the states and the sexes. As was true of the first arrangement of data, this second method also sets aside 400 computer words for storing the data. The advantages of four double-subscripted variables over eight single-subscripted ones are not immediately apparent but soon you will understand what they are.

As a third possible method of arranging these population data consider this statement:

```
INTEGER POPUL(50,4,2)
```

This time only one variable, POPUL, is used, but it is triple-subscripted such that the first subscript refers to state, the second subscript to race, and the third subscript to sex. Thus, if we assume that the ordering of data is in alphabetical order by name of state, that the fourth race refers to Caucasian, and the second sex is males, then the number of Caucasian males in Alabama would be stored in the computer memory location denoted by POPUL(1,4,2), and the number of Caucasian females would be stored in POPUL(1,4,1).

Obviously, all three of these methods for organizing such data are acceptable in the ForTran language. It is up to the programmer to select the arrangement that best suits the particular application being computerized. The knowledge as to which is best can be acquired most easily through experience.

Here are some more examples of array declarations.

```
LOGICAL FLAG(80)
REAL INDEX(20,3),WAGE(20)
CHARACTER NAME(12),CAPTAL(50)*12
```

The first example defines FLAG as a logical array of size 80. The second example defines INDEX as a doubly-subscripted array of size 60 and WAGE as a single-subscripted array of size 20. In the third example, NAME is defined as a character array of size 12 where each member of the array can store only one character. Also, in this example, CAPTAL is defined as a character array of size 50 where each member of the array can store up to twelve characters.

Refer back to Example 2 in this section and suppose instead of storing the *populations* of the state capitals we wanted to store their names. The following two statements could be substituted for the single DIMENSION statement of the previous example:

```
INTEGER STATE(50)
CHARACTER CAPTAL(50)*15
```

Since population figures are always integers it makes sense to define the array STATE as an integer array. And since the array CAPTAL is now to be used for storing *names* of capitals, it is appropriate to define it as a character array.

We suggest that at the time when an array is defined as a specific type, the type declaration statement should also be used to specify that it is an array by placing parentheses after the variable name and appropriate numbers within the parentheses rather than including a DIMENSION statement for that purpose. The array specification may appear in only one declaration statement.

Arrays in ForTran 77 (full set)

What we have presented in the previous sections on arrays (subscripted variables) is valid for many versions of ForTran and, in particular, for ForTran 77, subset. There are, however, some capabilities in ForTran 77, full set, that are related to the arrays that we discuss in this section.

The general forms for statements that declare a one-dimensional array in ForTran 77, full set, are

DIMENSION VAR1(n1:n2), . . . ,VARN(p1:p2)

or

Type VAR1(n1:n2), . . . ,VARN(p1:p2)

where "Type" is one of the declaration types INTEGER, REAL, CHARACTER, LOGICAL, and so on; VAR1, . . . ,VARN are any acceptable ForTran variable names; and n1, n2, p1, and p2 are integers (negative, zero, or positive) such that n1 is less than n2 and p1 is less than p2 and where each pair of integers defines the range of integer values for the subscript.

Here are some specific examples of the above general forms.

1.   `DIMENSION ACCT(5:10),PRICE(1979:1982)`
2.   `INTEGER DEMAND(1:10),AGE(-5:0)`
3.   `REAL NUM(-5:5)`
4.   `LOGICAL SWITCH(0:5),FLAG(-1:1)`

Example 1 declares ACCT as a one-dimensional array of six elements consisting of ACCT(5),ACCT(6),ACCT(7),ACCT(8),ACCT(9), and ACCT(10). Also declared in Example 1 is the one-dimensional array PRICE of which there are four members: PRICE(1979),PRICE (1980),PRICE(1981), and PRICE(1982).

Example 2 declares both DEMAND and AGE as one-dimensional integer arrays where DEMAND has ten members (subscripts of 1,2,3, . . . , 10) and AGE has six members (subscripts of -5, -4, -3, -2, -1, and 0).

Example 3 defines the real array NUM with eleven members whose subscripts are -5, -4, . . . , 0, 1, 2, . . . , 5. The last example above declares SWITCH and FLAG both as one-dimensional arrays of type LOGICAL. SWITCH has six members whose subscripts range from 0 to 5, and FLAG has three members whose subscripts are -1, 0, and 1.

Multidimensional arrays in ForTran 77 (full set)

In the full set of ForTran 77 one may specify up to seven subscripts for a given application. We refer to arrays of more than one dimension as multidimensional arrays. The general forms of the ForTran statements for declaring such arrays are

DIMENSION VAR1(m1:n1, . . . ,mk:nk), . . . ,VARN(il:jl, . . . ,ip:jp)

or

Type VAR1 (m1:n1, . . . ,mk:nk), . . . ,VARN(il:jl, . . . ,ip:jp)

with $k \leq 7$ and $p \leq 7$, where VAR1, . . . ,VARN are any valid ForTran variable names; ml,nl, . . . ,mk,nk are integers (negative, zero, or positive) such that ml is less than or equal to nl, . . . , and mk is less than or equal to nk; and similarly for $il \leq jl, . . . ,ip \leq jp$. In the case of the integers ml,nl, . . . , mk,nk,il,jl,. . . ,ip,jp, each pair defines the range of integer values allowed for the subscripts of the given array.

Here are some specific examples to illustrate the above general forms.

1. `DIMENSION YEARP(1979:1983,1:6),MONTH(1:12)`
2. `INTEGER POPUL(1:50,1:4,0:1,1980:1989)`
3. `CHARACTER NAME (1:100,0:1)*20`
4. `REAL SUPPLY (1980:1984,10:19,-1:1),PRICE(1980:1984,10:19)`

In Example 1, YEARP is declared a two-dimensional array whose row-subscript may range from 1979 to 1983 and whose column-subscript may range from 1 to 6. This DIMENSION statement also declares the one-dimensional array MONTH with twelve members whose subscripts may range from 1 to 12. The convenience of this kind of declaration statement is more obvious when one realizes that array YEARP could be used to store the prices of six different items for each of the years 1979, 1980, 1981, 1982, and 1983.

Example 2 declares POPUL to be a four-dimensional integer array whose first subscript ranges from 1 to 50, its second subscript from 1 to 4, the third from 0 to 1, and the fourth subscript ranges from 1980 to 1989. Array POPUL could be used to store population data for the fifty states of the United States, for each of four races, for either of the two sexes, and for each of the ten years 1980 through 1989.

Example 3 declares NAME to be a character array of two subscripts, the first ranging from 1 to 100 and the second from 0 to 1, thus giving the array 200 members, each of them able to store up to twenty characters. This array might be used to store the names (twenty letters at most) of 100 males and 100 females, where the second subscript designates one sex or the other.

The fourth example declares two real arrays, SUPPLY and PRICE, where SUPPLY is a three-dimensional array and PRICE is a two-dimensional array. These arrays might be used together such that array SUPPLY would indicate numbers of items 10 through 19 for the years 1980 through 1984 that were undersold (third subscript would be -1), sold as ordered (third subscript would be 0), or which were oversold (third subscript of 1). Array PRICE could be used to store the price of items 10 through 19 for each of the years 1980 through 1984.

As a summary to this section on arrays we state some rules for using arrays, and present a complete example demonstrating their use.

Rules for using arrays

1. A single DIMENSION or type-declaration statement may be used to declare more than one variable.

2. More than one DIMENSION or type-declaration statement may appear in any given program.
3. A given variable can be defined as an array only once during a given program.
4. A subscript may be any valid integer expression provided its value is within the range for that subscript.
5. Once a variable has been defined as an array of one or more subscripts, that variable should be used with the prescribed number of subscripts except that in certain situations to be discussed later it may appear without subscripts.
6. A subscripted variable must have appeared in a DIMENSION or type declaration statement previous to its being used in any other program statement.
7. A variable may be defined as an array by a DIMENSION statement or by a type declaration statement.
8. In ForTran 77 (full set) allowable *ranges* for subscripts may be specified in the DIMENSION statement or in the type declaration statement.

Let's now consider some applications of the preceding rules as related to a given DIMENSION statement. This DIMENSION statement is assumed to hold throughout *all* of the examples that follow it.

DIMENSION A(20), B(3,6), C(−5:0,0:3,20:22)

Example	Comment
A(15)	Valid.
A(NAME)	Valid provided NAME is an integer variable with value less than or equal to 20 and greater than 0.
A(I=J*K)	Invalid because assignment is not permitted in a subscript representation.
A(J*K)	Valid provided J*K is greater than or equal to 1 and less than or equal to 20.
A(30)	Invalid because 30 is greater than 20.
A(K+I*J)	Valid provided K+I*J is greater than or equal to 1 and less than or equal to 20.
B(0,1)	Invalid. Zero subscript not allowed.
B(6,2)	Invalid because the first subscript exceeds 3.
B(3*I)	Invalid because only one subscript is used whereas the DIMENSION statement specifies 2.
B(I,J)	Valid provided I is greater than or equal to 1 and less than or equal to 3, and J is greater than or equal to 1 and J is less than or equal to 6.
B(2,I*J)	Valid if I*J is between 1 and 6, inclusive.

B(2+K*J,K-I*J)	Valid provided 2+K*J is between 1 and 3, inclusive, and K−I*J is between 1 and 6, inclusive.
B(I-2,3*I)	Invalid. If I=2, first subscript is 0. If I is less than 2, the first subscript is negative (an invalid situation), and if I is greater than 2, the second subscript is greater than 6 (an invalid situation).
C(1,1,20)	Invalid because the first subscript is not in the range −5 to 0.
C(I,J,I+J)	Invalid because there are no integers for I and J such that I is in the range −5 to 0 and J is in the range 0 to 3, and I+J in the range 20 to 22.
C(I*J,J,J*20)	Valid provided I*J is in the range −5 to 0, and J is in the range 0 to 3, and J*20 is in the range 20 to 22.
C(-2,0,20)	Valid.
C(I,J,K)	Valid provided I, J, and K all fall within their respective ranges.
C(-2,20)	Invalid because only two subscripts appear whereas there must be three.

ARRAY INPUT AND OUTPUT

So far in our discussion of variables in the lists of READ, PRINT, and WRITE statements we have included only *simple* variables. We turn our attention now to the use of subscripted variables in such lists.

Whenever a single element of an array is to be input or output, it is handled in the same way as a simple variable except that specific subscripts identify the element of the array to be input or output. Consider these examples:

```
1    READ*,A(11)
2    READ*,B(2,3)
3    READ*,VALUE(K,J)
4    PRINT*,A(3)
5    WRITE(6,*)A(11),B(2,J)
```

Statement 1 reads one value and stores it in A(11). Statement 2 reads one value and stores it in B(2,3). Statement 3 reads one value and stores it in memory location VALUE(K,J), where K and J must have been previously defined within the boundary values specified in the statement that defines VALUE as an array. Statement 4 causes the printing of the contents of A(3). Statement 5 causes the printing of the contents of A(11) and B(2,J), where J must have been previously defined as an integer within the boundary values specified in the statement that defines B as an array.

Sometimes it is desirable to input or output an entire array. In such a case the array name only (without subscripts) should appear in the input or output statement. Consider the following program:

```
      PROGRAM TABLES
      INTEGER TABLE(10),I
      READ *,TABLE
      I =1
10    IF(TABLE(I) .LT. 0)TABLE(I) = 0
      I = I + 1
      IF(I .LE. 10)GOTO 10
      PRINT*,TABLE
      STOP
      END
```

The READ statement in this program causes the input of ten numbers and stores them in TABLE(1),TABLE(2),...,TABLE(10). The next four lines of this program result in any negative members of array TABLE being set to zero. The PRINT statement in the program causes all ten members of the integer array TABLE to be printed, including those that might have been set to 0.

An important consideration in this discussion is this: In what order does the computer store data when only the array name is given? In the previous example where array TABLE is single-subscripted, the answer is easy. The numbers are stored sequentially in TABLE(1),TABLE(2),...,TABLE(10).

What happens when an array has *two* subscripts? If one were to suggest orderings in which a double-subscripted array could be assigned values, quite likely two different methods would come to mind first. Let's consider a specific example to demonstrate these. Suppose array A is defined to have two subscripts as follows:

```
      INTEGER A(3,4)
```
or
```
      INTEGER A(1:3,1:4)
```

Suppose also that the statement

```
      READ*,A
```

appears and that the numbers 1,3,5,7,9,11,13,15,17,19,21, and 23 are provided, in that order. From the fact that the INTEGER statement specifies three rows and four columns, one possible order for storing the numbers would be to assume that they are entered by row with four elements per row. The result in computer memory would be as follows:

Row 1 $\begin{cases} A(1,1) = 1 \\ A(1,2) = 3 \\ A(1,3) = 5 \\ A(1,4) = 7 \end{cases}$

Row 2 $\begin{cases} A(2,1) = 9 \\ A(2,2) = 11 \\ A(2,3) = 13 \\ A(2,4) = 15 \end{cases}$

Row 3 $\begin{cases} A(3,1) = 17 \\ A(3,2) = 19 \\ A(3,3) = 21 \\ A(3,4) = 23 \end{cases}$

Another possible order for storing the numbers would result from assuming that they are entered by column, with three elements per column. The result in computer memory would be as follows:

Column 1 $\begin{cases} A(1,1) = 1 \\ A(2,1) = 3 \\ A(3,1) = 5 \end{cases}$

Column 2 $\begin{cases} A(1,2) = 7 \\ A(2,2) = 9 \\ A(3,2) = 11 \end{cases}$

Column 3 $\begin{cases} A(1,3) = 13 \\ A(2,3) = 15 \\ A(3,3) = 17 \end{cases}$

Column 4 $\begin{cases} A(1,4) = 19 \\ A(2,4) = 21 \\ A(3,4) = 23 \end{cases}$

Mathematicians, when referring to these two orderings of the elements of an array, assign the name *row-major* to the first method and *column-major* to the second method. Computer scientists use the same terminology. The ordering used by ForTran is the column-major ordering. Thus, given the ForTran program segment

```
INTEGER A(3,4)
READ*,A
```

and the input values 1,3,5,7,9,11,13,15,17,19,21,23, the values in array A would be

```
A(1,1) = 1
A(2,1) = 3
A(3,1) = 5
A(1,2) = 7
A(2,2) = 9
A(3,2) = 11
A(1,3) = 13
A(2,3) = 15
A(3,3) = 17
A(1,4) = 19
A(2,4) = 21
A(3,4) = 23
```

ForTran also uses column-major ordering in output. Thus, if we assume that A is the same 3 × 4 array as previously discussed, the statement

```
PRINT*,A
```

would result in the elements of array A being printed in the following order:

```
A(1,1),A(2,1),A(3,1),A(1,2),A(2,2),A(3,2),A(1,3),A(2,3),A(3,3),A(1,4),A(2,4),
A(3,4)
```

When triple-subscripted arrays are used in ForTran there is again the question of which ordering of the elements in such arrays is used by ForTran. We shall illustrate the method ForTran uses with the following example:

```
INTEGER A(2,3,2)   or   INTEGER A(1:2,1:3,1:2)
READ*,A
```

In this program segment array A is of size 12 and, therefore, the READ statement would require twelve numbers in order to fill array A. Suppose that the input data provided were 1,2,3,4,5,6,7,8,9,10,11,12. The above For-Tran program segment would result in storage as follows:

```
A(1,1,1) = 1
A(2,1,1) = 2
A(1,2,1) = 3
A(2,2,1) = 4
A(1,3,1) = 5
A(2,3,1) = 6
A(1,1,2) = 7
A(2,1,2) = 8
A(1,2,2) = 9
A(2,2,2) = 10
A(1,3,2) = 11
A(2,3,2) = 12
```

One of the very important uses made of arrays in developing programs occurs when repeated (or iterated) processes are necessary. In Chapter 3 we introduced a very simple case of an iteration structure without discussing the general concept of these useful program structures. At this point we shall proceed to such a discussion.

ITERATION STRUCTURES

We begin by considering iteration structures from the algorithmic viewpoint, to be followed later by a discussion of the iteration structures available in ForTran.

Any iteration structure consists of an entry point (including initialization of certain variables), a loop body, and an exit point, illustrated in diagram form as follows:

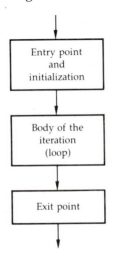

An iteration structure must always be entered at the entry point to ensure that appropriate initialization takes place. The exit point depends on the statements in the body of the iteration, although the most common and best-structured exit point is the first statement immediately following the body of the iteration. We shall refer to this as the *normal* or *default* exit point. Any other exit point must be specified in the body of the iteration.

This iteration body consists of those statements that would normally be executed at least once, the exact number depending on the initialization statements and on the statements in the body itself. We present now three examples of iteration structures appearing previously in this book.

Example 1

```
      PROGRAM VAR2
C***********************************************************************
C*        PROGRAM DEFINITION                                          *
C*              THIS IS THE SAME PROGRAM AS ON PAGE 181.  WE *
C*              INCLUDE IT HERE TO SHOW THE USE OF ITERATION *
C*              STRUCTURE INTRODUCED IN THIS SECTION.                 *
C***********************************************************************
      REAL NUMBER, MEAN, SUM, SUMSQ,  VAR
      INTEGER COUNT
C*
      COUNT=0
      SUM=0
      SUMSQ=0
      READ(5,*)NUMBER
C*****************************************************************
C*        COMPUTE MEAN AND VARIANCE                       *
C*****************************************************************
  3   IF(NUMBER .EQ. 99999)GOTO 5
        SUM = SUM + NUMBER
        SUMSQ = SUMSQ + NUMBER * NUMBER
        COUNT = COUNT + 1
        READ(5,*) NUMBER
      GOTO 3
C*
  5   MEAN = SUM / COUNT
      VAR = (SUMSQ - COUNT * MEAN * MEAN)/COUNT
      WRITE(6,*) ' THE VARIANCE IS ',VAR
      STOP
      END
```

In this example the entry point is where COUNT, SUM, SUMSQ and NUMBER are all initialized. The exit point is at statement 5. All statements between the first READ statement and statement 5 constitute the body of the iteration structure.

Example 2

This example is a program *segment,* so the usual statements appearing at the beginning of a program are purposely omitted.

```
        .
        .
        .

        READ*,N
100     IF (N .LE. 0) GO TO 110
          READ*,NUM
          SUM=SUM+NUM
          SUMSQ=SUMSQ+NUM*NUM
          N=N-1
        GO TO 100
110     PRINT*,SUM,SUMSQ

        .
        .
        .
```

In this iteration structure the entry point is at the first READ statement and its exit point is the PRINT statement. All other statements are in the body of the loop.

Example 3

Here is another program segment illustrating an iteration structure.

```
        READ*,N
100     IF(N .LE. 0) GO TO 110
          READ*,NUM
          IF(NUM .LT. 0) GO TO 150
          SUM=SUM+NUM
          SUMSQ=SUMSQ+NUM*NUM
          N=N-1
        GO TO 100
110     PRINT*,SUM,SUMSQ

          . .
          .
          .
```

```
150  PRINT*,'ERROR IN INPUT DATA.'
         .
         .
         .
```

The entry point to this loop is at the first READ statement where N is initialized. There are two exit points. The normal exit is the first PRINT statement where the value of SUM and SUMSQ are printed. The second exit point is at statement 150, which is executed only if input data are negative, an abnormal situation. Thus an exit like this is often called an abnormal exit.

Let's consider these three examples again with respect to the number of repetitions of the body statements. In Example 1 the number of repetitions is unknown in advance of entering the loop and is dependent on the condition that the value input for NUM be 99999. In the second and third examples, the number of repetitions, N, is known in advance and in both cases this value is input to be used later in determining when exit from the loop occurs. In Example 3, the abnormal exit conditions might reduce the number of repetitions.

In general there are two forms of iteration structures. In one of these, repetitions are controlled by a counter and we shall refer to this type as a *counter-controlled iteration structure.* Repetitions in the other type of iteration structure are controlled by the occurrence of a given condition; hence we shall refer to this type as *condition-controlled* iteration structures.

ForTran has some special capabilities for handling counter-controlled iteration structures, which we shall discuss in the next section.

Counter-controlled iteration structures

In any counter-controlled iteration there are the following features:

- Index: the counter of the number of repetitions (iterations).
- Initial value: the initial value of the index.
- Final value: the final value of the index.
- Step value: the amount of increment or decrement in the index after the completion of each iteration.

Referring again to the last two examples in the previous section, we note that in both examples the initial value of the index is N, its final value is 0, and the step value is -1.

As you can tell from these last three examples and from other examples given earlier in this book, a counter-controlled iteration structure may be created by using appropriate initialization statements, an IF statement, and a GO TO statement. ForTran provides a very convenient method for handling counter-controlled iterations by a single statement called the DO statement.

DO statement.   There are two forms of the DO statement as specified in ForTran 77 Standard (full set); these are designated by the type of the index. One is called the *integer* DO statement and the other the *real* DO statement. The integer DO statement has been available since the very beginning of ForTran and has this general form:

DO *n* INDEX=INIT,FINAL,STEP

where *n* is the reference number of the last program statement in the loop, INDEX may be any integer variable, INIT is the initial value of the index, FINAL is the final value of the index, and STEP is the value of the increment (or decrement). Note that in the *integer* DO all variables are integers. INIT, FINAL, and STEP may be integer constants, variables, or arithmetic expressions. If the value of STEP is the constant 1, it may be omitted together with the comma preceding it. Thus the statement

```
DO 50,INDEX=1,10
```

is an acceptable DO statement where the implied step value is 1, hence is not specified. Although the comma to the right of 50 in the preceding DO statement is optional in ForTran 77, many versions of ForTran compilers do not permit a comma there and if one is included it causes an error message. Therefore, we shall take the position of omitting the comma to the right of the reference number in any DO statement.

The general form of the *real* DO statement is the same as for the integer DO except that in the real DO statement, INDEX, INIT, FINAL, and STEP are all *real* constants, variables, or arithmetic expressions. Note that the real DO statement is included only in ForTran 77, full set. Other versions are apt to allow only the integer DO.

Execution of the DO statement.   Consider the statement

```
DO n INDEX=INIT,FINAL,STEP
```

When this statement is executed the following steps describe the process that is performed:

1. The ForTran 77 compiler uses the three values INIT, FINAL, and STEP to calculate the number of times the iteration occurs. This number is an integer—say, I—computed as follows:

$$I = ((FINAL - INIT)/STEP) + 1$$

If I is less than or equal to 0, then the loop is not executed even once. However, if I is greater than zero, then each time the loop is executed, I is decreased by 1 until one of the following occurs:
   (a) A RETURN statement (to be discussed in Chapter 7) is encountered (an abnormal exit).
   (b) A transfer out of the loop is executed (an abnormal exit).
   (c) A STOP statement (a statement that stops further processing) is encountered in the loop.
   (d) I becomes zero.
Thus if the loop is not ended by an abnormal exit or by a STOP statement, the iteration occurs I times. We shall refer to I as the *iteration counter*.
2. The value stored in variable INDEX is initially equal to whatever is stored in the location designated by INIT.
3. All statements following the DO are executed up to and including the statement whose reference number is *n*. (This is the reference number in the example.)
4. After the execution of statement *n*, the value stored in INDEX is replaced by the sum of STEP and the previous value in INDEX.

   Examples of valid and invalid DO statements. Now we present six DO statements and indicate which are valid and which are not.

Statement	Comment
1. DO 10 I=1,10	Valid.
2. DO 65 X=2,5, .1	Valid for full ForTran 77.
3. DO 17 I=1,5,1	Valid but the rightmost 1 and preceding comma are not required.
4. DO 10 I=2,1	Valid but since final value (1) is less than the initial value (2), the iteration counter is negative and the loop is not executed even once.
5. DO 50 I=N,N+K,J	Valid if values have been assigned to N, K, and J before execution of the DO. Note the effect (as in Example 4) if N>N+K.
6. DO 25,I=-5,5,2	Valid, but the comma between 25 and I is not valid in some versions of ForTran, although it is optional in ForTran 77.

Examples of complete DO loops.   Following are some complete DO loops with appropriate comments. As you study these examples, in each case determine the number of times the READ statement is executed.

1.
```
        DO 20 I=1,5
           READ*,X
           Y=X*I
    20    PRINT*,Y
```

2.
```
        DO 20 I=10,1
           READ*,X
    20    PRINT*,X
```

3.
```
        DO 20 I=10,1, −1
    20    PRINT*,I
```

4.
```
        N=2
        DO 50 I=1,N
           N=N+1
    50    PRINT*,N
```

5.
```
        DO 60 X=.1,1, .01
    60    PRINT*,X
```

6.
```
        DO 30 I=1,10,2
           READ*,X
           IF(X .LT. 0)STOP
    30    SUM=SUM+X
```

7.
```
        DO 30 I=1,10,2
           READ*,X
           IF(X .LT. 0) GO TO 50
    30    SUM=SUM+X
              .
              .
              .
    50...
```

Now let's review the action resulting from each of the seven DO loops. In Example 1 the iteration counter is $((5 - 1)/1)+1$, which is 5, so the READ statement is executed five times as are the two statements following it. Five different values are input for X, five different values are used for I in the assignment statement, and five different values for Y are printed by line 20.

In Example 2, the steps in the loop are not executed at all because the iteration counter is $((1-10)/1)+1$, which is -8. A negative iteration counter prevents any processing of the loop.

In Example 3, the iteration counter is computed as $((1-10)/(-1))+1$, which is 10, therefore the PRINT statement is executed ten times. The index, I, goes from an initial value of 10 to a final value of 1 in steps of -1.

In Example 4, at the entry of the iteration structure, N has the value 2, so the iteration counter is computed as $((2-1)/1)+1$, which is 2. Therefore, the body of the loop consisting of the statements $N=N+1$ and PRINT*,N are executed twice. The values output for N are 3 and 4. If this program segment were processed by a compiler other than a ForTran 77 compiler, the results would be radically different. In fact, the execution of the loop would go on endlessly because the final value, N, of the counter is being continuously increased and is, therefore, never equaled by the current value of the index I, a condition that must be satisfied in order to stop an iteration in most ForTran compilers other than ForTran 77.

Example 5 illustrates the possibility of minor errors in numbers owing to the finite representation in a binary computer of a decimal number. The iteration counter is $((1-.1)/.01)+1$, which is 91, so the loop should be executed 91 times. The output produced should be

```
.10000000
.11000000
.12000000
.13000000

        .

        .

        .

.98000000
.99000000
1.00000000
```

We suggest you run the following complete program version of Example 5:

```
PROGRAM XAMPL5
REAL X
DO 60 X = .1,1, .01
   PRINT*, X
60   CONTINUE
END
```

Compare the results of your program run with the expected output shown above. Not only is it possible that the output of your run may differ on some of the lines from the expected output but you may also find that the loop counter is not 91. Any such unexpected results are due to a truncated representation of .01 in the computer system.

In Example 6 the iteration counter is 5, the largest integer less than or equal to $((10-1)/2)+1$. In this loop the value input for X is tested for being less than 0. If it is, the program stops any further processing. However, if no negative number is input, the body of the iteration structure is executed five times. Similarly, in Example 7, if a negative value is input for X, control transfers out of the loop to the statement whose reference number is 50. If all numbers input are positive, the loop is executed five times, as in Example 6.

Note: Readers using a ForTran compiler other than ForTran 77 should be aware that most such compilers process DO loops as follows where the DO statement is

DO *n* INDEX=INIT,FINAL,STEP

The value stored at INIT is assigned to INDEX as the initial value and the loop is performed the first time. At the end of the body of the loop the value stored at STEP is added to the current value of INDEX and the sum is stored back in INDEX. The new value of INDEX is compared with the value at FINAL. Depending on whether STEP is greater than 0 or less than 0, the results are:

1. If STEP is greater than 0 the loop terminates when INDEX becomes greater than the value at FINAL.
2. If STEP is less than 0 the loop terminates when INDEX becomes less than the value at FINAL.

You are urged to find out exactly how the compiler available to you processes DO loops, because the results may differ quite a bit. (Recall the discussion for Example 4.)

CONTINUE statement

Provision is made in ForTran for a statement whose only function is to serve as a reference point in the program. This statement is the CONTINUE statement. Here is its form:

```
CONTINUE
```

As you can see, it is trivially simple and consists of only the single word CONTINUE. Typically this statement is given a reference number because that is its main reason for existing—that is, to serve as a reference point in the program. This statement is most frequently used as the last statement in a DO loop and is, therefore, referenced by the DO statement. Here is an example of such an application:

```
DO 20 I=1,5
    READ*,X
    Y=X*I
    PRINT*,X,Y,I
20  CONTINUE
```

You will notice in this program segment that the CONTINUE statement is the last statement in the loop that begins with DO. As previously stated, this is by far the most common use made of the CONTINUE statement. Of course, a CONTINUE statement may appear at *any* point in a ForTran program, but when not used as the end of a DO loop, it is a do-nothing statement that really has no purpose, though it is not incorrect.

THE DATA STATEMENT

A useful statement available in ForTran for accomplishing initialization is the DATA statement. Its general form is as follows:

DATA VAR1,VAR2, . . . , VARN/data-list/

where VAR1,VAR2, . . . , VARN represent any set of acceptable ForTran variable names (including arrays) and where "data-list" represents values (one for each variable) to be stored in the variables specified to the left of the first slash. The data in the data list are separated by commas and the entire data list is enclosed within a pair of slashes.

The variable list and the data list must match both in the number of items in each list, and in the type (integer, real, and so forth) item for item. Any mismatches in either of these situations causes some ForTran compilers to produce a warning type of error message. If there are more data items than variables, the message will be something like the following:

```
DATA CONSTANT LIST TOO LONG
```

This type of mismatch may not cause any problem but since it normally should not occur, the programmer is reminded of the possibility of providing incorrect data constants.

If there are more variables than data items, the warning will be something like

```
DATA VARIABLE LIST TOO LONG
```

Although this mismatch does not cause a fatal error in compiling the program, it does result in the excess variables being uninitialized. Note that some ForTran compilers automatically initialize every numeric variable to 0 and every character variable to blank spaces.

Here are some examples of DATA statements:

1. `DATA A,B/20,10/`
   Here A is assigned the value 20 and B the value 10.
2. `DATA A,B,C/15,2*1.0/`
   A is assigned the value 15 and both B and C are assigned the value 1.0. Note the use of the asterisk to indicate repetition of a value, where the number of repetitions is indicated by the appropriate integer at the left of the asterisk.
3. `DATA X,Y,Z/'XY', 'XYZ', 'XY/YZ'`
   X is assigned the character 'XY', Y is assigned the character string 'XYZ', and Z is assigned the six-character string 'XY/YZ.'
4. `DATA A/20/B,C/2HAB,25.6/`
   A is assigned the value 20, B is assigned the string AB, and C is assigned the value 25.6.
5. `DATA A,B,C,D/4*0,3/`
   All four of the variables A, B, C, and D are assigned the value 0 and the remaining data item, 3, is left unassigned. This situation would produce a warning message telling about the discrepancy between the variable list and the data list but would not result in a fatal error.
6. `DIMENSION A(1:100),B(0:9,-10:9)`
   `DATA A,B/100*1.0,200*0/`
   In this example A is a one-dimensional array of size 100 and B is a two-dimensional array of size 200 (ten rows and twenty columns). When just the variable names A and B appear in the DATA statement, that means that all 100 locations of array A and all 200 locations of array B are to be assigned values. Therefore, all 100 locations of array A are assigned the value 1.0 and all 200 locations of array B are assigned the value 0.
7. `INTEGER A(20,10),B(5)`
   `DATA B,A(10,10)/1,2,3,4,0,1/`

Here A and B appear in an INTEGER statement with subscripts. Therefore, A is defined as an integer array of twenty rows and ten columns for a total of 200 memory locations. Similarly, B is an integer array of five locations. In the DATA statement, since B appears without subscripts, all five of its memory locations are assigned values: $B(1)=1, B(2)=2, B(3)=3, B(4)=4$, and $B(5)=0$. Then one of the elements of array A—namely $A(10,10)$—is assigned the value 1.

The DATA statement is a nonexecutable statement in ForTran, although it does result in the assignment of values to specified variables. As is the case with all nonexecutable statements (except FORMAT statements), DATA statements must precede all executable statements in the program (or subprogram).

TWO PROBLEMS COMPLETE WITH SOLUTIONS

Now let's consider two complete examples in which we apply these latest programming concepts.

Example 1

Problem: A set of numbers (no more than 100 of them) is to be input in which sentinel datum of -999.999 is provided. Using these data, perform the following:

1. Determine the number of numbers; call it N.
2. Print the N numbers in an order reverse to that in which they are provided.
3. Change every negative number in the set to 0 and determine the number of negative numbers.
4. Print the new set of numbers and the count of negative numbers.

Algorithm:

1. Set aside an array, ANUM, for storing up to 100 numbers and initialize a counter, N, for counting the number of numbers to be input until sentinel datum is detected. Also initialize a counter, NEG, for counting negative numbers that are input.
2. Read a number, X. If it is sentinel datum proceed to step 4.
3. Increase N by 1 and assign to the Nth position in the array the number X. Then repeat the process from step 2.
4. Print the N numbers stored in array ANUM beginning with the number at ANUM (N) then ANUM $(N-1)$, and so on to ANUM (1).

5. Check each number in array ANUM for being negative. If it is, set it to 0 and increase NEG by 1.
6. Print each number in array ANUM as now modified to contain 0's rather than negative numbers.
7. Print the number of negative numbers.
8. End.

Algorithm with steps detailed.

1.0   Set aside an array, ANUM, for storing up to 100 numbers.

1.1   Initialize a counter, N, for counting the number of numbers input until sentinel datum is detected.

1.2   Initialize a counter, NEG, for counting negative numbers that are input.

1.3   Read a number X.

2.0   If X=−999.999 proceed to step 4.0.

3.0   Increase N by 1.

3.1   Assign to the Nth position in array ANUM the value of X and read the next number.

3.2   Repeat the process from step 2.0.

4.0   Set up a loop for printing ANUM(N) through ANUM(1).

5.0   Set up a loop with index I for examining numbers in array ANUM.

5.1   If ANUM(I) is negative, increase NEG by 1 and set ANUM(I) to zero.

6.0   Print ANUM(I).

7.0   Print NEG.

8.0   End.

Program

```
      PROGRAM EXAMP1
C*****************************************************************
C*          PROGRAM DEFINITION                                  *
C*                 SEE THE TEXT FOR COMPLETE DESCRIPTION OF      *
C*                 THE PROBLEM.                                  *
C*****************************************************************
C*          VARIABLE DEFINITION                                 *
C*                 NUM  IS THE ARRAY OF NUMBERS READ.           *
C*                 N   IS THE COUNT OF TOTAL NUMBERS.           *
C*                 NEG  IS THE COUNT OF NEGATIVE NUMBER.        *
C*                 TEMP  IS A TEMPRARY VARIABLE.                *
C*                 I  IS THE LOOP INDEX.                        *
C*****************************************************************
C*
      REAL NUM(100), TEMP
      INTEGER N, NEG, I
```

```
C*
        N = 0
        NEG = 0
        READ*, TEMP
C*
C*****************************************************************
C*              HERE A TOTAL OF N NUMBERS ARE READ              *
C*              ONE AT A TIME. IF THE NUMBER READ               *
C*              IS NOT -999.999 IT IS STORED IN  ARRAY NUM.     *
C*****************************************************************
C*
  10    IF(TEMP .EQ. -999.999 .OR. N .GE. 100)GOTO 20
           N = N +1
           NUM(N) = TEMP
           READ*,TEMP
        GOTO 10
C*
  20    PRINT*, ' A TOTAL OF ',N,' NUMBERS WERE READ.'
        PRINT*, ' THE LIST IN REVERSE ORDER IS:'
C*
C*              REPORT ARRAY NUM IN REVERSE ORDER.
C*
        DO 30 I = N,1, -1
           PRINT*,NUM(I)
  30    CONTINUE
C*
C*****************************************************************
C*              SET ANY NEGATIVE NUMBER IN ARRAY NUM TO         *
C*              ZERO AND COUNT NUMBER OF NEGATIVE NUMBERS.      *
C*****************************************************************
C*
        PRINT*,' THE LIST WITH NO NEGATIVE IS:'
C*
        DO 40 I = 1,N
          IF(NUM(I) .LT. 0)THEN
            NEG = NEG + 1
            NUM(I) = 0
          ENDIF
          PRINT*,NUM(I)
  40    CONTINUE
C*
        PRINT*,' THERE ARE ', NEG, ' NEGATIVE NUMBERS.'
        STOP
        END
```

Example 2

Problem: Read a positive integer, N, and compute the following sums:

(a) $1 + 2 + 3 + \cdots + N$.
(b) $1^2 + 2^2 + 3^2 + \cdots + N^2$.
(c) $1^3 + 2^3 + 3^3 + \cdots + N^3$.
(d) $1 + 3 + 5 + \cdots + M$ where M is the largest odd integer less than or equal to N.
(e) $2 + 4 + 6 + \cdots + K$ where K is the largest even integer less than or equal to N.

Include a test to select positive N not greater than 10,000. The output should be the value obtained for each of the sums together with appropriate identifying information.

Algorithm:

1. Initialize five variables (SUMA, SUMB, SUMC, SUMD, SUME) for accumulating sums.
2. Read an integer, N, and test for its being positive and no greater than 10,000. If either condition is not satisfied, output an appropriate error message and stop further processing.
3. Set up a loop in which the first three sums specified (SUMA, SUMB, SUMC) are computed.
4. Set up a loop for computing the fourth sum specified, SUMD.
5. Compute the fifth sum specified, SUME, by subtracting SUMD from SUMA.
6. Print each of the five sums with appropriate comments.
7. End.

Algorithm with detailed steps.

1.0   Set SUMA=SUMB=SUMC=SUMD=SUME=0.
2.0   Read N.
2.1   If N is less than or equal to 0 or greater than 10,000, print an error message and stop further processing.
3.0   Set up a loop with index I beginning at 1 and continuing through N. (Steps 3.1, 3.2, and 3.3 are the body of this loop.)
3.1   Assign to SUMA the value SUMA+I.
3.2   Assign to SUMB the value SUMB+$I*I$.
3.3   Assign to SUMC the value SUMC+$I*I*I$.
4.0   Set up a loop with index I and increment 2 such that I goes from 1 through the greatest odd integer less than or equal to N. (Step 4.1 is the body of this loop.)

4.1   Assign to SUMD the value SUMD+I.
5.0   Compute SUME=SUMA−SUMD.
6.0   Print SUMA with appropriate identification.
6.1   Print SUMB with appropriate identification.
6.2   Print SUMC with appropriate identification.
6.3   Print SUMD with appropriate identification.
6.4   Print SUME with appropriate identification.
7.0   End.

Program

```
      PROGRAM  SUMS
C******************************************************************
C*          PROGRAM DEFINITION                                    *
C*             SEE THE TEXT FOR COMPLETE DESCRIPTION OF           *
C*                  THE PROBLEM.                                  *
C*          VARIABLE DEFINITION                                   *
C******************************************************************
C*          SUMA   IS THE SUM OF ALL INTEGERS FROM 1 TO N.        *
C*          SUMB   IS THE SUM OF SQUARES OF INTEGERS FROM 1 TO N.*
C*          SUMC   IS THE SUM OF CUBES OF INTEGERS FROM 1 TO N.   *
C*          SUMD   IS THE SUM OF ODD INTEGERS FROM 1 TO N.        *
C*          SUME   IS THE SUM OF EVEN INTEGERS FROM 1 TO N.       *
C*          I   IS THE LOOP INDEX.                                *
C******************************************************************
      INTEGER SUMA, SUMB, SUMC, SUMD, SUME, N, I
C*
C*          INITIALIZATION
C*
      DATA SUMA, SUMB, SUMC, SUMD, SUME /5*0/
C*
C*        INPUT N AND VERIFY IT
C*
      READ*, N
      IF(N .LT. 0 .OR. N .GT. 10000)THEN
        PRINT*,' ERROR IN INPUT ', N
        STOP
      ENDIF
C*
C*          COMPUTE THE 5 SUMS.
C*
      DO 10 I = 1, N
        SUMA = SUMA + I
        SUMB = SUMB + I * I
        SUMC = SUMC + I * I * I
 10      CONTINUE
C*
      DO 20 I = 1, N, 2
        SUMD = SUMD + I
 20      CONTINUE
```

```
C*
       SUME = SUMA - SUMD
C*
       PRINT*, ' THE NUMBER N IS ', N
       PRINT*, ' SUM OF INTEGERS FROM 1 TO N IS ',SUMA
       PRINT*, ' SUM OF SQUARES OF INTEGERS FROM 1 TO N IS ',SUMB
       PRINT*, ' SUM OF CUBES OF INTEGERS FROM 1 TO N IS ',SUMC
       PRINT*, ' SUM OF ODD INTEGERS FROM 1 TO N IS ',SUMD
       PRINT*, ' SUM OF EVEN INTEGERS FROM 1 TO N IS ',SUME
       STOP
       END
```

In the preceding program, the computation of four of the required sums is done in two separate loops. As an exercise, try to combine the two loops into one while retaining the understandability of the program.

NESTED DO LOOP

Often situations will occur in solving problems where it is necessary to have a loop within a loop. One example might be the printing of an interest table for the simple interest formula $I = PRT$ where P, the principal, ranges from 4000 to 5000 in increments of 100; R, the annual interest rate, ranges from .08 to .15 in increments of .01; and T, the term of the loan, ranges from 20 to 30 in steps of 1. Consider the following program involving DO loops:

```
       PROGRAM   NEXT
       INTEGER PRINC, RATE, TERM
       REAL   INT
       DO 50 PRINC = 4000, 5000,100
         DO 40 RATE = 8, 15
           DO 30 TERM = 20, 30
             INT = PRINC * TERM * RATE/100.0
             PRINT*, PRINC, RATE, TERM, INT
30         CONTINUE
40       CONTINUE
50     CONTINUE
       STOP
       END
```

The preceding program would produce 968 lines of output and is an example of what a *nested loop* is: one loop within another. Before we try to analyze how the program produces that much output, let's use a diagram to consider nested DO loops in a general sense.

```
                              DO n₁ INDEX1=n₂,n₃,n₄
                                      .
   outer loop                         .
                                      .
                              DO m₁ INDEX2=m₂,m₃,m₄
                                      .
   middle loop                        .
                                      .
                              DO k₁ INDEX3=k₂,k₃,k₄
                                      .
   inner loop                         .
                                      .
      k₁                      CONTINUE
                                      .
                                      .
                                      .
      m₁                      CONTINUE
                                      .
                                      .
                                      .
      n₁                      CONTINUE
```

It is important to note some characteristics of this diagram:

1. Each loop is completely contained within another one.
2. Each loop has its own unique variable name for index.
3. Each loop ends in its own CONTINUE statement.

Now refer back to the example of nested loops for computing and printing simple interest. Each loop has a unique index name. Although the parameters for initial value, final value, and increment are different in each DO statement, they could have been the same. Those values, of course, depend on how many times each loop is to be executed. Now let's analyze what happens as these loops are executed.

When the first DO is executed, the value of PRINC is set at 4000 and the iteration counter for the loop is computed as $((5000-4000)/100) +1$, which is 11. Then the second DO statement is executed and RATE is assigned the value 8, and the iteration counter for that loop is computed as $((15-8)/1)+1$, which is 8. Next the third DO statement is executed and TERM is set at 20 and the third iteration counter is computed as $((30-20)/1)+1$, which is 11. Following that, INT is computed as shown and the PRINT statement causes a line of four numbers to be output. The CONTINUE statement with reference number 30 signals the end of the innermost loop, so its index, TERM, is increased by 1 since the step is not specified in the corresponding DO statement. The iteration counter is decreased by 1 and since that result is not 0, control returns to the line where INT is computed followed by the PRINT statement, after which line 30 is encountered again. The value of index TERM is incremented and the iteration counter is decreased by 1. These steps are repeated until the iteration counter of this loop has a value of 0, after which line 40 is executed. Since line 40 is a CONTINUE statement signalling the end of the loop whose index is RATE, this index (RATE) is increased by 1 and its corresponding iteration counter is decreased by 1. If the iteration counter is positive, program control returns to the first executable statement following the DO statement that starts this loop. The execution of that first executable statement means that once more the loop that starts with the line

```
DO 30 TERM = 20,30
```

becomes active; its index, TERM, is set to 20 and the loop is executed repeatedly until its iteration counter becomes 0. When that happens, line 40 is again executed and index RATE is incremented, and program control returns to the statement "DO 30 . . ." as long as the iteration counter is positive.

Eventually, the iteration counter for the RATE loop will be 0 and the CONTINUE statement whose reference number is 50 is executed. Now the index PRINC is incremented in the usual manner, after which program control returns to the statement "DO 40 . . ." and all of the steps described above are repeated as before. This process continues until finally the iteration counter of the "DO 50 . . ." loop becomes 0, at which time program control transfers to the statement immediately following line 50.

As stated previously, the PRINT statement is executed 968 times. Here's how it works. Each time the inner loop (DO 30 . . .) is completely satisfied, eleven lines of output are produced because the iteration counter of this loop is initially 11. Similarly, the middle loop (DO 40 . . .)

is executed eight times because the initial value of the iteration counter was 8. Since for each time the middle loop is executed the inner loop is executed eleven times, a total of eighty-eight lines of output results from complete satisfaction of both the inner loop and the middle loop.

The outer loop (DO 50 . . .) is executed eleven times because the initial value of its interaction counter was 11. For each execution of the outer loop, eighty-eight lines of output are produced as described above, thus 11 times 88, or 968 lines of output are produced when all three loops are completely satisfied.

The pattern of execution for nested loops is always the same as the one just described—that is, innermost loop is satisfied first, then the next outer loop, followed by subsequent outer loops until the outermost loop is satisfied. For each iteration of an outer loop, the inner loop goes through all required iterations to be completely satisfied again.

Now let's look at some other examples of nested DO loops.

Example 1

```
      DO 20 I=1,5
        K=I*I
        DO 10 J=1,5
          PRINT,K*J
10      CONTINUE
20    CONTINUE
```

The loop with index J is the innermost loop, so the PRINT statement within that loop is executed five times for each value of index I in the outer loop. Since I assumes five values and for each one of those, J assumes five values, a total of twenty-five lines of output are produced with each line consisting of one number.

Example 2

```
      DATA ISUM,JSUM,KSUM/3*0/
      DO 30 I=1,10
        ISUM=ISUM+1
        DO 20 J=1,10,2
          JSUM=JSUM+J
          DO 10 K=2,10,2
            KSUM=KSUM+K
10        CONTINUE
          PRINT*,KSUM
20      CONTINUE
        PRINT*,JSUM
30    CONTINUE
      PRINT*,ISUM
```

In this example, ISUM is printed once, JSUM is printed ten times, and KSUM is printed fifty times (five times for each JSUM).

Example 3

```
DO 20 I=1,15,2
    DO 10 J=I,10
        PRINT*,I*J
10      CONTINUE
20  CONTINUE
```

The inner loop (index is J) has a different initial value for its index each time because the initial value is I which is being changed in value from 1 to 3 to 5 to 7 to 9 to 11 to 13 to 15. As long as I is less than 10 (the final value for J), the inner loop is executed more than once. When I exceeds 10, then the inner loop is not executed at all. Verify that the PRINT statement is executed a total of thirty times.

Rules concerning DO loops

1. The terminating statement of a DO loop must be an executable statement but cannot be any transfer type statement like GO TO or IF, nor can it be a DO statement.
2. A DO loop must always be entered through its DO statement so as to properly initialize the index and compute the iteration counter.
3. Transfer *out* of a DO loop from a point other than the CONTINUE statement is permissible but transfer *into* a DO loop can only be done through its DO statement.
4. The initial value, final value, and increment may be constants, variables, or arithmetic expressions.
5. In nested loops there must be no overlapping of statements except that the same statement may be the terminating statement for more than one loop. However, this is not good programming style and should be avoided.

Implied DO loop

In using READ, WRITE, and PRINT statements whose data lists include subscripted variables, ForTran has a provision that results in processing very much like a DO loop but the loop is not explicitly specified. That structure is called an *implied DO loop*.

Consider this program segment:

```
    REAL A(20)
    DO 15 I=1,N
        READ*,A(I)
 15 CONTINUE
```

As you know, the execution of this program segment would require N lines (cards) of input where each line (card) consists of a single number.

Now study this line of ForTran code:

```
READ*, (A(I),I=1,N)
```

This READ statement, like the loop above, requires N values to be input into A(1), A(2), ..., A(N). However, using this statement makes it possible to input all N values from the same line (card) provided there is enough space on the line (card). The structure in this READ statement is called an *implied DO loop* because its effect is the same as that of a DO loop with the same parameters, except that new lines (cards) of data are not needed for each new subscript value as they are when a READ statement appears in an explicit DO loop.

The general forms for implied DO loops are as follows:

1. $(A(I), I=n_1,n_2,n_3)$ for a single-subscripted variable.
2. $((B(I,J),I=n_1,n_2,n_3),J=m_1,m_2,m_3)$ for a double-subscripted variable where the order of I and J depends on the desired order of the input or output.
3. $(((C(I,J,K),I=n_1,n_2,n_3),J=m_1,m_2,m_3),K=l_1,l_2,l_3)$ for a triple-subscripted variable where the order of I, J, and K depends on the desired order of the input or output.

In each of these general forms, the indices may be any acceptable integer variable and the parameters may be any integer constants, variables, or expressions, as is the situation for explicit DO loops. Furthermore, for two or three indices, the inner implied DO is executed first, then each succeeding outer one. Note that it is possible to have more than one item in an implied loop. For example, the statement

```
READ*,((A(I), B(I)),I=n₁,n₂,n₃)
```

is a correct ForTran statement.

Implied DO loops are used only to input, output, or initialize values for subscripted variables of one, two, or three subscripts. Following are some examples with explanatory comments:

```
10   INTEGER A(-1:3),B(2,10),C(3,4,10)
20   READ*,(A(I),I=0,2)
30   READ*,(((B(I,J),J=1,10),A(I)),I=1,2)
40   READ*,(((C(I,J,K),J=1,2),I=2,3),K=1,10,2)
```

Note that if the increment is 1 it need not be specified as a parameter, just as for explicit DO loops. Statement 20 would input three numbers and store them in A(0), A(1), and A(2), respectively.

Statement 30 would input twenty-two numbers and, in the order entered, they would be stored in

```
B(1,1),B(1,2),B(1,3), . . . ,B(1,10),A(1)
B(2,1),B(2,2),B(2,3), . . . ,B(2,10),A(2)
```

Finally, statement 40 would input 20 numbers and, in the order received, would store them as follows, in order from left to right, then top to bottom:

```
C(2,1,1) C(2,2,1) C(3,1,1) C(3,2,1)
C(2,1,3) C(2,2,3) C(3,1,3) C(3,2,3)
C(2,1,5) C(2,2,5) C(3,1,5) C(3,2,5)
C(2,1,7) C(2,2,7) C(3,1,7) C(3,2,7)
C(2,1,9) C(2,2,9) C(3,1,9) C(3,2,9)
```

The numbers could be entered on a single line (card) if there were space for them, or on more than one line (card).

The use of an implied DO loop with DATA statements produces identical results with those just described for READ statements. Consider these examples:

```
10   INTEGER A(-1:3),B(2,10),C(3,4,10)
20   DATA A,(B(1,J),J=1,10)/5*0,10*1/
30   DATA(((C(I,J,K),J=1,2),I=2,3),K=1,10,2)/5*1,5*2,5*3,5*4/
```

Statement 20, above, initializes all five elements of array A to 0 and the first row of array B to 1. Statement 30 will initialize 20 of the 120 elements of array C as follows (reading down the first column before proceeding to the second):

```
C(2,1,1)=1        C(3,1,5)=3
C(2,2,1)=1        C(3,2,5)=3
C(3,1,1)=1        C(2,1,7)=3
C(3,2,1)=1        C(2,2,7)=3
C(2,1,3)=1        C(3,1,7)=3
C(2,2,3)=2        C(3,2,7)=4
C(3,1,3)=2        C(2,1,9)=4
C(3,2,3)=2        C(2,2,9)=4
C(2,1,5)=2        C(3,1,9)=4
C(2,2,5)=2        C(3,2,9)=4
```

The same implied DO loops given earlier could be used with PRINT rather than READ. In such cases as many numbers would be printed (using unformatted PRINT) per line as the system dictates (usually five).

Another example of an implied DO loop will illustrate further. Consider this array declaration:

```
INTEGER C(0:2,4:7,−5:4)
```

Suppose this INTEGER statement appears in the same program as the following three READ statements:

(a) `READ*,((C(0,I,J),J=−5,4),I=4,7)`
(b) `READ*,((C(I,J,K),I=0,2),J=4,7)`
(c) `READ*,(((C(I,J,K),K=−5,0),I=0,2),J=4,7)`

When the READ in (a) is executed, forty values are read and stored in the following order:

```
C(0,4,−5),C(0,4,−4),C(0,4,−3),....,C(0,4,3),C(0,4,4)
C(0,5,−5),C(0,5,−4),C(0,5,−3),....,C(0,5,3),C(0,5,4)
C(0,6,−5),C(0,6,−4),C(0,6,−3),....,C(0,6,3),C(0,6,4)
C(0,7,−5),C(0,7,−4),C(0,7,−3),....,C(0,7,3),C(0,7,4)
```

The READ statement in (b) results in twelve values being read and stored as follows (reading left to right before proceeding to the next line):

```
C(0,4,K),C(1,4,K),C(2,4,K)
C(0,5,K),C(1,5,K),C(2,5,K)
C(0,6,K),C(1,6,K),C(2,6,K)
C(0,7,K),C(1,7,K),C(2,7,K)
```

In this case the value of K would have to be defined in some way previous to the execution of the READ and must be in the range from −5 to 4 (see the accompanying INTEGER statement).

Finally, in Example (c) above, that READ statement would cause the reading of 72 values which would be stored as follows (reading left to right):

```
C(0,4,−5),C(0,4,−4),C(0,4,−3),C(0,4,−2),C(0,4,−1),C(0,4,0)
C(1,4,−5),C(1,4,−4),C(1,4,−3),C(1,4,−2),C(1,4,−1),C(1,4,0)
C(2,4,−5),C(2,4,−4),C(2,4,−3),C(2,4,−2),C(2,4,−1),C(2,4,0)
C(0,5,−5),C(0,5,−4),C(0,5,−3),C(0,5,−2),C(0,5,−1),C(0,5,0)
C(1,5,−5),C(1,5,−4),C(1,5,−3),C(1,5,−2),C(1,5,−1),C(1,5,0)
C(2,5,−5),C(2,5,−4),C(2,5,−3),C(2,5,−2),C(2,5,−1),C(2,5,0)
C(0,6,−5),C(0,6,−4),C(0,6,−3),C(0,6,−2),C(0,6,−1),C(0,6,0)
C(1,6,−5),C(1,6,−4),C(1,6,−3),C(1,6,−2),C(1,6,−1),C(1,6,0)
C(2,6,−5),C(2,6,−4),C(2,6,−3),C(2,6,−2),C(2,6,−1),C(2,6,0)
C(0,7,−5),C(0,7,−4),C(0,7,−3),C(0,7,−2),C(0,7,−1),C(0,7,0)
C(1,7,−5),C(1,7,−4),C(1,7,−3),C(1,7,−2),C(1,7,−1),C(1,7,0)
C(2,7,−5),C(2,7,−4),C(2,7,−3),C(2,7,−2),C(2,7,−1),C(2,7,0)
```

As we conclude this section we call your attention again to the fact that the use of an array name without subscript designations in an input or output statement assigns or retrieves values to (from) the *entire* array. If only *some* of the array locations are to be used, methods described earlier in this chapter must be applied.

We have now completed our discussion of counter-controlled iteration structures. Next we consider condition-controlled iteration structures.

Condition-controlled iteration structures

This type of iteration structure is used in situations where the number of times the loop body is executed cannot be determined in advance of writing the program. To control the number of repetitions, a condition is built into the loop which, when satisfied, terminates the loop. This condition may be placed (a) at the beginning of the loop body, (b) within the loop body, or (c) at the end of the loop body. The following diagrams illustrate the three placements of the condition:

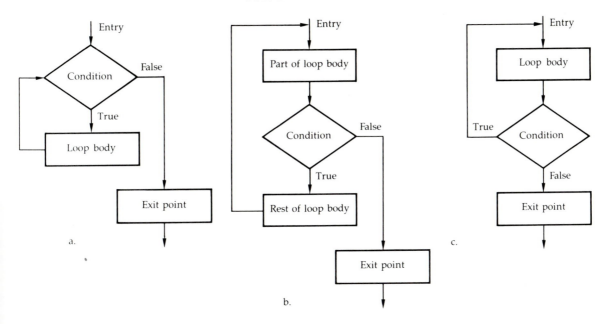

a.

b.

c.

All three of these cases may be implemented in ForTran 77 by using the IF and GO TO statements. We demonstrate this with three program segments.

(a)
```
        DATA COUNT,SUM,SUMSQ/3*0/
        READ*,X
   5    IF(X .LT. 0)GO TO 10                    ←Condition
            COUNT=COUNT+1
            SUM=SUM+X
            SUMSQ=SUMSQ+X*X                      } Loop body
            READ*,X
        GO TO 5
  10    AVG=SUM/COUNT
        VAR=(SUMSQ-COUNT*AVG*AVG)/COUNT
        PRINT*,AVG,VAR
                .
                .
                .
```

In this example the loop is bounded by the IF statement with reference number 5 and the GO TO 5 statement. The condition is tested *before* the loop body so it is possible to skip the loop execution completely.

(b)

```
      DATA COUNT,SUM,SUMSQ/3*0/
  5   READ*,X                          ←Part of loop body
        IF(X .LT. 0)GO TO 10           ←Condition
        COUNT=COUNT+1
        SUM=SUM+X                       } Rest of loop body
        SUMSQ=SUMSQ+X*X
      GO TO 5
 10   AVG=SUM/COUNT
      VAR=(SUMSQ-COUNT*AVG*AVG)/COUNT
      PRINT*,AVG,VAR
         .
         .
         .
```

Example (b) accomplishes the same results as Example (a) but this time the condition is tested within the loop body.

(c)

```
      DATA COUNT,SUM,SUMSQ/3*0/
  5   READ*,X
        COUNT=COUNT+1
        SUM=SUM+X                       } Loop body
        SUMSQ=SUMSQ+X*X
      IF(X .GE. 0)GO TO 5              ←Condition
      AVG=SUM/COUNT
      VAR=(SUMSQ-COUNT*AVG*AVG)/COUNT
      PRINT*,AVG,VAR
         .
         .
         .
```

This time the condition is checked *after* the loop body, so the loop is executed at least once. Although this program segment seems to solve the same problem as Examples (a) and (b) it has a logic error in it. When the sentinel datum (intended to stop the iteration) is input at statement 5, it is processed by the loop body as if it were valid data. Thus this placement of the test for the condition could not be used to solve the problem. We have included it to demonstrate all three placements of the condition test using essentially the same program statements.

WHILE-DO statement. Although we have just seen that various forms of condition-controlled structures can be developed by using IF and GO TO statements, some ForTran compilers provide for specific instructions designed to write condition-controlled iteration structures. At the time of this writing standard ForTran 77 does *not* have these features. We include them here because they are very useful and because a number of the commonly used ForTran compilers make provision for them. Furthermore, the use of these features makes for a better-structured program.

Two new ForTran statements will now be introduced: the WHILE-DO and the ENDWHILE. The general format of these statements is as follows:

```
WHILE (Condition) DO
                        }  Statements: loop body
ENDWHILE
```

The word DO is optional.

The entry to the loop is the WHILE-DO statement where the condition is tested. If the condition is true all statements between the WHILE-DO and the ENDWHILE are executed, after which the condition is tested again. As long as the condition is true, this process continues. The iteration is terminated when

(a) The condition tested in the WHILE-DO becomes false.
(b) A STOP statement within the loop body is executed.
(c) A RETURN statement (discussed later in this book) within the loop body is executed.
(d) A transfer out of the loop body occurs during the execution of the loop body.

The normal exit from a WHILE-DO loop is the statement immediately following the ENDWHILE. Abnormal exits may be the result of a RETURN, a STOP, or a GO TO statement.

Because ForTran 77 does not support the WHILE-DO statement we shall not place much emphasis on the program structure made possible through its use. Nevertheless, if it is available on the ForTran compiler to which you have access we urge you to use it. To help you learn this loop structure we present some examples.

Example 1

```
PROGRAM  AVERAG
INTEGER  COUNT, SUM, NUM, AVG
DATA COUNT, SUM, AVG/3*0/
READ*, NUM
WHILE(NUM .GT. 0)DO
   SUM = SUM + NUM
   COUNT = COUNT + 1
   READ*, NUM
ENDWHILE
AVG = SUM/COUNT
PRINT*, AVG, COUNT
STOP
END
```

This program reads and processes a set of numbers until a 0 or negative number is encountered. Thus, the program is designed to read a set of positive numbers, compute the average of the set and print the count of positive numbers and their average. The reason for the READ statement preceding the loop is to handle the unlikely situation where there are no positive numbers, only the sentinel datum. As written, the program does not correctly process this unlikely case to its conclusion because the statement immediately following the ENDWHILE would result in an attempt to divide 0 by 0. To avoid that attempted division we could replace the statement

```
AVG=SUM/COUNT
```

by the statement

```
IF(COUNT .GT. 0)AVG=SUM/COUNT
```

This statement would allow division to occur only if COUNT were greater than 0.

Example 2

```
      PROGRAM VAR3
C*********************************************************
C*              THIS IS A SECOND VERSION OF PROGRAM *
C*              VAR2 OF PAGE 181.  IN THIS VERSION  *
C*              THE WHILE-DO STRUCTURE IS USED.     *
C*********************************************************
      INTEGER COUNT
      REAL SUM, SUMSQ, NUMBER, MEAN, VAR
      DATA COUNT, SUM, SUMSQ /3*0/
C*
      READ*, NUMBER
C*
      WHILE (NUMBER .NE. 99999) DO
         COUNT = COUNT + 1
         SUM = SUM + NUMBER
         SUMSQ = SUMSQ + NUMBER * NUMBER
         READ*, NUMBER
      ENDWHILE
C*
      MEAN = SUM/COUNT
      VAR = (SUMSQ - COUNT * MEAN * MEAN)/COUNT
      PRINT*, VAR
      STOP
      END
```

PROBLEMS WITH COMPLETE SOLUTIONS

We conclude this chapter by presenting two problems, with algorithms for solving them, and the associated ForTran programs.

Problem 1

Develop an algorithm and write a computer program that provides practice problems in the skill of finding the sum of two 2-digit integers. The program should allow only one attempt for each practice problem and should keep track of the number of wrong and correct responses. Output should tell the user (1) the number of practice problems attempted, (2) the number answered correctly, (3) the number answered incorrectly, and (4) the percent answered correctly.

Algorithm:

1. Initialize necessary variables.
2. Ask user to specify number of problems wanted, then read that number.
3. Produce two randomly-generated two-digit integers and increase by 1 the count of problems generated.
4. Print the problem using the two integers.
5. Read the answer.
6. If answer is correct, increase the count of correct answers and output a message that answer is correct. If answer is incorrect, increase the count of incorrect answers and print a message that answer is incorrect.
7. If there are more problems to be printed, then repeat the process from step 3.
8. Print summary of number of problems attempted, number answered correctly, and number answered incorrectly and the percentage of correctly answered problems.
9. End.

Program

```
      PROGRAM   MATDRL
C***************************************************************
C*               PROVIDE THE DESCRIPTION OF   PROGRAM           *
C*               AND VARIABLES AS AN EXERCISE.                  *
C***************************************************************
C*
      INTEGER ANSWER, CORANS, TOTAL, SCORE
      INTEGER OPRND1, OPRND2, RESULT, INCANS, I
C*
      DATA CORANS, INCANS /2*0/
      PRINT*, ' HOW MANY PROBLEMS DO YOU WANT'
      READ*, TOTAL
C*
C***************************************************************
C*               VARIABLE TOTAL HOLDS THE NUMBER OF PROBLEMS*
C*               TO BE GENERATED. A DO LOOP IS SET UP TO        *
C*               GENERATE EXACTLY THAT MANY PROBLEMS. THE       *
C*               FUNCTION RANDOM IS USED TO GENERATE A RAN-     *
C*               DOM NUMBER BETWEEN 0 AND .999999,INCLUSIVE.    *
C*               YOU SHOULD CHANGE THIS FUNCTION TO THE         *
C*               RANDOM FUNCTION NAME OF YOUR LOCAL SYSTEM      *
C*               BEFORE EXECUTING THIS PROGRAM.                 *
```

```
C*                                                          *
C*            THE ANSWER TO EACH PROBLEM IS READ AND IS     *
C*            TESTED. IF THE ANSWER GIVEN IS INCORRECT,     *
C*            THE CORRECT ANSWER IS PRINTED.                *
C*            A SUMMARY REPORT IS PREPARED AT THE END.      *
C*******************************************************************
C*
          DO 50 I = 1, TOTAL
            OPRND1 = RANDOM(1) * 100 + 1
            OPRND2 = RANDOM(1) * 100 + 1
            RESULT = OPRND1 + OPRND2
            PRINT*, OPRND1, ' + ',OPRND2, ' = '
            PRINT*, ' ENTER YOUR ANSWER'
            READ*,ANSWER
            IF(ANSWER .EQ. RESULT)THEN
              CORANS = CORANS + 1
              PRINT*,' CORRECT'
            ELSE
              INCANS = INCANS + 1
              PRINT*, ' SORRY. THE ANSWER IS ',RESULT
            ENDIF
  50      CONTINUE
C*
          SCORE = 100 * CORANS / TOTAL
          PRINT*,' YOU WERE GIVEN ',TOTAL,' PROBLEMS.'
          PRINT*,' YOU ANSWERED ',CORANS,' CORRECTLY.'
          PRINT*,INCANS, ' OF YOUR ANSWERS WERE WRONG. '
          PRINT*,' YOUR PERCENT SCORE IS ',SCORE
          STOP
          END
```

Problem 2

Input a set of N numbers (N not greater than 100) and store them in an array—say, ANUM. Then sort this set of numbers in ascending order such that the sorted numbers are stored back in array ANUM. The output of the program is to be the sorted numbers.

Algorithm:

1. Read the numbers into array ANUM, counting them as they are read, stopping when sentinel datum of −999,999 is encountered.
2. Sort the numbers, storing them back in array ANUM in sorted order.
3. Print array ANUM.

Algorithm with detailed steps.

1.0   Define an array ANUM of size 100.
1.1   Initialize a number counter, N, to 0.
1.2   Input numbers into array ANUM until the sentinel datum −999,999 is reached.
2.0   Set loop index, I, to 1.
2.1   Find the smallest of the numbers at ANUM(I),ANUM(I+1), . . . , ANUM(N). Assume this smallest number is at ANUM(K).
2.2   Interchange ANUM(I) and ANUM(K).
2.3   Increase the index I by 1. If I is less than N, then repeat the process from step 2.1.
3.0   Output the N numbers in array ANUM.

The only step in the previous algorithm that has any complexity is step 2.1, so we explain a process that could be used. One method for finding the smallest of a set of N numbers ANUM(1), ANUM (2), . . . , ANUM(N), proceeds as follows: Beginning with I=1, assume that ANUM(I) is the smallest number in array ANUM, and store a copy of ANUM(I) in location SMALL. Compare the entire array from ANUM(I) to ANUM(N), one at a time, with the value at SMALL. If a number is found smaller than the current value of SMALL, the value at SMALL is replaced by that value and the location of that smaller number in array ANUM is stored at K. When all members of array ANUM from ANUM(I) to ANUM(N) have been compared with the number at SMALL by means of this process, K will contain the position number of the smallest of ANUM(I) to ANUM(N). Now ANUM(I) and ANUM(K) are interchanged, I is increased by 1, and the process is repeated with the new value for I. By this process, when I reaches the value N−1, the whole array will be sorted in ascending order.

Here is a ForTran program developed from the algorithm presented above.

Program

```
PROGRAM   SORT
C***********************************************************************
C*         PROGRAM DEFINITION                                         *
C*                 THIS PROGRAM INPUTS UP TO 100 NUMBERS,             *
C*                 SORTS THEM IN ASCENDING ORDER AND                 *
C*                 OUTPUTS THE SORTED NUMBERS.                        *
C***********************************************************************
```

```
C*          VARIABLE DEFINITION                               *
C*              NUM   IS AN ARRAY OF SIZE 100 TO STORE        *
C*                    THE SET OF NUMBERS.                     *
C*              SMALL IS TO STORE THE SMALLEST NUMBER         *
C*                    EACH TIME THROUGH THE LIST.             *
C*              TEMP  IS A TEMPORARY VARIABLE TO STORE        *
C*                    NUM(I).                                 *
C*              COUNT   IS TO HOLD THE NUMBER OF NUMBERS *
C*                    READ INTO ARRAY NUM.                    *
C*              K   IS THE POSITION OF SMALL IN   ARRAY       *
C*                    NUM.                                    *
C*              I,J  ARE LOOP INDEXES.                        *
C********************************************************************
C*
        REAL   NUM(100), SMALL
        INTEGER   COUNT,K
        COUNT = 0
C*
C*        READ NUMBERS UNTIL THE DUMMY NUMBER -999999
C*        IS ENTERED. AT THIS POINT COUNT IS THE NUMBER
C*        OF NUMBERS READ.
C*
        DO 10 I = 1, 100
          READ*, NUM(I)
          IF(NUM(I) .EQ. -999999)GOTO 12
          COUNT = COUNT + 1
  10      CONTINUE
C*
C********************************************************************
C*        AT THIS POINT A TOTAL OF COUNT NUMBERS ARE   *
C*        READ INTO ARRAY NUM. NEXT WE SORT THESE      *
C*        NUMBERS USING THE ALGORITHM DESCRIBED HERE.  *
C*        THIS IS KNOWN AS THE EXCHANGE ALGORITHM.     *
C*        SOME AUTHORS CALL IT THE  LINEAR SORT        *
C*        ALGORITHM.                                   *
C*                                                     *
C*        NOTE THAT SMALL HAS NO EFFECT ON THE PROGRAM *
C*        AND CAN BE REPLACED BY NUM(K). ITS PURPOSE   *
C*        IS TO HELP THE UNDERSTANDING OF THE PROGRAM. *
C********************************************************************
```

```
C*
 12        DO 15 I = 1, COUNT - 1
             SMALL = NUM(I)
             K = I
             DO 14 J = I, COUNT
               IF(NUM(J) .LT. SMALL)THEN
                 SMALL = NUM(J)
                 K = J
               ENDIF
 14          CONTINUE
             NUM(K) = NUM(I)
             NUM(I) = SMALL
 15        CONTINUE
C*
           PRINT*, (NUM(I),I=1,COUNT)
           STOP
           END
```

SUMMARY

In this chapter we have presented concepts related to arrays and iteration structures. The DATA statement was introduced as a convenient method for initializing variables. We summarize here some of the more important concepts.

Arrays

An array is a collection of computer memory locations all having the same name but distinguished from one another by the use of subscripts: single, double, or triple. ForTran statements available for defining a variable as an array are the DIMENSION statement and any of the type-declaration statements. Here are some examples of general forms of array-definition statements:

```
DIMENSION VAR1(n1), VAR2(n2), . . . ,VARN(nn)
INTEGER VAR1(n1), VAR2(n2), . . . ,VARK(nk)
REAL VAR1 (n1), VAR2(n2), . . . ,VARJ(nj)
CHARACTER VAR1(n1)*k1,VAR2(n2):k2, . . . ,VARN(nn)*kn
DIMENSION VAR1(m1:n1),VAR2(m1:n1,m2:n2),
  +   VAR3(m1:n1,m2:n2,m3:n3)
   REAL VAR1(m1:n1,m2:n2)
   INTEGER VAR2(m1:n1,m2:n2,m3:n3,m4:n4)
```

In the above examples, the capitalized combinations of letters and digits represent variable names and the small letters and digits represent integer constants that define the sizes of the arrays. In the case of the CHARACTER statement, k1,k2,..., kn are integer constants specifying the maximum number of characters each member of the array is to hold. If this is not specified the number of characters is assumed to be 1. Double- and triple-subscripted arrays are specified similar to any of the first four examples above as well as those shown in the last three examples.

Rules for using arrays

1. A single array-declaration statement may be used to define more than one array.
2. More than one array-definition statement may appear in any given program.
3. A given array name may appear in an array-declaration statement only once during a given program.
4. A subscript may be any valid constant, variable, or arithmetic expression leading to an integer within the range specified for that subscript in the array-declaration statement.
5. Once a variable name has appeared in an array-declaration statement and defined with one, two, or more subscripts, that variable must be used with the prescribed number of subscripts, except that in output, input, DATA statements or subprogram CALL statements (described in Chapter 7) it may appear without subscripts.
6. An array name must have appeared in an array-declaration statement prior to its being used as a subscripted variable.

The DATA statement

1. The general form of the DATA statement is

    DATA VAR1,VAR2, ... ,VARN/data1,data2, ... ,datan/

    where VAR1, ... , VARN are variable names and data1,data2, ... ,datan are the actual data to be stored.
2. If a given datum is to be used more than once, such use can be indicated by placing an asterisk to the left of the datum, preceded by the number of times it is to be used, as in this example:

```
DATA V1,V2,V3,V4/4*0/
```

3. If an array name appears without subscripts in a DATA statement, this means that *all* elements of the array are to be assigned values by the DATA statement, and this number of data values must then be accounted for between the pair of slashes following the array name.

DO loops

A DO loop is a collection of program statements repeated some finite number of times under the control of a DO statement and another statement identified in the DO statement. The general form of the DO statement follows:

DO *r* INDEX = exp1,exp2,exp3

where *r* is an integer constant specifying the reference number of the last statement in the loop; INDEX is any integer variable (or real variable in the case of a real DO loop); and exp1, exp2, and exp3 represent any constants, variables, or expressions resulting in integer (or real, in the case of a real DO loop) values. The first of these three, namely, exp1, is the initial value for INDEX, exp2 is the final value for INDEX, and exp3 is the increment to be used in going from the value of exp1 to the value of exp2.

Rules related to DO loops

1. The terminating statement of a DO loop must be an executable (preferably CONTINUE) statement but cannot be any transfer-type statement like GO TO or IF, nor can it be a DO statement.
2. A DO loop must always be entered through its DO statement so as to properly initialize the index and compute the iteration counter for the loop.
3. Transfer *out* of a DO loop is permissible but transfer *into* a DO loop can only be done through its DO statement.
4. The initial value, final value, and increment may be constants, variables, or arithmetic expressions.
5. In nested loops there must be no overlapping of statements except that the same statement may be the terminating statement for more than one loop.

EXERCISES

1. Indicate which of the following statements are true or false. (Note that a statement must be considered false if any part of it is false.) Be able to make changes in those that are false so that they will be true.

   (a) An implied loop may be used in an assignment statement.

   (b) ForTran 77 allows at most two subscripts for any given variable.

   (c) A ForTran compiler stores a two-dimensional array in row-major order.

   (d) If a DIMENSION statement is to appear in a program, it must be the first statement in the program.

   (e) If an INTEGER statement is to appear in a program, it must immediately follow any DIMENSION statements, or if there are none, the INTEGER statement must come first.

   (f) The REAL statement may be anywhere in the program as long as any variables declared in it are not used until after its appearance.

   (g) There may be at most three levels of nesting in nested DO loops in ForTran 77.

   (h) Implied DO loops may be used only in READ statements.

   (i) ForTran stores matrices in column-major order.

   (j) When using DO loops, it is permissible to transfer into and out of the body of a loop.

   (k) Transferring into a DO loop is permitted if the programmer sets the index with an assignment statement before so doing.

   (l) Real-indexed DO loops are available in almost every version of ForTran.

   (m) The following is an incorrect statement:

       DO I=1,1,1

   (n) One may use no more than ten DO loops in any given program.

   (o) It is not permissible to use the index of a DO loop in any statement that appears in the body of a DO loop.

   (p) Two or more nested loops can have the same variable as their indices.

   (q) Two or more nested loops may terminate in the same statement.

   (r) Every DO loop *must* terminate in a CONTINUE statement or else it is incorrect.

   (s) There must be at least one statement between a DO statement and a CONTINUE statement in order to have a DO loop.

(t) The terminating statement in a DO loop may be any executable statement.

(u) In ForTran 77, it is permissible to use a variable as the subscript in a DIMENSION statement.

(v) DIMENSION statements not only specify the *size* of subscripted variables but also cause values to be stored in the variables.

(w) A subscript may be any valid arithmetic expression leading to a positive integer value less than or equal to the original amount defined in a DIMENSION statement.

(x) The CONTINUE statement is an executable statement.

(y) When using nested DO loops, it is permissible for an inner loop to have its terminating statement later in the program than the terminating statement of an outer loop.

(z) Two or more DO loops may terminate in the same CONTINUE statement.

2. Determine which of the following ForTran statements are correct ForTran 77 statements and which contain errors. For those that are incorrect, make the changes needed to correct them.

(a) `DIMENSION A(10) B(20)`

(b) `DIMENSION A,B(10)`

(c) `DIMENSION A(7:10),B(10)`

(d) `DIMENSION A(10,20),B(5,6,10),`

(e) `DIMENSION IF(10),GOTO(20)`

(f) `DIMENSION X(1000,1000),Y(100,200,20)`

(g) `DIMENSION X(2,3,5,5)`

(h) `INTEGER VAR1,VAR(2)`

(i) `INTEGER X(1),X(2),X(3)`

(j) `CHARACTER A(20),*10`

(k) `INTEGER OUT(-10:10),IN(-5:0,5:10)`

(l) `REAL IN(10,20),A(50)`

(m) `CHARACTER*10,A(5),B,C(10)`

(n) `REAL A,B,C(20)`

(o) `REAL NUM(1),NUM(5)`

(p) `READ*,(I(A),A=1,5)`

(q) `READ*,(X(I),Y(I),I=1,5`

(r) `READ*,A,(A(J),J=1,5)`

(s) `READ*,A,IN,OUT(I),X(5)`

(t) `READ*,(A(I),I=1,3),(B(I),I(-1,10)`

(u) `READ*,(A(I,J,K),K=1,5)`

(v) `READ(3,*,END=3)(A(I,J,1),J=1,5)`

(w) `10 READ *,(A(I,1),I=1,5)`

(x) `READ(1,*,END=6) B(2),(B(I),I=3,5)`

(y) READ(2,*,END=10) A(1,2),A(I,J,K)

(z) PRINT*,A,B(10)

(aa) PRINT*,A(10,20),A(5)

(bb) PRINT*,B(1),B(2,6)

(cc) PRINT*,B(1)+B(2)

(dd) PRINT *,A,(B(I)+A,I=1,5)

(ee) PRINT (2,*),(A(I),I=1,5)

(ff) WRITE(6,*),(A(I,J),I=10,1)

(gg) WRITE(1,*)A,(B(I,J),I=J=1,5)

(hh) WRITE(2,*,END=15)X,Y(1)

(ii) WRITE(1,*)N,A(N),A(N+1)

(jj) DATA A,B(5)/6*2/C/3./

(kk) DATA A(1,J),J=1,10/10*0/

(ll) DATA (B(I,J),I=1,10)/10*1/

(mm) DATA A,B,C/20*1,3,7/

(nn) DATA A/10./B/20./

(oo) DATA ((A(I,1,K),I=0,4),K=2,3)/10*1./

(pp) 10 CONTINUE

(qq) DO 20 I=1,10,−1

(rr) DO 50 I=10,1

(ss) DO 9 J=1.5

(tt) DO 10 I=1,5

(uu) DO 20 J=1,1,2

3. In the following segments of ForTran programs, identify any syntactical errors. Correct those in which you find errors and use appropriate indentation where applicable.

(a)   DIMENSION A(10)
        READ*,(A(I),I=1,5)
        DO 10 I=1,10
     10 A=A+A(I)

(b)   INTEGER A(10)
        READ*,A(2)

(c)   DIMENSION A(−8:8)
        INTEGER A
        A(X)=X*X+1
        READ*,X
        PRINT*,A(X)

(d)   DIMENSION A(10,2)
        READ*,A
        DO 10 I=1,10
     10 A(I,1)=A(I−1,2)

(e)
```
      DO 20 I=1,10
      K=K+I
   20 DO 30 J=K,100
   30 PRINT*,J
```

(f)
```
      DIMENSION A(20)
      DO 20 I=1,20
      READ(1,*) A(I)
      T=A(I)+A(I+1)
      Q=A(I-1)+A(I)
      PRINT*,T,Q
   20 CONTINUE
      END
```

(g)
```
      DIMENSION A(0:1,0:5),B(1:12)
      INTEGER A,B
C*
C* THIS PROGRAM SHOULD COPY A(0,0) THROUGH A(0,5) INTO B(1) THROUGH B(6)
C* AND A(1,0) THROUGH A(1,5) INTO B(7) THROUGH B(12)
C*
      READ*,A
      DO 2 I=0,5
      B(I)=A(0,I)
      B(I+6)=A(1,I)
    2 CONTINUE
```

(h)
```
      DIMENSION A(100),B(10,10)
      READ *,A
      DO 10 I=1,100
   10 B(I,I)=A(I)
```

(i)
```
      INTEGER B(10,10)
      DO 10 I= 1,10
      DO 10 J= 1,10
      B(I,J)='*'
   10 CONTINUE
      PRINT*,(B(I,J),J=1,10)
   10 CONTINUE
```

(j)
```
      INTEGER B(10,10)
      READ*,B
      DO 10 I=1,10
      PRINT*,B(I,I),B(I,I-1),B(I-1,I)
   10 CONTINUE
```

(k)
```
      INTEGER A(50),B(5,10)
      READ*,A
      DO 20,J=1,5
      DO 20,I=1,10
   20 B(I,J)=A(I*J)
```

4. Specify the output produced by each of the following programs:

(a)
```
      PROGRAM A
      DIMENSION INT(10)
      DO 10 I=1,10
         DO 20 J=1,10
            INT (J)=I*J
20       CONTINUE
         PRINT*,INT
10 CONTINUE
      END
```

(b)
```
      PROGRAM B
      CHARACTER A(-30:30)
      DO 10 I=-30,30
         A(I)='$'
10 CONTINUE
      DO 20 I=-30,30,10
         PRINT *,A(I)
         DO 30 J=1,6
            PRINT *,(A(K),K=-30,30)
30       CONTINUE
20 CONTINUE
      END
```

(c)
```
      PROGRAM C
      CHARACTER A(10,10)
      DO 10 I=1,10
         A(I,I)='$'
         A(I,11-I)='*'
10 CONTINUE
      DO 20 I=1,10
         DO 20 J=1,10
            IF(A(I,J) .NE. '$' .OR. A(I,J) .NE. '*')A(I,J)=' '
20 CONTINUE
      PRINT *,A
      END
```

(d)
```
      PROGRAM D
      READ*,N
      IF(N .LT. 1 .OR. N .GT. 15)GO TO 30
      K=1
      DO 10 I=1,N
         K=K*I
         PRINT*,I,'FACTORIAL=',K
10 CONTINUE
30 PRINT*,'N MUST BE BETWEEN 1 AND 15.'
      END
```

```
(e)    PROGRAM E
       INTEGER N,SUM,K,I
       REAL X
       READ*,N
       SUM=0
       DO 10 I=1,N−1
          X=I
          K=N−(N/X)*I
          IF(K .EQ. 0) SUM= SUM+I
    10 CONTINUE
       IF(SUM .EQ. N)THEN
          PRINT*,N,'IS A COMPLETE NUMBER.'
       ELSE
          PRINT*,N,'IS NOT A COMPLETE NUMBER.'
       ENDIF
       END
```

5. In the following programs, look for errors in program logic. As you identify them, make the changes necessary to correct any such errors.
   (a) This program is intended to compute and print the sum of all odd integers between J and K.

```
       PROGRAM ODD
       DATA ISUM,JSUM/2*0/
       READ*,J,K
       IF(K−J .LE. 0)THEN
          PRINT*,'NO INTEGERS BETWEEN',J,'AND',K
          STOP
       ENDIF
       DO 10 I=1,J,2
          JSUM=JSUM+I
    10 CONTINUE
       DO 20 I=1,K,2
          KSUM=KSUM+I
    20 CONTINUE
       ISUM=KSUM−JSUM
       PRINT*,ISUM
       END
```

(b) A *perfect number* is a positive integer such that the sum of all its factors equals the number itself. The smallest perfect number is 6, whose factors are 1, 2, and 3. The sum of 1, 2, and 3 equals 6. This program is intended to find and print the second smallest perfect number.

```
      PROGRAM PERFEC
      INTEGER N,SUM,K
      DO 30 N=7,10000
          SUM=0
          DO 20 J=1,N-1
             IF(N/J .EQ. N/FLOAT(J)) SUM=SUM+1
20        CONTINUE
          IF (SUM .EQ. N)THEN
          PRINT*,'THE SECOND PERFECT NUMBER IS',SUM
          STOP
          ENDIF
30    CONTINUE
      END
```

(c) This program should read into an array, A, a set of 100 numbers that is already sorted so that its members are in ascending order. Then a number X is to be read and the array A searched to determine the position of X in array A.

```
      PROGRAM SEARCH
      REAL A(100),X
      READ*,A
      DO 10 I=1,100
         READ *,X
         IF(X .EQ. A(I))THEN
            PRINT*,X,'IS IN POSITION',I
            STOP
         ELSEIF(X .GT. A(I))THEN
            PRINT*,X,'IS NOT IN THIS ARRAY.'
            STOP
         ENDIF
10    CONTINUE
      END
```

(d) This program is intended to read an integer, then print each digit of the integer on a separate line beginning with the most significant digit.

```
    PROGRAM DIGIT
    READ*,N
    DO 10 I=1,10
       X=N/10.0
       K=(X−N/10)*10
       PRINT*,K
 10 CONTINUE
    END
```

6. (General)
   Given two positive integers, M and N, write a program to compute the sum of all integers from M to N, inclusive; the sum of the even integers from M to N; and the sum of the odd integers from M to N. For the last two sums, include the integers M and N as appropriate.

7. (General)
   Write a program that, for all positive integers I, J, and K from 1 to 1000, inclusive, finds all combinations of I, J, and K such that

   $$I + J = K$$

8. (General)
   Write a program that for all positive integers I, J, K, and L from 1 to 1000, inclusive, finds all combinations of I, J, K, and L such that

   $$I + J + K = L$$

9. (Mathematics, General)
   The *transpose* of a matrix, A, is the matrix obtained by interchanging the rows and columns of A. Write a program that will read the elements of a 10 × 10 array, A, and store its transpose in array B. Then write array B in matrix format.

10. (Mathematics, General)
    A square matrix (one having the same number of rows and columns) is called a *diagonal matrix* if its only nonzero elements are on the diagonal from upper left to lower right. It is called *upper triangular* if all elements below the diagonal are 0's, and *lower triangular* if all elements above the diagonal are 0's. Write a program that reads a square matrix (no larger than 10 × 10) and determines if it is one of these three special matrices.

11. (Mathematics, General)

Two matrices having the same number of rows and the same number of columns may be added, the *sum matrix* being found by adding corresponding elements in the two matrices. Similarly, the *difference matrix* is obtained by subtracting corresponding elements. Write a program that reads two matrices of the same dimensions and finds the sum and difference matrices. Output should be properly identified.

12. (Mathematics, General)

Write a program to find all integer solutions to the equation

$$X + Y = 2(X - Y)$$

where X is greater than or equal to Y and both X and Y are less than 1000.

13. (Mathematics, General)

Write programs to compute the values of each of the following series:

(a) $1 + \left(\frac{1}{8}\right) + \left(\frac{1}{27}\right) + \cdots + \left(\frac{1}{100}\right)$

(b) $1 - \left(\frac{1}{2}\right) + \left(\frac{1}{3}\right) - \left(\frac{1}{4}\right) + \cdots - \left(\frac{1}{1000}\right)$

(c) $1 + \left(\frac{X}{1}\right) + \left(\frac{X}{2!}\right) + \left(\frac{X}{3!}\right) + \cdots + \left(\frac{X}{10!}\right)$

where

$$2! = 2 \cdot 1 = 2$$
$$3! = 3 \cdot 2 \cdot 1 = 6$$
$$4! = 4 \cdot 3 \cdot 2 \cdot 1 = 24$$
$$k! = k \cdot (k-1) \cdot (k-2) \ldots 1$$

Compute values of this series for $X = 1,2,3,\ldots,10$.

14. (General)

Write a program that reads any Roman numeral from I to XX and prints the equivalent decimal number. The twenty Roman numerals are I,II,III,IV,V,VI,VII,VIII,IX,X,XI,XII,XIII,XIV,XV,XVI,XVII, XVIII,IX,XXX.

15. (General)

Consider five adjacent squares arranged as follows:

Write a program to find all possible ways of placing the digits 1 through 8 in the above pattern so that no two squares having a common side contain consecutive digits. One such arrangement is

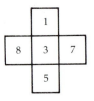

16. (Statistics, General)

A rectangular room has a floor consisting of one-inch tiles. The dimensions of the room are $m \times n$ inches. Suppose a cockroach is placed on any given tile located at row i and column j, where i is less than or equal to m and j is less than or equal to n. Assume that the cockroach crawls from tile to tile at the rate of one tile per second, and that, from any single tile, it crawls randomly with equal probability to any of the eight tiles surrounding it (unless it is against the wall). Write a program to compute the time needed for the cockroach to touch every tile on the floor at least once. Assume that $2 < m < 40$ and $2 < n < 40$ and that the cockroach cannot enter any given tile more than 50,000 times. (If this maximum is reached before all tiles are touched, restart the cockroach at a different tile.)

The output for each trial should be (1) the total number of seconds required, (2) the number of entries onto each tile, and (3) a list of the tiles by position from the one(s) entered most often to the one(s) entered just once.

Here are some hints for solving this problem:

1. Establish an array COUNT(M,N) to keep track of the number of entries to the tile whose position in the room is row M and column N.

2. If the cockroach is on tile (I,J), then the position of the next tile to which it moves is given by

```
(I+IMOVE(K),J+JMOVE(K))
```

where K is a random number from 1 through 8 and

```
IMOVE(1) — IMOVE(7) = IMOVE(8) = −1
IMOVE(2) = IMOVE(6) = 0
IMOVE(3) = IMOVE(4) = IMOVE(5) = 1
```

and if we assume the arrangement of tiles with respect to the cockroach as follows:

3	4	5
2	0	6
1	8	7

A little thought will tell you the values to use for

```
JMOVE(1)...JMOVE(8)
```

17. (Mathematics, General)

The Magic Square Problem. A magic square is an n by n matrix of integers from 1 to n^2 such that the sums of every row, column, and diagonal are equal. In fact, that sum will always be the value of

$$(n/2) * (n^2 + 1)$$

Mathematicians have shown that for even values of n no magic square exists.

For $n = 3$, here is a magic square:

6	1	8
7	5	3
2	9	4

Here is a magic square for $n = 5$:

15	8	1	24	17
16	14	7	5	23
22	20	13	6	4
3	21	19	12	10
9	2	25	18	11

One algorithm for creating a magic square once n is given is as follows:

1. Place 1 in the center of the top row. (Remember, n is odd.)
2. Subsequent placement of successive numbers is done by moving diagonally upward to the left unless the number just placed is in the top in its column.
3. If the number just placed is the top one in its column, the next number is placed in the bottom square of the adjacent column to the left.
4. If there is no column to the left when applying step 2 or step 3, place the next number in succession in the appropriate row (as defined in steps 2 and 3) of the rightmost column of the matrix.
5. Every time the number just placed is a multiple of n, the next number is placed just below it in the same column, after which step 2 or step 3 is applied.
6. Continue until the last number, n, has been placed.

Write a program to read an odd integer and write a magic square.

CHAPTER 6
Additional Capabilities for Input and Output

In Chapter 2 you were introduced to input and output statements. You were also introduced to unformatted READ, PRINT, and WRITE statements. Recall that these were called unformatted because there was no way for the programmer to specify columns in which data were to be read from cards, or to prescribe the columns across the page in which output was to appear. The layout of both input and output was what the computer system specified rather than what the programmer would like. There was no specified format, hence the adjective *unformatted*.

A major benefit of unformatted input/output is that the programmer need not be concerned about the details of precisely organizing either the input or the output, thus being relieved of some of the drudgery of programming.

Of course, there are a great many situations where the organization of information can mean the difference between its being useful or not. It might be essential for understanding some reports that they be printed in precise formats. It may be that cards have already been punched with specific information in certain fields, and we want to be able to read those fields without looking at *all* the columns of the cards.

It is for situations like these, and others, that ForTran makes available formatted input and output. We begin with formatted input.

FORMATTED INPUT

The general form of a formatted READ is either of the following:

1. `READ` f,V1,V2, . . . ,VN
2. `READ`(u,f,END=n)V1,V2, . . . ,VN

In both of the preceding, *f* represents either the reference number of a statement (called the FORMAT statement) having the function of specifying the precise layout of information to be input or the format specification string itself. V1,V2, . . . ,VN represent the list of variables into which data are being read. In form 2, *u* stands for the logical-unit number (see, in Chapter 2, the section on unformatted READ), and "END=n" means that a program statement having reference number *n* is the next one to be executed if an end-of-file mark is encountered. The "END=n" is an optional feature in a READ statement.

The first form of the formatted READ is used when input takes place from a card reader (batch system) or terminal keyboard (interactive system). The second form is used, as we learned in Chapter 2, when input occurs from an electronically stored file such as magnetic tape or magnetic disk.

We have said nothing yet about the FORMAT statement except to point out that *f* in the two general forms of the formatted READ was the reference number of such a statement. We shall examine the FORMAT statement more closely after the next section.

FORMATTED OUTPUT

Formatted PRINT

The general form is as follows:

`PRINT` f, data-list

where *f* represents the format specification string or the reference number of the associated FORMAT statement, and data-list represents the list of output to be produced. Note the use of commas to separate the number *f* from the data list and, though they do not appear in the given PRINT statement, the entries in the data list from each other. This use of commas is required here as it is in the READ statements discussed previously.

As we learned in Chapter 2 for unformatted output, the data list in a PRINT statement may consist of variables, numeric constants, string constants, or arithmetic expressions. The PRINT statement always causes output to occur on the printer of a batch computer system or on the terminal of an interactive system. Here is an example of a formatted PRINT statement:

```
PRINT 10,A,B,A+B, 'THIS IS OK.'
```

The number 10 is the reference number of the associated FORMAT statement elsewhere in the program, A and B are variables assumed to have been previously defined, A+B is an arithmetic expression, and 'THIS IS OK.' is a character string.

Formatted WRITE

Sometimes it is desirable or even necessary to produce output on a device other than a printer—for example, on magnetic tape or on a magnetic disk. As with the unformatted WRITE discussed in Chapter 2, that capability is available by using the formatted WRITE statement. Here is its general form:

```
WRITE(u,f) data-list
```

As with the formatted READ, u represents the number of the logical unit where output is to occur. Logical unit 6 is commonly used to designate the printer. The letter f represents the format specification string or the reference number of the associated FORMAT statement, and data-list represents the list of outputs to be produced. It is possible to have a blank data list, in which case the output statement produces one blank record of output.

THE FORMAT STATEMENT

We are now ready to consider the FORMAT statement mentioned several times in the preceding paragraphs. Because there are at least five different specifications permitted in a FORMAT statement or format specification string, it is not possible to give an example of a completely general FORMAT statement. A format specification may be provided in either of two ways: (1) by means of a FORMAT statement whose reference number appears in an associated input or output statement, and (2) by means of a character string bounded by a pair of parentheses. Here are some examples:

FORMAT statements	FORMAT strings
10 FORMAT(I2,I7)	'(I2,I7)'
20 FORMAT(I4,1X,I6)	'(I4,1X,I6)'

As in these examples, every FORMAT statement must have a reference number which, you recall, appears in the READ, PRINT, or WRITE statement specified by it. Of course, this reference number must not be assigned to any other statement. More than one READ, PRINT, or WRITE statement may reference the same FORMAT, however. Following the word FORMAT is at least one pair of parentheses inside of which appear the specifications that determine the spacing and form of the information being input or output.

The FORMAT statement is a nonexecutable statement since it causes no action by the computer. It specifies or modifies the action resulting from another statement such as the READ, PRINT, or WRITE. Here is a simple program to illustrate the use of the FORMAT and the format string.

```
      PROGRAM  SIMPLE
      INTEGER  M,N
      READ '(I3,I7)', M, N
      WRITE(6,20)M, N, M + N
 20   FORMAT(I3,1X,I7,1X,I9)
      STOP
      END
```

The READ statement reads two integers N and M in the form and spacing specified in the format string immediately following the word READ. A look at this format string reveals that I3 and I7 appear inside parentheses. The letter "I" means that input numbers are integers and must be stored internally as integers. The number 3 to the right of the first I specifies that the integer input for N occupies three columns on a punched card or three spaces when entered from a terminal. Similarly, the number 7 indicates seven columns on a punched card if the computer is a batch system or seven spaces if an interactive terminal is being used. Note that the READ statement has two items (N and M) in its data list and the associated format string has two specifications, I3 and I7.

The WRITE statement shows 20 as the reference number of the associated FORMAT. The FORMAT whose reference number is 20 has three I-type specifications and two X-type specifications. We have briefly mentioned some of the properties of the I-type specification. When used in conjunction with an output statement, the I-format causes the output information to be in the form of an integer. The X-format specification is new; it designates blank spaces. In particular, then, FORMAT 20 in the preceding program results in an integer (including sign if it is negative) being printed in positions 1 through 3, one blank space in position 4, an integer in positions 5 through 11, one blank space in position 12, and an integer in positions 13 through 21.

Now we discuss each of several possible format specifications permitted in FORMAT statements or format strings. As we do so you will realize how much control the programmer gains over the form and spacing of input and output through the use of formatting. In fact, the control is complete. Therefore it is most important to learn how to apply FORMAT statements appropriately.

I-format specification

The form of the I format is as follows:

In

where n is a positive integer specifying the number of spaces allowed for input or output, and I designates that the information is an integer. If there is a sign associated with the integer it must be included in the designated spaces. Whenever an I format is used, the integer being input is always assumed to be right-justified in its field. That is, it must be positioned so that the units digit is in the rightmost space of the field. If by mistake this is not done, the spaces to the right of the integer are assumed by the system to be 0's. Consider this example:

```
READ'(I5)',N
```

Suppose that the card in which the value of N is punched looks like this:

Then the value input to variable N is 50000 since the blanks in columns 3, 4, and 5 are interpreted as 0's in conjunction with the I5 format specification.

Similarly, information *printed* under an I format is positioned at the right of its designated field.

Here is a complete program to further illustrate the I format.

```
1          PROGRAM  TEST
2          INTEGER NUM1, NUM2, NUM3
3          READ(5,10)NUM1, NUM2, NUM3
4          WRITE(6,15)NUM1, NUM2, NUM3
5 10       FORMAT(I2,I5,I4)
6 15       FORMAT(I5,I8,I7)
7          STOP
8          END
```

Line 3 causes the reading of three pieces of information from logical unit 5 (card reader or interactive terminal) in compliance with FORMAT 10 (line 5), the FORMAT that specifies the spacing to be used for the three integers being read. NUM1 is read from positions 1 and 2, NUM2 from positions 3 through 7, and NUM3 from positions 8 through 11. Line 4 causes printing on logical unit 6 (printer or interactive terminal) in compliance with FORMAT 15, which reserves positions 1–5 for NUM1, the next eight spaces for NUM2, and the next seven spaces for NUM3. Line 7 signals the end of processing.

Suppose the preceding program were processed on a batch system and that the data were punched on a card as follows:

```
|  | 7 |  | - | 2 | 1 | 0 |  |  | 1 | 3 |  |  |  |
```

The READ statement together with the FORMAT statement causes the value 7 to be stored in NUM1, the value −210 to be stored in NUM2, and the value 13 to be stored in NUM3. Note the following fields on the card: columns 1 and 2, columns 3 through 7, and columns 8 through 11. These fields exactly match the I-format specifications in the FORMAT statement whose reference number is 10—that is, I2, I5, and I4.

Assume that the preceding program is executed on an interactive system and that the data entered are as follows:

```
| 1 | 7 | 2 | 3 | 4 | 1 | 5 | 2 |  |  |  |  |
```

The value stored in NUM1 is determined by the specification I2, so the first two digits constitute this value—namely, 17. NUM2 is read according to the specification I5, so the next five digits, 23415, make up the value stored in NUM2. Finally, NUM3 is read with the specification I4 so the next four digits would constitute the value stored in NUM3. Since only one digit remains of the eight entered as input, the program assumes the other three digits of the four needed to satisfy the I4 specification will be 0's. Therefore, the value stored in NUM3 will be 2000. If the user of the program had intended NUM3 to have the value 2, then the input should have been

| 1 | 7 | 2 | 3 | 4 | 1 | 5 | | | | 2 |

in which case the digit 2 is right-adjusted in its four-space field.

Suppose the program is executed with this most recent input data. In this case the output is as follows:

| | | | 1 | 7 | | | 2 | 3 | 4 | 1 | 5 | | | | | | | 2 |

Thus the number to the right of the I in an I-format output-specification specifies the total number of columns set aside in the output for the value to be printed from a given memory location. That value is always right-adjusted in the output field, with blank spaces filled in to the left of the value, thus completing the specified field.

X-format specification

The form of the X format is

nX

where n is a positive integer indicating a number of spaces to be *skipped*, as is specified by the letter X. As with the I format, the X format can be used for both input and output. When used in conjunction with a formatted READ, the X format causes the designated number of characters to be ignored. In an output situation the X format causes the designated number of blank spaces to be produced on the output medium.

Consider this sample program:

```
1              PROGRAM EXAMPL
2              INTEGER   N, M
3              READ 100, N, M
4  100    FORMAT(2X,I3,1X,I4)
5              PRINT 100, N, M
6              STOP
7              END
```

Suppose this program were executed and suppose the input keyed in were

23567,2381

In this case the output would be

567 2381

Let's review both the input and output processes in this program. First the input. Format specification 2X in line 4 causes the first two characters of input to be ignored. The format specification I3 causes the next three digits, 567, to be stored in variable N. The format specification 1X causes the next character input (the comma) to be ignored. Finally, the specification I4 causes the four digits 2381 to be stored as the value of M.

The use of the same FORMAT statement for output will have these results:

1. The specification 2X causes two blank spaces.
2. The specification I3 causes the contents of N—namely, 567—to be printed.
3. Next, 1X causes one blank space.
4. Finally, I4 causes the contents of M—namely, 2381—to be printed. Since the number of positions allotted in I4 is exactly the same as the number of digits in M, no blank spaces are output to the left of the value of M (2381) except for the one space produced by specification 1X.

For practice in the use of I and X formats, consider the following different input data to the preceding program:

1. 3681275100
2.   812.51
3. 2-812*51

Verify that in all three cases the output is

```
812 5100
```

F-format specification

The form of the F format is

Fw.d

where F indicates that the value to be read or printed is a real number. (Real numbers were called *floating-point* numbers when ForTran was first developed, hence the letter F for this specification.) The *w* specifies the *width* of the field (how many spaces to be used, including sign and decimal point), and *d* specifies the number of decimal digits to the right of the decimal point in the input or output. The period, as shown in this general form, must be there.

Consider these three programs and realize that they all perform the same tasks and produce the same output. They differ only in the handling of format specifications.

```
      PROGRAM F1          PROGRAM F2                 PROGRAM F3
      REAL A, B           REAL A, B                  REAL A, B
      A = 26.812          A = 26.812                 CHARACTER FMT*9
      B = 17              B = 17                     A = 26.812
      PRINT 15, A, B      PRINT '(3X,F6.2)', A, B    B = 17
15    FORMAT(3X,F6.2)     STOP                       FMT = '(3X,F6.2)'
      STOP                END                        PRINT FMT, A, B
      END                                            STOP
                                                     END
```

Program F1 uses a FORMAT statement to specify the form of the output, program F2 uses a format string in the PRINT statement, and program F3 uses a character variable, FMT, to specify the format. All three programs produce this output:

```
26.81
17.00
```

Let's analyze program F1. (Analysis for the others would be very similar.) The value 26.812 is stored in variable A and the value 17 is stored in variable B. The PRINT statement in conjunction with FORMAT 15 causes three blank spaces (3X) followed by a field of six positions (F6.2) to contain the value of A. Only two fractional digits (F6.$\underline{2}$) are to be output in the value of A. Since the value stored in A is 26.812, and since only two fractional digits are to be output, the value that is printed for A is 26.81, a total of five characters including the decimal point. (*Note that output is correctly rounded.*) Six spaces were designated as the field width, so that leaves one blank space plus the five characters 26.81 to fill out the field. Any blank spaces resulting from an F-format specification are inserted at the left of the nonblank information. The effect of this is to right-adjust information that is output by means of the format.

You probably noticed in the output produced by the preceding three programs that the two numbers printed appeared on separate lines. This is the result of FORMAT 15 having only *one* F specification while the PRINT statement designates *two* variables to be output, A and B. When this condition occurs—that is, when there are more items in the data list of a PRINT or WRITE than there are format specifications in the accompanying FORMAT—the program is compiled so that output format specifications are repeated from the point where the rightmost right–facing parenthesis appears. Thus, in this example, the specification in FORMAT 15 is used twice by the PRINT statement. Also note that the numbers that are output appear on two separate lines. This is because in applying FORMAT 15 to the PRINT statement, the closing parenthesis in the FORMAT statement was encountered before all items in the PRINT statement had been output. Whenever the end of a FORMAT statement is encountered like that, a new line is produced before any more output occurs. Repetition of the FORMAT specification begins from the rightmost right-facing parenthesis, which appears in FORMAT 15 just to the left of 3X. A similar situation occurs when a formatted READ statement is used. When there are more items in the data list of the READ statement than there are format specifications in the accompanying FORMAT statement, then the format specifications are repeated from the point where the rightmost right-facing parenthesis occurs. In a *batch system*, this situation results in a new card being processed. That is, the repetition of formats from the rightmost right-facing parenthesis causes the card reader to take information from a new card. In the case of most *interactive systems*, this situation results in additional input data being entered on a new line.

F format with input.   When the F-format specification is used in conjunction with a READ statement, the following situations hold:

1. If the data as entered from the terminal or input on a punched card include no decimal points, then the F specification causes the data to be stored with as many fractional digits as indicated in the F format.

Example

```
    READ 100,A,B
100 FORMAT(F6.2,F8.4)
```

Suppose the preceding lines where executed as part of a complete program and that input data were entered as follows:

```
12345612345678
```

The value stored in variable A is 1234.56 because the specification F6.2 causes six characters to be read with the two rightmost digits being the fractional digits. Similarly, the number in variable B is 1234.5678 because F8.4 causes the next eight characters to be read with four fractional digits.

2. If the data entered from the terminal or input on a punched card include decimal points, then the placement of the decimal point in a given datum determines the number of fractional digits in the value stored.

Example

```
    READ 100,A,B
100 FORMAT(F6.3,F8.0)
```

This time suppose the preceding program segment were executed with the following input data:

```
123.45123.4567
```

The value stored in variable A is 123.45 because the specification F6.3 causes six characters to be input, and since the decimal point is included in those six characters, *it* determines that *two* fractional digits are to be stored, not three digits as indicated by F6.3. In other words, the decimal point included in the input data overrides the specification in the F format. Similarly, the value stored in variable B is 123.4567, the F8.0 specifying eight characters but since the decimal point is included in the input data, it overrides the specification of 0 fractional digits in F8.0.

E-format specification

This specification is used when data are to be read or printed in exponential form. Its general form is

E *w.d*

where w represents the total width of the input or output field and d represents the number of fractional digits in the coefficient that appears to the left of the E. For example, if the number 13562E05 is read with format specification E8.4, the value stored would be 1.3562×10^5, which is the same value as 135620. Note again that in the specification E$w.d$, d specifies the number of fractional digits in the real number to the left of E when the number is in exponential form, and w is the number of positions in the entire field in which the number appears.

When using E-format specification for input, if the real part of the number in exponential form contains a decimal point, then the d part of the format specification has no purpose, because the placement of the decimal point in the datum overrides the specified placement in the format specification. For example, if the number 1.3562E5 is input with format specification E8.3, the value stored is $1.3562 \times 10^5 = 135620$. The fact that a decimal point appears in the datum, so that there are four fractional digits, takes precedence over the specification of three fractional digits. This is similar to the situation with F format described above.

In connection with output, the use of specification E$w.d$ causes a total of w spaces to be available for the number to be output, and specifies that d fractional digits will appear in the real part of the exponential number being output. The form of the output is always

$$\pm 0.X_1 X_2 \ldots X_d E(\text{exp})$$

where a preceding sign is printed if it is negative, otherwise not; where the *integer* portion of the real coefficient is always 0 and is printed; and where the exponent (exp) appears with sign if it is negative and with a leading space if it is positive. Because of this output form, whenever E-format specification is used with data of unknown size, one must allow for a position for a possible leading negative sign, a 0, a decimal point, the letter E, a possible negative sign or blank space for the exponent, and two digits for the exponent. This means that to be sure to have enough space for the number to be printed, the value of w (in the specification E$w.d$) must be at least seven greater than the value of d.

Consider this program as an example of correct use of E-format specifications:

```
        PROGRAM EXAMPL
        REAL A, B
        READ 10, A, B
        PRINT 20, A, B
10      FORMAT(E7.4,F9.2)
```

```
20      FORMAT(E13.6, E9.2)
        STOP
        END
```

Check carefully the relationship between input data and output as they relate to the format specifications used.

A-format specification

The four format specifications discussed so far are used either for numeric data or for blank spaces. We turn our attention now to a specification used when the input or output data are character-type data. The format specification is the A format and has the form

$$A n$$

where n specifies the number of characters being read or printed by a given use of the A format. The n in this format specification must be considered in two different contexts.

Whenever the variable being read or printed has been declared as a character variable it is not necessary to include n in the A specification. When n is omitted it means that the input or output operation is to process the full number of characters specified for that variable in the CHARACTER statement in which it was declared. Note that some versions of ForTran *require* the use of n in the A-format specification. Furthermore, it is simply a good programming practice to specify n.

If the n is included in the A-format specification and the variable to which it applies has been declared in a CHARACTER statement the following situations hold (assume m is the length of the variable as declared in the CHARACTER statement):

1. n equals m.
   In this case the situation is as described in the preceding paragraph.
2. n is greater than m.
   In this case only the m leftmost characters are read and the remaining $n - m$ characters are truncated from the right end of the string being read. Where the variable is to be *printed*, $n - m$ blank spaces are output at the left of the m characters of the string.
3. n is less than m.
   In this case only the n leftmost characters are processed out of the m characters available. In the case of reading, the rightmost $m-n$ characters are made blanks.

Here are two programs to demonstrate these cases.

```
PROGRAM ONE
CHARACTER ST*10
ST = 'ABCDEFGHIJ'
PRINT '(A)',ST
PRINT '(A2)',ST
PRINT '(A12)',ST
END
```

The output is:

```
ABCDEFGHIJ
AB
  ABCDEFGHIJ
```

The first PRINT statement has no *n* with the A-format specification, so exactly all ten characters of ST are printed. The second PRINT statement (case 3 above) results in only the leftmost two characters of ST being printed. Finally, the last PRINT statement calls for twelve characters to be printed, but there are only ten available, so two blank spaces are printed at the left of the ten-character string. This last PRINT statement is an example of case 2 above.

```
      PROGRAM EXTWO
      CHARACTER  ST*10
      READ 10, ST
10    FORMAT(A)
      PRINT 11, ST
11    FORMAT(A2)
      PRINT 12, ST
12    FORMAT(A12)
      READ 11, ST
      PRINT 10, ST
      PRINT 11, ST
      PRINT 12, ST
      READ 12, ST
      PRINT 10, ST
      PRINT 11, ST
      PRINT 12, ST
      STOP
      END
```

There are three READ statements, so assume the following three lines of input data:

```
ABCDEFGHIJ
UVWXYZ
UVABCDEFGHIJ
```

The output resulting from the preceding input run by PROGRAM EXTWO is

```
AB
  ABCDEFGHIJ
UV
UV
   UV
ABCDEFGHIJ
AB
  ABCDEFGHIJ
```

Let's analyze these results. The first READ statement causes as many characters to be input as are declared for variable ST in the CHARACTER statement. The two PRINT statements that come next are similar to the last two PRINT statements in PROGRAM ONE. The second READ statement uses FORMAT statement 11, which calls for two characters to be read from the left end of the second input line. Thus UV are the characters read into variable ST and the remaining eight characters of ST are made blanks. The next three PRINT statements produce lines 3, 4, and 5 of the output. Note that line 5 is the result of using an A12-format specification to output the ten-character string, ST, consisting of two letters and eight blanks. Thus two blank spaces appear to the left of the first character in the string.

The third line of input contains 12 characters and FORMAT 12 allows for twelve characters. However, the string ST is declared to have only ten characters. Thus the leftmost two characters of input data are omitted. If the format specification in FORMAT 12 had been A10, then the rightmost two characters of the input line would have been truncated.

If the variable associated with the "An" specification has *not* been declared in a CHARACTER statement, the situation now to be discussed prevails. (We strongly recommend that alphanumeric variables be declared in CHARACTER statements, but also believe the student should know what happens if they are not.) As you know, information is stored in a computer as combinations of electronic representations for 1's and 0's, and such combinations are in some code, a very common one requiring eight bits per character. Thus if the computer word length is sixteen bits, such a word could store at most two characters and A2 would be the maximum A-format specification. If the word length were thirty-two bits, the maximum A-format specification would be A4, and so on.

To help clarify the use of this format specification if the CHARACTER statement is not used, we shall consider three cases in which the computer word length allows for j characters, the input or output item contains w characters, and n is the number of characters specified in the A format.

Case 1: Assume $w = n$ and that both are less than or equal to j. All w characters of the input (output) item are processed.

Case 2: Assume $w > n$ and that n is less than or equal to j. There are more characters seeking input (output) than are specified in the A format, so only the n leftmost characters in the data item are input (output), and the rightmost $w - n$ characters are ignored.

Case 3: Assume $w < n$ and that n is less than or equal to j. All w characters in the input (output) item are processed as the leftmost characters and $j - w$ blanks are made the rightmost characters of the item being processed.

Single quotation marks in FORMAT statements

Recall that in the unformatted PRINT or WRITE we are able to use a string constant as a data element to be printed. It is also possible to include string constants in a format specification, and such string constants will be printed in the same manner as when they appear in the PRINT or WRITE. Here are some examples:

Example 1

```
     PRINT 10, N
10   FORMAT (1X,'TOTAL=',I3)
```

When this is executed the string consisting of the six characters TOTAL= is printed, followed by the value stored at N printed in I3 format.

Example 2

```
     PRINT 20, A, '+', B, A+B
20   FORMAT(F6.2,A1,F6.2,'=',F7.2)
```

When this PRINT statement is executed the output list consists of the value of A, the character "+", the value of B, and the sum of A and B. The FORMAT statement has an F6.2 specification for the value of A, an A1 specification for the character "+", an F6.2 specification for the value of B, and an F7.2 specification for the sum, A+B. Thus, each item in the output list has a corresponding format specification. There is, however, the character '=' that appears in the FORMAT statement. Its effect is to cause an equal sign to be output between the value for B and the value for A+B. Any character constant appearing between a pair of single quotation marks in a FORMAT statement will be output in the order it appears in the FORMAT.

Example 3

```
      WRITE (6,10)
10    FORMAT(5X,'NAME',5X,'NUMBER',1X,'CODE')
```

This is a typical method for printing a heading. There is no output list in the WRITE statement. Everything to be printed is specified in the FORMAT, including spacing (five spaces at the beginning of the line, five spaces between NAME and NUMBER, and one space between NUMBER and CODE).

H-format specification

In older versions of ForTran, character (string) constants were output with the H format, which has this form:

nHstring

where "string" represents a character string of exactly n characters in length, n being a positive integer. Although this older method for treating character strings is not as convenient as the single quotations we have just discussed, the H-format specification is still accepted by ForTran 77. Here is an example:

```
      PRINT 10,AVG
10    FORMAT(20HTHE AVERAGE WAGE IS ,F6.2)
      PRINT 15, A,B,A+B
15    FORMAT(F6.2,1H+,F6.2,1H=,F8.2)
```

The FORMAT with statement number 10, causes twenty characters immediately following the letter H to be printed first in the output line. Just to their right the value of AVG is printed in F6.2 format. In FORMAT 15, F6.2 specifies the output form for the value of A. To the right of that value a plus sign is printed followed by the value of B in F6.2 format. Next is printed the single character "=" after which the sum of A and B is output in F8.2 format. A distinct disadvantage of the H format is that the count of characters must be exact. If it is not, a fatal error is apt to result.

CONTROLLING VERTICAL SPACING

As you know from examples that we have used, the execution of each PRINT or WRITE statement causes one line of output. Thus it has the effect of vertically spacing the output. Sometimes it is desirable to be able to space immediately to a new page, or to double space the output, or even to cause no vertical spacing at all. Such controlling of vertical spacing is referred to as *carriage control*. This comes from the fact that on batch computer systems, the major output device is usually a line printer. A feature of the line printer is the *paper carriage*—the mechanism that positions the paper for the device that actually does the printing of characters on the paper. The paper carriage contains a device called the *carriage control*, which responds to special signals from the program, causing vertical movement of the paper corresponding to those signals. We turn our attention now to the programming method for controlling movement of the paper carriage.

Carriage control

On batch computer systems, the first character of every line of formatted output controls the movement of the paper carriage. Therefore, the first character specified for output by a FORMAT statement is the carriage-control character. The allowable carriage-control characters are given in Table 6.1. Since the first character specified for output in a FORMAT statement is used by the computer system for carriage control, it is important that the programmer provide the proper character in this place.

TABLE 6.1   Carriage control characters

Character	Effect on paper carriage
Blank	Single space forward. This is the most common character used in FORMAT statements.
0	Double space forward. This causes a one-line skip after each printed line.
1	Move to the top of a new page.
+	Print on the same line as the previous output. This character has the effect of suppressing the normal spacing to a new line before printing.

Furthermore, the first character is not printed on the output page. It is the *second* specified character that is printed in the first space on the output page. The first character has the sole function of controlling the paper carriage. If that character is not one of the special ones reserved for carriage control (the ones in Table 6.1) the system tries to interpret it as one of the special characters and, of course, does not print it.

Here is an example to demonstrate the concept of controlling vertical spacing.

Example
Problem.   Input these data for each employee of a company:

1. Name, in the first twenty-two positions of input.
2. Identification number, in the next six positions.
3. Hours worked, in the next five positions (hours worked is a number between 0.00 and 99.99).
4. Hourly rate of pay, in the next five positions (a number between 0.00 and 99.99).

Input is to continue until sentinel data are encountered such that the characters END occur as the first three positions of input data. Output is to be as follows:

1. Print the headings at the beginning of a page.
2. Double space the individual employee output.
3. Print twenty-five lines of individual employee output per page.
4. Print the summary output on a separate page.

Program

```
        PROGRAM EMPLOY
C*****************************************************************
C*          PROGRAM DEFINITION                              *
C*              THIS PROGRAM READS EMPLOYEE'S NAME          *
C*              IDENTIFICATION NUMBER, TOTAL HOURS          *
C*              WORKED AND HOURLY PAY RATE. IT              *
C*              COMPUTES WAGES FOR EACH EMPLOYEE,           *
C*              TOTAL HOURS WORKED, TOTAL PAYMENT AND       *
C*              AVERAGE HOURLY RATE. FOR EACH               *
C*              EMPLOYEE IT REPORTS NAME, ID NUMBER,        *
C*              HOURS WORKED AND WAGES EARNED. THE          *
C*              INPUT DATA ARE ENDED WITH A CARD            *
C*              CONTANING 'END' IN ITS FIRST 3 COLUMNS.*
C*              AT THE END OF DATA THE PROGRAM REPORTS      *
C*              TOTAL PAYMENTS, TOTAL HOURS WORKED AND      *
C*              AVERAGE HOURLY PAY ON A SEPARATE PAGE.      *
C*****************************************************************
C*          VARIABLE DEFINITION                            *
C*              NAME = NAME OF EMPLOYEE                     *
C*              EMPLID = EMPLOYEE ID NUMBER                 *
C*              HOURS = HOURS WORKED                        *
C*              RATE = HOURLY PAY RATE                      *
C*              WAGE = HOURS * RATE                         *
C*              TWAGES = TOTAL WAGES FOR ALL EMPLOYEES*
C*              THOURS = TOTAL HOURS WORKED BY ALL          *
C*                      EMPLOYEES                           *
C*              AVG = AVERAGE RATE OF PAY PER HOUR          *
C*              LINE = LINE COUNTER TO CONTROL PAGING       *
C*****************************************************************
C*
        REAL HOURS, RATE, WAGE, TWAGES, THOURS, AVG
        INTEGER   LINE
        CHARACTER NAME*22, EMPLID*6
C*
C*          INITIALIZATION
C*
        DATA TWAGES, THOURS, LINE/3*0/
        PRINT20
        READ 10, NAME,EMPLID, HOURS, RATE
C*
C*                  START OF READING AND PROCESSING DATA
```

```
C*
  4        IF(NAME .EQ. 'END')GOTO 8
             WAGE = HOURS * RATE
             LINE = LINE + 1
             IF(LINE .GT. 25)THEN
                PRINT 20
                LINE = 0
             ENDIF
             PRINT 30, NAME, EMPLID, WAGE
             THOURS = THOURS + HOURS
             TWAGES = TWAGES + WAGE
             READ 10, NAME, EMPLID, HOURS, RATE
           GOTO 4
C*
C*         PREPARING FINAL RESULTS
C*
  8        IF(THOURS .GT. 0)THEN
             AVG = TWAGES/THOURS
             PRINT 40, THOURS,TWAGES
             PRINT 50, AVG
           ENDIF
           STOP
C*
 10        FORMAT(A22,A6,F5.2,F5.2)
 20        FORMAT('1', 10X, 'NAME',10X,'ID # ',1X, 'WAGE')
 30        FORMAT('0',2X,A22,1X,A6,1X,F5.2)
 40        FORMAT('1','FOR A TOTAL OF ',F7.2,' HOURS WE PAID $',F7.2)
 50        FORMAT('0', ' THE AVERAGE HOURLY RATE IS $',F5.2)
           END
```

The slash (/) in FORMAT statements

When performing input and output operations, it is often use-ful to consider a collection of information as a single unit—for example, the information on *one* punched card, or the information on *one* printed line. Processing of such a unit of information occurs so often that a phrase is used to identify it—namely, *physical record*. A *physical record* in computer science, then, is a collection of information that is to be input or output at approximately the same time by the appropriate hardware device. All the information on one punched card is input or output at nearly the same time by the card reader/punch. An entire line of print-ing is done almost instantly by a line printer. On magnetic tape such a collection of information is identified by an end-of-record mark, and in-put occurs until the end-of-record mark is encountered. These are all examples of processing physical records by input/output devices.

Before we conclude this discussion of the concept of a *record*, it should be noted that there is a related notion in computer programming called the *logical record*. A *logical record* is a collection of related information stored in computer memory in such a way that it can be processed as a unit, not necessarily on one card or on one line as with the physical record, but, nevertheless, as a single unit. For example, in the data processing systems used by a college registrar it would be common to organize information about students in such a way that the information related to any given student is stored as one logical record. Thus, when computer processing of student information occurs, it is done by student record.

There is in ForTran the capability of causing input or output to proceed to the next physical record, and the symbol in FORMAT statements that causes such processing to occur is the slash (/). Since the most common output device you are apt to use is a printer, the following discussion of the use of the slash will be in that context.

When used in a FORMAT statement referenced by a PRINT or WRITE, the slash indicates that the next character of output specified to the right of the slash is to appear on a new line. If a slash is followed by a slash, each slash specifies a new line on which subsequent output is to appear. Here is a simple example:

Suppose values stored at A, B, and C are 20.575, 30.5, and .752, respectively. Consider these program statements for printing those values.

```
      PRINT 20,A,B,C
20    FORMAT (//1X,F10.2,/1X,F10.2,1X,F10.2)
```

The output produced by the preceding two lines of program is as follows:

```
20.58
30.50     .75
```

The first two slashes cause two blank lines and positioning of the print mechanism to be ready for printing on the third line. The 1X to the right of the second slash produces the carriage control character while F10.2 causes the value of A (20.575) to be rounded to two decimal places and printed with two fractional digits right-justified in the leftmost ten-space field. The third slash causes a new line to begin; again 1X produces the carriage control character; F10.2 causes the value of B to be printed with two fractional digits right-justified in the leftmost ten-space field; 1X causes one blank space; and the last F10.2 causes the value of C to be printed with two fractional digits (rounded) in the next ten-space field.

If a slash appears in a FORMAT referenced by a READ statement, it causes the current record of input to be skipped. Consider the following example of a partial program for a batch computer system:

```
       READ 10,A,B,C
10     FORMAT(F8.3//F8.2,F6.1)
```

Assume these input cards are provided:

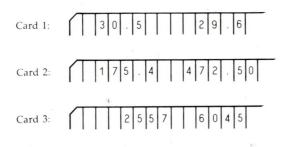

The value stored in A would be 30.5 (from the leftmost eight columns of card 1). The rest of card 1 and all of card 2 would be skipped because of the two slashes in FORMAT 10. The leftmost eight columns of card 3 are used to provide a value for B and since two decimal digits are specified (F8.2), the value stored in B is 25.57. The next six columns of card 3 are used to provide a value for C and since one decimal digit is specified (F6.1) the value stored in C is 604.5.

Here are the contents of the three memory locations after the execution of the above READ with its accompanying FORMAT:

A $\boxed{\text{30.5}}$ from the leftmost eight columns of card 1 with F8.3 format.
B $\boxed{\text{25.57}}$ from the leftmost eight columns of card 3 with F8.2 format.
C $\boxed{\text{604.5}}$ from columns nine through fourteen of card 3 with F6.1 format.

Here is a short but complete program that also illustrates the use of slashes in FORMAT statements.

```
        PROGRAM EXAMPL
        REAL A,B,C
        READ 15,A, B, C
15      FORMAT(F10.4/F10.4/F10.4)
        PRINT 25,A, B, C
25      FORMAT(1X,F10.4,F10.4,F10.4,//)
        PRINT 35,A+B, A+C, B+C
35      FORMAT(1X,F10.4,F10.4,/,1X,F10.4)
        STOP
        END
```

Suppose the above program is executed on an interactive system. The input provided and output produced follow:

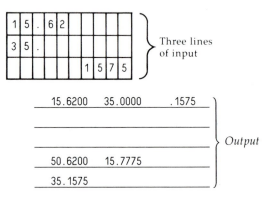

Note that the two slashes in FORMAT 15 specify that each value is taken from the first ten spaces of each of three different lines. When decimal points are included in the input, as on the first two lines, their placement determines the number of fractional digits in the number being input. When no decimal point appears in the input (as in the third line) the format specification determines the number of fractional digits.

The two slashes in FORMAT 25 produce two blank lines. The slash in FORMAT 35 causes the last value to be output on a new line.

PARENTHESES IN FORMATS

Sometimes format specifications are repeated in a given FORMAT statement as, for example, in FORMAT 25 of the previous example. ForTran allows such repeated specifications to be written once and then preceded by an integer that indicates the number of repetitions. Therefore, FORMAT 25 could have been written as follows:

```
25    FORMAT(1X,3F10.4//)
```

This statement causes exactly the same results as the one shown in the program.

When repeated *patterns* of format specifications occur, rather than rewriting the pattern as often as it occurs, you may enclose the pattern in parentheses just once and precede the parentheses by an integer indicating the number of repetitions. For example, suppose this FORMAT statement occurs in a program:

```
10    FORMAT(1X,F10.4,1X,F10.4,1X,F10.4)
```

Obviously, the repeated pattern is 1X,F10.4. Therefore, the preceding statement could be written as

```
10    FORMAT(3(1X,F10.4))
```

Here is another example.

```
20    FORMAT(1X,I10,I10,I10,A2,F10.2,F10.2)
```

The preceding statement could be written as follows:

```
20    FORMAT(1X,3I10,A2,2F10.2)
```

Notice that the repeated I10's and F10.2's need be written only once and are preceded by an integer indicating the number of times each is to be repeated.

FORMAT SPECIFICATIONS AND INPUT/OUTPUT LIST

We have discussed before the need for making sure that the items in the list of a READ, PRINT, or WRITE statement correspond in type to the format specifications in the associated FORMAT statement. It is also important that there be the same *number* of format specifications as there are items in the associated data list.

The general rule for the processing of a case where the number of items in a data list is not the same as the number of format specifications is as follows:

1. The items in the input/output list will be corresponded to the format specifications until all items in the data list have been processed.
2. If there are more items in the list than there are format specifications, format specifications are repeated, beginning with the replication number, if any, in front of the rightmost right-facing parenthesis, until condition 1 occurs.

The programmer must take note of the fact that when repeated use of format specifications occur according to part 2 of the preceding rule, it can easily happen that the mode of the data item does not match that of the specification. Unless such use is carefully designed and is intentional, this can be a source of difficulty.

We present some examples to clarify the preceding discussion.

```
    REAL A,B,X
    INTEGER K
    READ 10,A,B,K,X
10  FORMAT(F10.5,F6.2,I3)
```

In this case the first three items in the data list of the READ correspond correctly to the first three format specifications in the FORMAT. The fourth item in the READ data list is associated with specification F10.5 since repeated use of format specifications begins with the rightmost right-facing parenthesis. Thus, in this example, no mismatch of mode occurs between data list items and format specifications, so no error message occurs. Note, however, that the value for X will be taken from the first ten columns of a second input card (or line).

In this next example, we present just one READ statement with four different possible FORMAT statements and a discussion of the effect of each combination.

```
        REAL A,X
        INTEGER K,J
        READ 10,A,K,X,J
(a)  10  FORMAT(F10.2,I6)
(b)  10  FORMAT(2(F10.2,I6))
(c)  10  FORMAT(F10.2,I6,F10.2,I6)
(d)  10  FORMAT(F10.2,I6,F10.2,I6,2A5)
```

If (a) is used, variable A is read with the specification F10.2 and variable K with specification I6. Since there are no more format specifications, an automatic return to the rightmost right-facing parenthesis occurs and format specifications are repeated from that point in the FORMAT statement, and a new line is then read. Therefore, variable X is read with F10.2 specification and variable J is input with specification I6, from a second card (line) of input.

If the FORMAT in (b) is used, the specifications F10.2 followed by I6 are repeated; thus variable A is read with F10.2, variable K with format I6, variable X with format F10.2, and variable J with I6, where the input data are all on the same card (line).

The FORMAT in (c) has exactly the same effect as for (b). The FORMAT in (d) would result in variable A being read according to format F10.2, variable K according to format I6, variable X according to format F10.2, and variable J according to format I6. The remaining format specifications (2A5) are not used by the given READ statement but will cause no errors.

There are many more intricacies that occur in the use of FORMAT statements. However, there is probably no satisfactory method of discussing most of them until they are encountered in actual situations. That will be the approach taken from this point on. That is, as special cases arise in the use of FORMATS, we shall explain the effects.

PROBLEMS WITH COMPLETE SOLUTIONS

We conclude Chapter 6 by discussing four complete programming examples with the objective that your studying them will help you acquire problem-solving and programming techniques.

Problem 1

Problem: A computerized accounts receivable file exists which contains two items of information on each customer: (1) six-digit customer number, and (2) account balance. This file is a sequential file sorted in ascending order of customer number. A ForTran program is needed which accepts as input any number of triplets of data as follows:

1. A single letter (A, U, D, or any other letter) where
   A means a new customer is to be added to the file,
   U means an existing customer record is to be updated,
   D means an existing customer record is to be deleted,
   any other letter means end of data.
2. A six-digit customer number.
3. A positive or negative number, where positive indicates a customer purchase, and negative indicates customer payment on account.

The input data are to be entered in ascending order of customer number. For each triplet of input data the program must output the following information:

1. If the letter input is A, the output is:

   ```
   NEW CUSTOMER ADDED XXXXXX  XXXXX.XX
   ```

   where the first six X's represent customer number and the last X's represent new account balance.
2. If the letter input is U, the output is either

   ```
   CUSTOMER # XXXXXX NEW PURCHASE XXXXX.XX BALANCE XXXXX.XX
   ```

   or

   ```
   CUSTOMER # XXXXXX PAYMENT XXXXX.XX BALANCE XXXXX.XX
   ```

3. If the letter input is D, the output is

   ```
   CUSTOMER # XXXXXX HAS BEEN DELETED.
   ```

4. If the letter input is anything other than A, U, or D, output the entire file in two columns with appropriate headings.

To solve this problem we must modify a data file (called the master file) in accordance with three items of data. The first of the three items is the code specifying the type of modification to be made to the data file, the second item identifies the customer account to be modified, and the third item indicates the amount of change in the account balance. In order to modify the master file, it must be read one record at a time, and each record tested to determine if it is the one to be modified. If it is, then appropriate modification takes place. Otherwise, records continue to be read from the master file until the one to be modified is encountered. These steps are repeated until the code is encountered that indicates no more modifications are to be made. At that time, the master file is repositioned so that it can be output from the beginning so that the user may be aware of the most recent data in the data file.

Here is an algorithm to accomplish these tasks.

Algorithm:

1.0 Prepare MASTFL as logical unit 1 for input, and a scratch file as logical unit 2 for temporary input and output. Declare all variables.

2.0 Input a single record from logical unit 1 consisting of NUMF and BALANC.

3.0 Input current transaction data consisting of CODE, NUMI, and TRANS.

3.1 If CODE indicates the adding of a new record to the file, proceed to step 5.0 to do this.

3.2 If CODE indicates the updating of a record, proceed to step 6.0 to do this.

3.3 If CODE indicates the deleting of this record, proceed to step 7.0 to do this.

4.0 (Note that this step with its substeps are executed only when the end of all transactions is reached or when the end of MASTFL is encountered or if a CODE is keyed incorrectly.) Copy the rest of MASTFL (unit 1) to the scratch file (unit 2).

4.1 Initialize unit 2 to its beginning.

4.2 Prepare headings for the output.

4.3 Copy unit 2 to output and then back to unit 1.

4.4 Save unit 1 as the new MASTFL and stop further processing.

5.0 (At this step a new record is added to the file.) Find the proper location in the file for the new record.

5.1 Add the new record to the file.

5.2 Repeat the process from step 3.0.

6.0 (At this step an existing record is updated.) Locate the record which must be updated.

6.1 Update the record.

6.2 Repeat the process from step 3.0.

7.0 (At this step a record must be deleted.) Locate the record which must be deleted.

7.1 Delete the record.

7.2 Repeat the process from step 3.0.

Some of the steps in this algorithm such as "Locate the record," "Update the record," and so forth, need more detailed description if they are to be translated one-for-one into program statements. We describe here one of these steps ("Locate the record") and leave the others as exercises. The reader will gain some insight into the writing of these algorithm steps by examining the program segments that perform the associated processes.

To "locate a record" we consider two factors: (1) the unit 1 file is sorted in ascending order of customer number; and (2) transactions are entered in ascending order of customer number. Thus, when a record of transaction data is input, we have the customer number available as NUMI and we read record after record of the file designated as unit 1 and copy it to unit 2 until the account number, NUMF, of unit 1 matches NUMI. The CODE specifies whether the located record of the file on unit 1 is to be updated or deleted. If the record is to be updated, such action is then taken, followed by copying the updated record to the file on unit 2. If CODE specifies deletion of the record on unit 1, then no copying to unit 2 occurs. In other words, when processing is complete, unit 2 will be a copy of unit 1 with appropriate updates and deletions. Similar handling of new records to be added will ensure that unit 2 also contains such new records.

In order to run this program on any given computer system one must provide an appropriate set of statements to do the following:

1. Connect a permanent file named MASTFL to logical unit 1;
2. Connect a scratch file to logical unit 2;
3. Save the updated copy of the permanent file MASTFL.

In ForTran 77 these tasks are done by using the OPEN and CLOSE statements.

The program we present next was developed to implement the preceding algorithm. Study the program very carefully to see if there are certain cases of input data that the program does not process correctly.

Complete Program

```
      PROGRAM  UPDATE
C*******************************************************************
C*               Note that if you prepare your program by     *
C*               means other than keypunch machine, it is      *
C*               a good practice to prepare your comments      *
C*               using lower case letters.                     *
C*                                                             *
C*       Program definition                                   *
C*           For a complete description of the problem        *
C*           see the text.                                    *
C*                                                             *
C*         Variable Definition                                *
C*             NUMI= customer # read from transaction         *
C*             NUMF= customer # read from file MASTFL         *
C*             TRANS= Amount of transaction (purchase or      *
C*                    payment)                                *
C*             BALANC= amount read from MASTFL as account     *
C*                     balance                                *
C*             CODE= 'A' for adding new account               *
C*                   'U' for updating existing account        *
C*                   'D' for deleting an account              *
C*             any other thing indicates the end of data      *
C*******************************************************************
      CHARACTER CODE
      INTEGER NUMI,NUMF
      REAL BALANC,TRANS
C*               Initialization
C*
      OPEN(1,FILE='MASTFL',STATUS='OLD')
      REWIND 1
      OPEN(2,FILE='TEMPF',STATUS='NEW')
      READ(1,10,END=30)NUMF,BALANC
C*******************************************************************
C*               Read transactions one at a time and check    *
C*               for appropriate action based on the content*
C*               of the variable CODE.                        *
C*******************************************************************
C*
 3    READ 20, CODE, NUMI, TRANS
C*
C*******************************************************************
C*               In this section a record is to be added in   *
C*               its proper place in MASTFL.                  *
C*******************************************************************
```

```
C*
          IF(CODE.EQ.'A') THEN
  100        IF(NUMI .LE. NUMF) GOTO 150
              WRITE(2,10)NUMF, BALANC
              READ(1,10,END=21)NUMF,BALANC
             GOTO 100
  150        IF(NUMI .EQ.NUMF)THEN
              PRINT*, ' DUPLICATE RECORD IN MASTFL ',NUMI
             ELSE
              WRITE(2,10)NUMI,TRANS
              PRINT110, NUMI,TRANS
             ENDIF
C*
C****************************************************************
C*              In this section an existins record is beins*
C*              updated.                                    *
C****************************************************************
C*
          ELSEIF(CODE .EQ. 'U')THEN
  200        IF(NUMI .LE. NUMF) GOTO 250
              WRITE(2,10)NUMF, BALANC
              READ(1,10,END=30)NUMF,BALANC
             GOTO 200
  250        IF(NUMI .EQ. NUMF)THEN
              BALANC = BALANC + TRANS
              WRITE(2,10)NUMI,BALANC
              IF(TRANS .LT. 0)THEN
                PRINT220, NUMI,TRANS,BALANC
              ELSE
                PRINT210, NUMI,TRANS,BALANC
              ENDIF
             ELSE
              PRINT*, ' NONEXISTENT RECORD ',NUMI
             ENDIF
             READ(1,10,END=30)NUMF,BALANC
C*
C****************************************************************
C*              In this section a record is beins deleted  *
C*              from the file MASTFL.                       *
C****************************************************************
          ELSEIF(CODE .EQ. 'D') THEN
  300        IF(NUMI .LE. NUMF) GOTO 350
              WRITE(2,10)NUMF,BALANC
              READ(1,10,END=30)NUMF, BALANC
             GOTO 300
```

```
350       IF(NUMI .EQ. NUMF)THEN
            IF(BALANC .LE. .01)THEN
              PRINT 310, NUMI
            ELSE
              PRINT 320, NUMI,BALANC
              PRINT*, ' BUT IS DELETED ANYWAY '
            ENDIF
            READ(1,10,END=30)NUMF,BALANC
          ELSE
            PRINT*,' NO SUCH RECORD IN MASTFL ',NUMI
            PRINT*,' OR CUSTOMER # OUT OF SEQUENCE'
          ENDIF
        ELSE
C*
C*******************************************************************
C*              If code is not A , U or D that is the signal*
C*              for the end of transaction data. At this    *
C*              point the rest of the data from MASTFL      *
C*              should be copied to file 2.                 *
C*******************************************************************
C*
 13       WRITE(2,10)NUMF,BALANC
            READ(1,10,END=30)NUMF,BALANC
          GOTO 13
        ENDIF
        GOTO 3
C*
C*******************************************************************
C*              Here the end of MASTFL and the transaction  *
C*              data has occurred. File 2 is copied back to *
C*              file 1 and all data are reported.           *
C*******************************************************************
 21       WRITE(2,10)NUMI,TRANS
          PRINT 110, NUMI, TRANS
 30       ENDFILE 2
C*
C*         The above line puts an end-of-file mark
C*         at the end of the file associated with
C*         unit 2.
C*
        REWIND 1
        REWIND 2
        PRINT 18
 32     READ(2,10,END=50)NUMF,BALANC
        WRITE(1,10)NUMF,BALANC
        PRINT25,NUMF,BALANC
        GOTO 32
```

```
C*********************************************************************
C*              The following statement is to remove         *
C*              unit 2 from the system. The CLOSE            *
C*              statement has a parameter called STATUS.     *
C*              If STATUS is omitted or is KEEP the unit     *
C*              will be saved on the system. If STATUS       *
C*              is DELETE the unit is  purged.               *
C*********************************************************************
 50       CLOSE(2,STATUS='DELETE')
          CLOSE(1)
          STOP
 10       FORMAT(I6,F8.2)
 18       FORMAT('1',2X,'CUST #  BALANCE',//)
 20       FORMAT(A1,I6,F8.2)
 25       FORMAT(2X,I6,F9.2)
 109      FORMAT(' ERROR IN INPUT ',I6,1X,F8.2,' DATA IGNORED')
 110      FORMAT(' NEW ACCOUNT ADDED ',I6,1X,F8.2)
 210      FORMAT(' CUSTOMER # ',I6,' NEW PURCHESE ',F8.2,' BALANCE ',F8.2
 220      FORMAT(' CUSTOMER # ',I6,' PAYMENT ',F8.2,' BALANCE',F8.2)
 310      FORMAT(' CUSTOMER # ',I6,' HAS BEEN DELETED')
 320      FORMAT(' CUSTOMER # ',I6,' HAS A BALANCE OF ',F8.2)
          END
```

We have run the program with the following test data:

Contents of MASTFL

```
123456   -84.39
123457   239.60
123458   260.80
123460    50.75
123464   100.00
123466   100.00
```

Input data

```
U123456   -12.20
U123457   100.00
A123460    50.75
D123462
U123464   100.00
D123466
```

Program output

```
CUSTOMER # 123456 PAYMENT    -12.20 BALANCE    -96.59
CUSTOMER # 123457 NEW PURCHASE    100.00 BALANCE    339.60
NEW ACCOUNT ADDED 123460    50.75
   NO SUCH RECORD IN MASTFL    123462
   OR CUSTOMER # OUT OF SEQUENCE
CUSTOMER # 123464 NEW PURCHASE    100.00 BALANCE    200.00
CUSTOMER # 123466 HAS A BALANCE OF    100.00
   BUT IS DELETED ANYWAY



   CUST #  BALANCE
   123456   -96.59
   123457   339.60
   123458   260.80
   123460    50.75
   123460    50.75
   123464   200.00
```

If you have concluded that the preceding program works correctly for all sets of correct input data, you are wrong. Consider this set of input data assuming the contents of MASTFL the same as previously:

```
A123460    50.00
A123468    50.00
A123470   100.00
```

If you run the program with these input data you will find that only the first line of data (as shown) is processed correctly while the other two lines seem to be ignored.

Another inadequacy of this program is that it performs incorrectly, without any notice, when input data are not entered as expected by the program. For example, suppose these were the input data:

```
U123461    -20.10
D123470
A123460    50.00
A123472   100.00
```

The result would be the addition of customer number 123460 to MASTFL when that number is already in the file. Furthermore, no output is given to help us know how the other data were processed. The problem, of course, is that the input data were not entered in ascending order of customer number as we had assumed when the algorithm was developed. A well-written program should check for such inconsistencies and provide for appropriate messages, so that the user will know what is happening.

Problem 2

As mentioned, the program for Problem 1 had two weaknesses in particular: (1) it made no checks on the validity of input data, and (2) it could not add new accounts with account numbers greater than the largest one already in the master file. These are significant weaknesses, although the program does all that the statement of the problem requires.

Problem: Modify the program of Problem 1 so that the input data are checked for (1) correct sorting in ascending order of customer number and (2) the existence on the master file of all customer numbers for which update or deletion are to occur. If the input do not meet these requirements, the *update* data are not to be processed but, instead, an appropriate message should be printed. The *add* data are to be processed even though the customer number of the account to be added is larger than the last one on the master file.

Analysis and algorithm changes. Since the algorithm presented in Problem 1 will be changed in just a few places, we will consider only those portions of it as identified by step number. First we make the changes needed to check for correct order of input data. We need a variable to store the account number read immediately preceding the current one being processed. Let's name it NUMPRV, to identify it as the number previously input. We initialize NUMPRV = 0. Then the following step should be added after Step 3.0 of the algorithm of Problem 1.

3.01 If NUMPRV is greater than or equal to NUMI, print an appropriate message about out-of-order data. Replace the current value of NUMPRV by the value of NUMI just input, then input the next set of transaction data, if any.

Next we consider the changes needed to be able to add accounts to the master file if the account number to be added is larger than any of those previously on the master file. The first change must be made so that if end-of-file is reached, we do not transfer to the final tasks in the solution but add the new account, then proceed to reading any remaining transaction data. A second change must be made in order to continue processing data after the end of MASTFL has been reached. There may be a number of records that must be added as new ones to MASTFL. These records are those whose customer numbers are all greater than the last customer number on MASTFL.

Here is a ForTran program that is a modification of the one in Problem 1. It incorporates the new features just discussed.

Modified Program

```
        PROGRAM  UPDAT2
C*****************************************************************
C*          Program definition                                  *
C*              For a complete description of the problem        *
C*              see the text.                                    *
C*****************************************************************
C*          Variable Definition                                 *
C*              NUMI= customer # read from transaction           *
C*              NUMF= customer # read from file MASTFL           *
C*              TRANS= Amount of transaction (purchase or        *
C*                   payment)                                    *
C*              BALANC= amount read from MASTFL as account       *
C*                   balance                                     *
C*              CODE= 'A' for adding new account                 *
C*                    'U' for updating existing account          *
C*                    'D' for deleting an account                *
C*              Any thing else indicates the end of data.        *
C*              NUMPRV  Previous transaction number              *
C*              EOF     To indicate the end of MASTFL            *
C*****************************************************************
        CHARACTER CODE
        INTEGER NUMI, NUMF, NUMPRV
        REAL BALANC,TRANS
        LOGICAL EOF
C*              Initialization
C*
        NUMPRV = 0
        EOF = .FALSE.
        OPEN(1,FILE='MASTFL',STATUS='OLD')
        REWIND 1
        OPEN(2,FILE='TEMPF',STATUS='NEW')
        READ(1,10,END=410)NUMF,BALANC
C*****************************************************************
C*              Read transactions one at a time and check       *
C*              for appropriate action based on the content     *
C*              of the variable CODE.                           *
C*****************************************************************
C*
 3      READ 20, CODE, NUMI, TRANS
        IF(CODE .EQ. 'A' .OR. CODE .EQ. 'U' .OR. CODE .EQ. 'D')THEN
           IF(NUMPRV .LT. NUMI) GOTO 8
           PRINT 109, NUMI,TRANS
           PRINT *, 'TRANSACTION NUMBER OUT OF SEQUENCE'
        ELSEIF(.NOT. EOF) THEN
```

```
C*
C*****************************************************************
C*               If code is not A , U or D that is the signal *
C*               for the end of transaction data. At this     *
C*               point the rest of the data from MASTFL ,     *
C*               if any, should be copied to file 2.          *
C*****************************************************************
C*
  13        WRITE(2,10)NUMF,BALANC
              READ(1,10,END=30)NUMF,BALANC
            GOTO 13
          ELSE
  30        ENDFILE 2
C*
C*               The line "ENDFILE 2" puts an end-of-file
C*               mark at the end of unit 2.
C*
          REWIND 1
          REWIND 2
          PRINT 18
  32        READ(2,10,END=50)NUMF,BALANC
              WRITE(1,10)NUMF,BALANC
              PRINT25,NUMF,BALANC
            GOTO 32
C*****************************************************************
C*               The following statement is to remove         *
C*               unit 2 from the system. The CLOSE            *
C*               statement has a parameter called STATUS.     *
C*               If STATUS is omitted or is KEEP the unit     *
C*               will be saved on the system. If STATUS       *
C*               is DELETE the unit is  purged.               *
C*****************************************************************
  50        CLOSE(2,STATUS='DELETE')
            CLOSE(1)
            STOP
          ENDIF
          GOTO 3
   8      NUMPRV = NUMI
C*
C*               If MASTFL is all read we can only add to
C*               MASTFL.
```

```
C*
          IF(EOF)THEN
            IF(CODE .EQ. 'A') THEN
              WRITE(2,10)NUMI,TRANS
              PRINT 110, NUMI,TRANS
            ELSE
              PRINT 109, NUMI,TRANS
            ENDIF
C*
C***************************************************************
C*            In this section a record is to be added in   *
C*            its proper place in MASTFL.                   *
C***************************************************************
C*
          ELSEIF(CODE.EQ.'A') THEN
  100       IF(NUMI .LE. NUMF) GOTO 150
              WRITE(2,10)NUMF, BALANC
              READ(1,10,END=400)NUMF,BALANC
            GOTO 100
  150       IF(NUMI .EQ.NUMF)THEN
              PRINT*, ' DUPLICATE RECORD IN MASTFL ',NUMI
            ELSE
              WRITE(2,10)NUMI,TRANS
              PRINT110, NUMI,TRANS
            ENDIF
C*
C***************************************************************
C*            In this section an existing record is being  *
C*            updated.                                      *
C***************************************************************
C*
          ELSEIF(CODE .EQ. 'U')THEN
  200       IF(NUMI .LE. NUMF) GOTO 250
              WRITE(2,10)NUMF, BALANC
              READ(1,10,END=400)NUMF,BALANC
            GOTO 200
  250       IF(NUMI .EQ. NUMF)THEN
              BALANC = BALANC + TRANS
              WRITE(2,10)NUMI,BALANC
              IF(TRANS .LT. 0)THEN
                PRINT220, NUMI,TRANS,BALANC
              ELSE
                PRINT210, NUMI,TRANS,BALANC
              ENDIF
            ELSE
              PRINT*, ' NONEXISTENT RECORD ',NUMI
            ENDIF
            READ(1,10,END=410)NUMF,BALANC
```

```
C*
C******************************************************************
C*              In this section a record is being deleted   *
C*              from the file MASTFL.                        *
C******************************************************************
        ELSEIF(CODE .EQ. 'D') THEN
  300      IF(NUMI .LE. NUMF) GOTO 350
             WRITE(2,10)NUMF,BALANC
             READ(1,10,END=400)NUMF, BALANC
          GOTO 300
  350      IF(NUMI .EQ. NUMF)THEN
             IF(BALANC .LE. .01)THEN
               PRINT 310, NUMI
             ELSE
               PRINT 320, NUMI,BALANC
               PRINT*, ' BUT IS DELETED ANYWAY '
             ENDIF
             READ(1,10,END=410)NUMF,BALANC
           ELSE
             PRINT*,' NO SUCH RECORD IN MASTFL ',NUMI
             PRINT*,' OR CUSTOMER # OUT OF SEQUENCE'
           ENDIF
         ENDIF
         GOTO 3
C*
C******************************************************************
C*              Here the end of MASTFL and the transaction  *
C*              data has occurred. File 2 is copied back to  *
C*              file 1 and all data are reported.            *
C******************************************************************
  400     IF(CODE .EQ. 'A' .AND. NUMI .NE. NUMF)THEN
             WRITE(2,10)NUMI, TRANS
             PRINT 110, NUMI,TRANS
          ELSEIF(EOF)THEN
             PRINT 109, NUMI, TRANS
          ENDIF
  410     EOF = .TRUE.
          GOTO 3
C*
  10      FORMAT(I6,F8.2)
  18      FORMAT('1',2X,'CUST #  BALANCE',//)
  20      FORMAT(A1,I6,F8.2)
  25      FORMAT(2X,I6,F9.2)
  109     FORMAT(' ERROR IN INPUT ',I6,1X,F8.2,' DATA IGNORED')
  110     FORMAT(' NEW ACCOUNT ADDED ',I6,1X,F8.2)
```

```
210     FORMAT(' CUSTOMER # ',I6,' NEW PURCHASE ',F8.2,' BALANCE ',F8.2)
220     FORMAT(' CUSTOMER # ',I6,' PAYMENT ',F8.2,' BALANCE',F8.2)
310     FORMAT(' CUSTOMER # ',I6,' HAS BEEN DELETED')
320     FORMAT(' CUSTOMER # ',I6,' HAS A BALANCE OF ',F8.2)
        END
```

Sample run 1:   The content of MASTFL follows:

```
123456   -84.39
123457   239.60
123458   260.80
123460   600.00
123464   200.00
222222     3.33
```

The input data are as follows:

```
A111111    100.00
A123456   1000.00
A222222    200.00
A222223    300.00
A322222   4000.00
D222222
U222223   -100.00
A332222    400.00
S
```

Program output

```
NEW ACCOUNT ADDED 111111    100.00
DUPLICATE RECORD IN MASTFL   123456
   CUSTOMER #    123456    IS IN MASTFL
   CUSTOMER #    222222    IS IN MASTFL
NEW ACCOUNT ADDED 222223   300.00
NEW ACCOUNT ADDED 322222  4000.00
ERROR IN INPUT 222222      0.   DATA IGNORED
   TRANSACTION NUMBER OUT OF SEQUENCE
ERROR IN INPUT 222223  -100.00 DATA IGNORED
   TRANSACTION NUMBER OUT OF SEQUENCE
NEW ACCOUNT ADDED 332222    400.00
```

```
      CUST #   BALANCE
      111111    100.00
      123456    -84.39
      123457    239.60
      123458    260.80
      123460    600.00
      123464    200.00
      222222      3.33
      222223    300.00
      322222   4000.00
      332222    400.00
```

Sample run 2:   The content of MASTFL is as reported at the end of run 1. The input data are as follows:

```
D111111
U123456    84.39
D222222
D332222
D324444
A344444    40.00
S
```

Program output

```
CUSTOMER # 111111 HAS A BALANCE OF    100.00
  BUT IS DELETED ANYWAY
CUSTOMER # 123456 NEW PURCHASE    84.39 BALANCE      0.
CUSTOMER # 222222 HAS A BALANCE OF      3.33
  BUT IS DELETED ANYWAY
CUSTOMER # 332222 HAS A BALANCE OF    400.00
  BUT IS DELETED ANYWAY
ERROR IN INPUT 324444     0.    DATA IGNORED
  TRANSACTION NUMBER OUT OF SEQUENCE
NEW ACCOUNT ADDED 344444     40.00
```

```
CUST #   BALANCE
123456      0.00
123457    239.60
123458    260.80
123460    600.00
123464    200.00
222223    300.00
322222   4000.00
344444     40.00
```

The previous program differs from the one in Problem 1 in two ways. First, it corrects the situations where, if input data are not in customer number order, such data are output with an appropriate error message and any additional data are processed.

Secondly, whenever the end of MASTFL is reached before all of the input data are processed, and the transaction code is one of the valid ones, the results of the transaction are added to the file on unit 2 if its code is "A" and are rejected if its code is not "A." All the leftover transactions are processed before the file on unit 2 is copied to unit 1 to be saved as the updated version of file MASTFL.

You are urged to run this program with a variety of input data and find answers to these questions:

1. What happens if the first input has a code other than A, D, and U?
2. What happens if file MASTFL doesn't exist when the program is run?
3. What happens if we try to update and delete the same record in the same run of the program?

In Chapter 7 we will present this program once more, taking care of the situation suggested in question 2 above.

Problem 3

Problem: Write a computer program to generate monthly bills for long distance telephone calls for an unspecified number of customers. For each customer the following input data are provided on punched cards:

Card 0 (Personal data):

    Columns 1–10: Customer telephone number including area code.
    Columns 11–28: Customer name.
    Columns 29–46: Street address.
    Columns 47–58: City.
    Columns 59–60: State abbreviation.
    Columns 61–65: Zip code.

Cards 1 through N (data about a given call): N is the number of calls completed during this billing period and for each call, there is a card in which is punched the following data:

    Columns 1–10: Customer telephone number (including area code).
    Columns 11–20: Telephone number dialed (including area code).
    Columns 21–23: Duration of the call, in minutes (called "connect time").
    Columns 24–27: Clock time (as on a 24-hour clock) at which the call was dialed (first two digits give the hour, last two give the minute).
    Columns 28–30: Rate of charge in cents per minute.

Card $N+1$:   (Blank card)

A blank card indicates the end of data for one customer.

The rule for computing the charge for a given call made at time T with connect time C and rate R is given by the following:

1. If $0800 \leq T \leq 1700$, the charge is the larger of the two products 3*R or C*R.
2. For any other value of T the charge is C * R * .5.

The program output should be one or more pages per customer, where each page is as follows:

```
CUSTOMER TELEPHONE:    (aaa)ddddddd
customer name
street address
city, state, zip


LONG-DISTANCE CALLS FOR THIS PERIOD
CALL      DIALING TIME    NUMBER DIALED     CONNECT TIME     CHARGE
----      ----            ----              ----             ----
----      ----            ----              ----             ----
----      ----            ----              ----             ----
----      ----            ----              ----             ----
          TOTAL CHARGE  XXX.XX
```

Algorithm:

1. Read personal data card.
2. If it is blank or if end of data is encountered, terminate the process.
3. Print headings for the first page of output for this customer.
4. Read data card for the next telephone call.
5. If the personal data card is blank proceed to step 9.
6. Compute the charge for this call.
7. Print a line on the customer's statement with billing information about this call.
8. Repeat the process from step 4.
9. Print a line on the customer's statement showing total charge for this customer.
10. Repeat the process from step 1.
11. End.

We shall omit further detailed steps of this algorithm but refer you instead to the comment statements included in the program.

Program

```
      PROGRAM  PHBILL
C********************************************************************
C*          PRGRAM DEFINITION                                       *
C*               THIS PROGRAM IS TO READ DATA ON LONG-DISTANCE      *
C*               CALLS FOR VARIOUS CUSTOMERS AND PROVIDE            *
C*               PRINTED OUTPUT FOR EACH CUSTOMER AS FOLLOWS:       *
C*                  A. CUSTOMER PHONE NUMBER                        *
C*                  B. CUSTOMER BILLING ADDRESS                     *
C*                  C. DETAIL ABOUT EACH CALL                       *
C*                  D. TOTAL CHARGE                                 *
C*                                                                  *
C*          INPUT DESCRIPTION                                       *
C*               FOR EACH CUSTOMER THERE IS ONE PERSONAL DATA       *
C*               CARD AS FOLLOWS:                                   *
C*                  AREA CODE  COLS 1 TO 3                          *
C*                  PHONE NUMBER COLS 4 TO 10                       *
C*                  NAME  COLS 11 TO 28                             *
C*                  STREET ADDRESS COLS 29 TO 46                    *
C*                  CITY  COLS 47 TO 58                             *
C*                  STATE ABBREVIATION  COLS 59 TO 60               *
C*                  ZIP CODE  COLS 61 tO 65                         *
C*                                                                  *
C*               NOTE THAT THE END OF INPUT COULD BE A              *
C*               ZERO CUSTOMER PHONE NUMBER OR JUST                 *
C*               THE REGULAR END-OF-FILE.                           *
C*                                                                  *
C*          INPUT FOR EACH CALL                                     *
C*               FOR EACH LONG-DISTANCE CALL THERE IS A DATA        *
C*               CARD AS DESCRIBED HERE.                            *
C*                  CUSTOMER AREA CODE COLS 1 TO 3                  *
C*                  CUSTOMER PHONE NUMBER COLS 4 TO 10              *
C*                  DIALED AREA CODE COLS 11 TO 13                  *
C*                  DIALED PHONE NUMBER COLS 14 TO 20               *
C*                  CONNECT TIME IN MINUTES  COLS 21 TO 23          *
C*                  CLOCK TIME (24-HOUR CLOCK) AT WHICH CALL WAS    *
C*                     DIALED  COLS 24 TO 27                        *
C*                                                                  *
C*               NOTE THAT THE LAST DATA CARD FOR EACH CUSTOMER     *
C*               MUST BE A BLANK CARD.                              *
C*               ALSO, NOTE THAT WE SEPARATE AREA CODE FROM         *
C*               PHONE NUMBER BECAUSE SOME COMPUTER SYSTEMS         *
C*               CAN'T STORE 10-DIGIT NUMBERS.                      *
```

```
C*          OUTPUT DESCRIPTION                                          *
C*                  THIS PROGRAM PRODUCES AT LEAST ONE PAGE OF          *
C*                  OUTPUT FOR EACH CUSTOMER WHERE THE FIRST            *
C*                  PAGE INCLUDES CUSTOMER PHONE NUMBER, ADDRESS,       *
C*                  DETAILS ON EACH PHONE CALL INCLUDING THE CHARGE,    *
C*                  AND TOTAL CHARGE. NOTE THAT THE NUMBER OF OUTPUT    *
C*                  PAGES DEPENDS ON THE NUMBER OF CALLS THE CUS-       *
C*                  TOMER HAS.                                          *
C********************************************************************************
C*              VARIABLE DEFINITION                                     *
C*              CAREA  = CUSTOMER AREA CODE                             *
C*              CPHONE = CUSTOMER PHONE NUMBER                          *
C*              NUMBER = CUSTOMER PHONE NUMBER AT THE                   *
C*                       BEGINING OF EACH CALL CARD                     *
C*              AREA   = CUSTOMER AREA CODE AT THE                      *
C*                       BEGINING OF EACH CALL CARD                     *
C*              DAREA  = DIALED AREA CODE                               *
C*              DPHONE = DIALED PHONE NUMBER                            *
C*              NAME   = CUSTOMER NAME                                  *
C*              STREET = STREET ADDRESS                                 *
C*              CITY   = CITY NAME                                      *
C*              STATE  = STATE ABBREVIATION                             *
C*              ZIP    = ZIP CODE                                       *
C*              CALLS  = TOTAL NUMBER OF CALLS                          *
C*              CHARGE = CHARGE FOR A SINGLE CALL                       *
C*              TCHARG = TOTAL CUSTOMER CHARGE                          *
C*              MCHARG = CHARGE FOR FIRST 3 MINUTES                     *
C*              RATE   = COST PER MINUTE IN CENTS                       *
C*              CTIME  = TOTAL CONNECT TIME IN MINUTES                  *
C*              DTIME  = CLOCK READING AT DIAL TIME                     *
C*                       IN FORM OF HHMM                                *
C*              HOUR   = THE HOUR PART OF THE DTIME                     *
C********************************************************************************
C*
        CHARACTER  *18 NAME, STREET
        CHARACTER CITY*12, STATE*2, ZIP*5
        INTEGER CPHONE, CALLS, CAREA, DAREA,NUMBER
        INTEGER CTIME, DTIME, RATE,HOUR,AREA, DPHONE
        REAL  CHARGE, MCHARG , TCHARG
C*
C*       READING THE CUSTOMER INFORMATION CARD AND
C*       INITIALIZING VARIABLES.
```

```
C*
 5        READ(5,10)CAREA,CPHONE, NAME,STREET,CITY,STATE,ZIP
          IF(CAREA .EQ. 0)STOP
          CALLS = 0
          TCHARG = 0
          PRINT 20,CAREA,CPHONE
          PRINT 30, NAME ,STREET,CITY,STATE,ZIP
          PRINT 40
C*
C*            START TO READ LONG-DISTANCE DATA
C*            AND PROCESS THE CALL.
C*
 7        READ(5,50)AREA,NUMBER,DAREA,DPHONE,CTIME,DTIME,RATE
          IF(AREA .EQ. 0) GOTO 8
            IF(NUMBER .NE. CPHONE .OR. AREA .NE. CAREA)THEN
              PRINT 65 ,AREA,NUMBER
            ELSE
              CALLS = CALLS + 1
              HOUR = DTIME/100
              IF(HOUR .GE. 8 .AND. HOUR .LT. 17)THEN
                CHARGE = CTIME * RATE /100.0
                MCHARG = 3 * RATE /100.0
                IF(MCHARG .GT. CHARGE)CHARGE = MCHARG
              ELSE
                CHARGE = .5 * RATE * CTIME /100
              ENDIF
              TCHARG = TCHARG + CHARGE
              PRINT 70, CALLS, DTIME,DAREA, DPHONE,CTIME, CHARGE
            ENDIF
            GOTO 7
 8        PRINT 60, TCHARG
          GOTO 5
C*
 10       FORMAT(I3,I7,A18,A18,A12,A2,A5)
 20       FORMAT('1',///,10X, 'CUSTOMER PHONE (',I3,')',I7)
 30       FORMAT(2(10X,A18,/),10X,A12,1X,A2,1X,A5,///)
 40       FORMAT(20X,'LONG DISTANCE CALLS FOR THIS PERIOD',/
     +    20X,'------------------------------------',/
     +    16X,'DIALING   NUMBER        CONNECT',/
     +     9X,'CALL      TIME      DIALED       TIME       CHARGE')
 50       FORMAT(2(I3,I7),I3,I4,I3)
 60       FORMAT(10X,40('-'),/10X,'TOTAL CHARGE  $',F8.2)
 65       FORMAT(' ERROR IN CUSTOMER NUMBER   (',I3,')',I7)
 70       FORMAT(10X,I3,4X,I4,3X,'(',I3,')',I7,2X,I3,4X,F6.2)
          END
```

Problem 4

Note that the solution to this problem is the only one of the four in this chapter that makes use of arrays and DO loops, so if Chapter 5 has not yet been covered, this problem should be postponed until after Chapter 5 has been studied.

Problem: Develop an algorithm and write a computer program to score multiple-choice examinations. Any given examination may have from one to eighty items in it, each item having a value of from zero to nine points and each item having as its correct answer either a single decimal digit or a single letter of the alphabet. The input consists of the following (the word *Card* may be read *Line* for an interactive system):

1. Instructor input:
   Card 1: Number of students taking the examination (N) and number of items in the exam (M).
   Card 2: Correct answers for all items in order, beginning with item 1, one answer per column. There are M items so the first M columns of this card will be used.
   Card 3: Number of points per item for each item in the exam.
   This number must be between 0 and 9, inclusive. Since there are M items, this card will have point-values in the first M columns.

2. Student input:
   Card 1:
      Columns 1–6: Student identification number.
      Columns 7–30: Student name.
   Card 2: Responses for all items in the exam in order, beginning with item 1, one response per column.
   The output produced by the program should be the following:

1. Instructor output:

```
STUDENT NO.        SCORE
    ---             ---

    ---             ---

    ---             ---
```

2. Student output:

```
######     NNNNNNNNNNNNNNNNNNNNNNNNNNNN
CORRECT ANSWERS AAAA...A
STUDENT ANSWERS SSSS...S
POINTS PER ITEM PPPP...P
```

In the above, # represents a digit of a student number, N represents a letter of a student name, A represents the correct answer for a given item, S represents the student response for a given item, and P represents the point-value for a given item.

Algorithm:

1. Read instructor data.
2. Print heading for instructor output.
3. Read the data for one student.
4. Compute score for this student.
5. Print student number and score for this student.
6. Output to a file the student number, name, and all test-item responses for this student.
7. If there are more student data, repeat process from step 3.
8. Produce the student output for all N students.
9. End.

Program

```
      PROGRAM SCORE
C***********************************************************************
C*           PROGRAM DEFINITION                                        *
C*               THIS PROGRAM READS THE NUMBER OF STUDENTS             *
C*               TAKING AN EXAM AND THE NUMBER OF QUESTIONS IN         *
C*               THE EXAM. IT ALSO READS THE CORRECT ANSWER            *
C*               AND THE POINT VALUE FOR EACH QUESTION.                *
C*               THEN THE DATA REQUIRED FROM EACH STUDENT IS           *
C*               STUDENT'S NUMBER , NAME AND RESPONSE TO EACH          *
C*               QUESTION. THE STUDENT SCORE IS COMPUTED AND           *
C*               APPROPRIATE OUTPUT IS PROVIDED, BOTH FOR THE          *
C*               STUDENT AND THE INSTRUCTOR.                           *
C***********************************************************************
C*
      CHARACTER NAME*20, NUM*6, CAN(80),SAN(80)
      INTEGER POINT(80),SCORE,STUDT,QUEST,I,J
```

```
C*
C********************************************************************
C*           VARIABLE DEFINITION                                   *
C*           CAN      AN ARRAY OF 80 TO STORE CORRECT ANSWERS.*
C*           SAN      SAME AS CAN TO STORE STUDENT'S ANSWERS. *
C*           POINT    TO STORE POINT VALUES FOR EACH QUESTION.*
C*           STUDT    NUMBER OF STUDENTS TAKING THE TEST.     *
C*           QUEST    NUMBER OF QUESTIONS IN THE TEST.        *
C*           NUM      STUDENT IDENTIFICATION NUMBER           *
C*           NAME     STUDENT NAME.                           *
C*           SCORE    TEST SCORE FOR A GIVEN STUDENT.         *
C*           I AND J ARE LOOP INDICES.                        *
C********************************************************************
C*                                                                 *
C*           DATA FILE USED                                        *
C*           UNIT 1 IS A TEMPORARY FILE STORING                    *
C*           INTERMEDIATE RESULTS NEEDED TO                        *
C*           PRODUCE STUDENT OUTPUT.                               *
C********************************************************************
C*                                                                 *
C*               ENTERING INSTRUCTOR DATA                          *
C********************************************************************
C*
      READ*,STUDT,QUEST
      READ 10,(CAN(I),I=1,QUEST)
      READ 20,(POINT(I),I=1,QUEST)
      PRINT 25
C*
      OPEN(1,FILE='TEMP',STATUS='NEW')
C*
      DO 70 I = 1, STUDT
        READ 40,NUM,NAME
        READ 10,(SAN(J),J=1,QUEST)
        SCORE = 0
        DO 50 J=1,QUEST
          IF(SAN(J) .EQ. CAN(J))SCORE = SCORE +POINT(J)
 50     CONTINUE
        PRINT 60,NUM,NAME,SCORE
        WRITE(1,40)NUM,NAME,SCORE
        WRITE(1,10)(SAN(J),J=1,QUEST)
 70   CONTINUE
      ENDFILE 1
      CLOSE (1)
```

```
C*
C*                    PREPARE STUDENT OUTPUT
C*
          PRINT 80
          OPEN(1,FILE='TEMP',STATUS='OLD')
          REWIND 1
          DO 90 J=1,STUDT
            READ(1,40)NUM,NAME,SCORE
            READ(1,10) (SAN(I),I=1,QUEST)
            PRINT 85, NUM,NAME,SCORE
            PRINT 86,(CAN(I),I=1,QUEST)
            PRINT 87,(SAN(I),I=1,QUEST)
            PRINT 88,(POINT(I),I=1,QUEST)
   90     CONTINUE
          CLOSE(1,STATUS='DELETE')
          STOP
C*
   10     FORMAT(80A1)
   20     FORMAT(80I1)
   25     FORMAT('1',10X,'STUDENT NO.      STUDENT NAME              SCORE')
   40     FORMAT(A6,A20,I4)
   60     FORMAT(13X,A6,5X,A20,I4)
   80     FORMAT('1',10X,'STUDENT OUTPUT',//)
   85     FORMAT(10X,A6,1X,A20,I4)
   86     FORMAT(1X,'**CORRECT ANSWERS**',2X,80A1)
   87     FORMAT(1X,'**STUDENT ANSWERS**',2X,80A1)
   88     FORMAT(1X,'**POINTS/QUESTION**',2X,80I1)
          END
```

SUMMARY

In this chapter we have discussed formatted input and output and data files.

1. Formatted input.
   General form of the two allowable statements:

   (a) READ f, V1, V2, . . . , VN

   where f represents the reference number of the associated FORMAT statement (or the format specification itself), and V1, V2, . . . , VN represent variables into which information is to be stored.

(b) READ (u,f,END=n) V1, V2, ..., VN

where u represents an integer specifying logical unit number, f represents the reference number of the associated FORMAT statement (or the format specification string itself), n represents the reference number of the statement to be executed when an end-of-file message is encountered, and V1, V2, ..., VN represent variables into which information is stored.

2. Formatted output.
   General form of the two allowable statements:

   (a) PRINT f, data-list

   where f represents the reference number of the associated FORMAT statement (or the format specification itself), and "data-list" represents the list of output items.

   (b) WRITE (u,f) data-list

   where u represents an integer specifying logical unit number, f represents the reference number of the associated FORMAT statement (or the format specification string itself), and "data-list" represents the list of output items.

3. FORMAT statement.
   General form:

   r   FORMAT (s1, s2, ..., sn)

   where r represents the reference number, and s1, s2, ..., sn represent format specifications.

4. Format specifications.
   (a) I format (for integer input/output).

   In

   where n represents the maximum number of digits allowed in the integer.
   (b) X format (blank spaces).

   nX

   where n specifies the number of blank spaces to be output or spaces to be skipped in the input record.

(c) F format (for real numbers).

$Fw.d$

where w specifies the width of the input or output field and d specifies the number of decimal digits to the right of the decimal point in the input or output.

(d) E format (exponential form for real numbers).

$Ew.d$

where w specifies the total width of the input or output field and d represents the number of fractional digits in the coefficient that appears to the left of the E.

(e) A format (character data).

An

where n specifies the number of characters being input or output.

5. Carriage control.

Vertical spacing is controlled in FORMAT statements in that the first character of the first specification inside the parentheses of the FORMAT is used for that purpose. Following are the allowable carriage-control characters:

Character	Effect
Blank	Single space.
0	Double space.
1	Move to the top of new page.
+	Stay on same line as previous output.

Vertical spacing may also be controlled by using the slash (/) in FORMAT statements. Whenever a slash occurs among the format specifications, it causes the input or output device to perform the next operation on the next physical record. For printed output, this means printing on the next line. For input, it means subsequent input occurs from the next card (or line). Slashes may appear in multiples, in which case each slash brings in a new physical record (card or line).

6. Repeating format specifications.

Patterns of format specifications may be repeated by preceding them with an integer that indicates the number of repetitions. In the case of a single specification no parentheses are required, but for more than one specification, they must be enclosed in parentheses. Following is an example:

Single specification repeated 3 times:

3F8.1

Multiple specification repeated 3 times:

3(I5, F6.0, F8.1)

EXERCISES

1. Indicate which of the following statements are true or false. (Note that a statement must be considered false if any part of it is false.) Be able to make changes in those that are false so that they will be true.
   (a) It is impossible to write a complete ForTran program without using a FORMAT statement.
   (b) A FORMAT statement must appear in the program immediately following the input or output statement that references it.
   (c) A FORMAT referenced by a READ statement cannot be referenced by a WRITE statement.
   (d) It is not permissible to use both the H-format specification and quotation marks in the same FORMAT statement.
   (e) Every FORMAT statement must have a reference number.
   (f) It is possible to have in a program a READ statement that results in no input taking place.
   (g) A single READ statement may cause the input of more than one record.
   (h) Every WRITE statement will cause the printing of at least some information on the printer.
   (i) Every READ statement causes the input of exactly one record of data.
   (j) A READ statement need not reference a FORMAT statement.
   (k) A single FORMAT statement may be referenced by more than one READ statement.
   (l) Whenever a slash occurs in a FORMAT referenced by a READ statement, reading will occur from the next record of data.

(m) If a slash occurs as the last character within the parentheses of a FORMAT that is referenced by a PRINT statement, the result is the printing of one blank line.

(n) One way of printing a blank line of output is to include a PRINT statement with no data list.

(o) The data list of a formatted PRINT statement cannot include any character strings.

(p) The following is a correct ForTran statement:

```
FORMAT(A10,2X)
```

(q) The following statement would cause the printing of the value of A on one line and the value of B on the next line:

```
PRINT*,A,'/',B
```

(r) The X-format specification can only be used in those FORMAT statements that are referenced by output statements.

(s) The following is a correct ForTran statement:

```
10 READ(1,10)A,B
```

(t) If in a READ statement data list a variable that has been defined as real is made to correspond to an I specification in a FORMAT statement, only the integer part of the data will be input.

(u) If a variable contains string data, that information can be printed using either an A or F format specification.

(v) FORMAT statement 10 as referenced in the PRINT statement

```
PRINT 10,A,B,'BE CAREFUL'
```

need only include format specifications for the variables A and B.

(w) The statement

```
READ 10,A
```

could never cause the input of more than one data card.

(x) The statement

```
READ*,A,B
```

will cause the input of exactly one data card on a batch computer system.

(y) If NAME is a variable into which data are read using I6 as the format specification, then it would be possible to output the contents of NAME correctly with A6 as the format specification, without creating an error message.

(z) The number of records input or output by a READ or PRINT statement depends on the referenced FORMAT statement.

2. Determine which of the following ForTran statements or groups of statements are correct and which contain errors. For those that are incorrect, make the changes needed to correct them. Assume the default data type based on the first letter of a variable name.

(a)   `READ 23`

(b)   `READ*,'A',A`
       `FORMAT(1X,F6.2)`

(c)   `READ*,PRINT,WRITE,READ`

(d)   `READ 10,READ 20`
    `10 FORMAT(F6.2)`
    `20 FORMAT(I6)`

(e)   `READ 10,A,B,C`
    `10 FORMAT('1',3F6.2)`

(f)   `READ,10,A,B,C`
    `10 FORMAT(3F6.2)`

(g)   `READ 10,A,I,C`
    `10 FORMAT(1X,F10.2,I6)`

(h)   `READ 10,A,I,C`
    `10 FORMAT(1X,2F10.6,I6)`

(i)   `READ 10,A,I,C`
    `10 FORMAT(1X,F4.6,I6,F6.4)`

(j)   `READ 20,THIS,IS,GOOD`
    `20 FORMAT(A4,A2,F6.2)`

(k)   `READ 20,THIS,IS,BAD`
    `20 FORMAT(A4,I6)`

(l)   `READ 20,THINK,BRIGHT`
    `20 FORMAT('NOT NOW',A5,A6)`

(m)   `READ(1,10,END=10)A,B`
    `10 FORMAT(2F6.2)`

(n)   `READ(1,10)A,B,C`
    `10 FORMAT(F6.2,2*(1X,F6.2))`

(o)   `READ(5,10,END=100)`
    `10 FORMAT(1X,/)`

(p)   `READ(5,*,END=100)PRINTS`
     `5 FORMAT(2X,F6.2)`

(q)   `READ(10,10,END=7)A,B`
    `10 FORMAT(F.2)`

```
(r)     READ(10,10)A1,A2,A3,A4
     10 FORMAT(1X,A6,2(A6,F6.2))
(s)     READ(1,100,END=100A,B
(t)     PRINT 20,READ 30
     20 FORMAT(2X,A8)
(u)     PRINT(6,20)A,B
     20 FORMAT(A6,2X,F6.2)
(v)     PRINT,20,A,B
     20 FORMAT('IS THIS A',A6,)
(w)     PRINT 20,A,B
     20 FORMAT(5X,A,F6.2)
(x)     PRINT,PRINT
(y)     PRINT,PRINT5,READ5
      5 FORMAT(A6,A6)
(z)     PRINT(6,20,END=10)A,B
     20 FORMAT(A6,'‡',A6
(aa)    X=10
        PRINT10,X
     10 FORMAT(I3)
(bb)    X=Y=10
        PRINT 10,X,Y,X+Y
     10 FORMAT(F6.2/F6.2/12X,F6.2)
(cc)    READ 10,A,B
        PRINT 10,A,B,A+B
     10 FORMAT(F6.2,F6.2,'A+B='F6.2)
(dd)    READ 10,FORMAT,NOT
     10 FORMAT(2A3,I6
        PRINT 10,NOT,FORMAT
(ee)    READ 10,ONE,TWO,THREE
     10 FORMAT(1X,2(A3,1X),A4)
        PRINT 10,ONE,TWO,THREE,ONE+TWO
(ff)    A=10.0
        B='AB'
        PRINT,A,A+B
(gg)    A=10.0
        B=A*2
        PRINT A,B,A/B
(hh)    READ(6,10,END=30)A,B
        WRITE(1,30)A,A+B,B
     10 FORMAT(2F10.4)
     30 FORMAT(3F10.4)
```

```
(ii)    WRITE(6,10)
     10 FORMAT(A,B,C)
        READ 10,A,B,C
(jj)    WRITE(6,10)A,A+B,'A*B'
     10 FORMAT(3F10.6)
(kk)    READ,10,A,B
        X=A/B
        PRINT,10,X,B
     10 FORMAT(1X,2F6.2)
(ll)    READ 20
        WRITE(6,20)
     20 FORMAT('THIS IS A HEADLINE',/)
(mm)    WRITE(6,20)
     20 FORMAT('0',200('*'))
(nn)    WRITE(6,20)
     20 FORMAT('1*',131(*)/' *',130' ')
(oo)    WRITE(6,20)
     20 FORMAT(1X,A6,//)
(pp)    WRITE(6,20,END=5)A,B
     20 FORMAT(2F6.2)
(qq)    I=2
        X=I*2.5
        WRITE(6,10)I,X,I
     10 FORMAT(I4,F6.2,A1)
```

3. Write an appropriate program segment for each of the following, using FORMAT statements.

   (a) Print the value of A at about the middle of the second line where each line is 132 spaces wide.

   (b) Print on eleven different lines the values for

   $$I, \; I*10, \; I*10^2, \; I*10^3, \; \ldots, \; I*10^{10}$$

   (c) Use FORMAT and PRINT statements (and maybe others) to draw a square with sixty asterisks along each of its sides.

   (d) Use FORMAT and PRINT statements (and maybe others) to draw a square of sixty asterisks and its diagonals.

   (e) Use FORMAT and PRINT statements to print your name with asterisks.

   (f) With FORMAT and PRINT statements (and maybe others) draw a circle with a radius of ten spaces and a horizontal diameter.

   (g) Print the following headings:
   ```
   NAME  STREET  CITY  STATE  TELEPHONE
   ```

(h) Print the following heading:

```
EXAMPLE OF HEADING
I DATA I DATA I DATA  I DATA I DATA I DATA I TOTAL I
I ONE  I TWO  I THREE I FOUR I FIVE I SIX  I DATA  I
```

4. Describe the output produced by each of the following program segments (assume default data types based on the first letter of a variable name):

(a)
```
     CHARACTER*6 B
     A=25
     B='TOTAL'
     PRINT 10,A,B
  10 FORMAT('1',2X,F4.0,1X,'IS THE',1X,A6)
```

(b)
```
     I=10
     J=I*2.98
     PRINT 10,I,J,I*J
  10 FORMAT('1',2X,I3/3X,I3/3X,'   *'/3X,I6)
```

(c)
```
     READ 10,A,B
     PRINT 20,A,B
  20 FORMAT('1',F6.2)
  10 FORMAT(2F6.3)
```

(d)
```
   5 READ 10,A,B
     IF(A .EQ. B)STOP
     PRINT 20,A,B,A/(A+B)
  20 FORMAT('1',3F6.2)
     GO TO 5
```

(e)
```
     A=0
     PRINT 10
  10 FORMAT('I')
  15 A=A+1
     PRINT 20,A, A*A, A*A*A
  20 FORMAT(2X,3(F6.2,2X))
     IF(A .LT. 100)GO TO 15
```

(f)
```
     CHARACTER *6 B,LNAME,FNAME,INIT
     B= 'NAME: '
     READ 10,LNAME,FNAME,INIT
  10 FORMAT(3A6)
     PRINT 10,B,LNAME,FNAME,INIT
     PRINT 20,B,LNAME,FNAME,INIT
  20 FORMAT(2X,A6,1X,3A6)
```
Use your own name as input data for the above program.

(g)
```
      CHARACTER *12 FNAME
      PRINT*,'WHAT IS YOUR FIRST NAME?'
      READ 10,FNAME
   10 FORMAT(A12)
      PRINT 20,FNAME
   20 FORMAT(1X,'WHAT IS YOUR AGE',A12,'?')
      READ*,IAGE
      PRINT 30,FNAME,IAGE
   30 FORMAT(1X,A12,1X,'YOU MUST BE',I3,1X,'YEARS OLD')
```

(h)
```
      CHARACTER *12 NAME1
      PRINT*,'HI, I AM JOE COMPUTER. WHO ARE YOU?'
      READ 10,NAME1
   10 FORMAT(A12)
      A=15.6
      B=2.3
      PRINT 20,NAME1,A,B
   20 FORMAT(1X,2A6,'CAN YOU ADD',F4.1,'AND',F4.1,'?')
      PRINT*,'WHAT IS YOUR ANSWER?'
      READ*,ANS
      IF(ANS .EQ. A+B)THEN
         PRINT,'CORRECT'
      ELSE
         PRINT*,'YOU ARE WRONG.'
      ENDIF
```

(i)
```
    5 X=RANF(Z)*100.
      IF(X .LE. 0.1)STOP
      Y=RANF(Z)*200.
      Z=X+Y
    7 PRINT*,X,'+',Y,'=?'
      PRINT*,'WHAT IS YOUR ANSWER?'
      READ*,ANS
      IF(ANS .EQ. Z)THEN
         IF(X.LE.50.0)PRINT *, 'GOOD'
         IF(X.GT.50.0)PRINT *, 'SUPER'
         GO TO 5
      ELSE
         PRINT*,'INCORRECT! TRY AGAIN.'
         GO TO 7
      ENDIF
```

5. Identify the compile-time errors and the run-time errors, if any, in the following program segments (assume default data types based on the first letter of a variable name):

(a)
```
    READ*,A,I
    PRINT 20,A,I,A*I
20 FORMAT(1X,F6.2,A6,F6.2)
```

(b)
```
    A=45.6
    B=A*10
    I=B
    IF(I .EQ. 456)THEN
        PRINT*,I
        STOP
    ENDIF
```

(c)
```
    I=1
    ISUM=0
10 ISUM=ISUM+I
    PRINT 10,I,ISUM
    I+I+3
    GO TO 10
10 FORMAT(1X,'THE SUM OF ODD INTEGERS FROM 1 TO',I8,'IS',I10)
```

(d)
```
    READ*,A,B,C
    DELTA=B*B-4.*A*C
    IF(DELTA .LT. 0)THEN
        PRINT*,'THERE IS NO ANSWER FOR'
        PRINT 10,A,B,C
10      FORMAT(1X,F6.2,'*X*X+(',F6.2,')*X+(',F6.2,')=0')
    ELSE IF(DELTA .EQ. 0)THEN
        X1=-B/2.
        PRINT*,X1
        PRINT*,'IS THE ONLY ANSWER FOR'
        PRINT 10,A,B,C
    ELSE IF(DELTA .GT. 0)THEN
        X1=-B/2+DELTA**.5
        X2=-B/2-DELTA**.5
        PRINT*,X1,X2
        PRINT*,'ARE THE TWO ANSWERS FOR'
        PRINT 10,A,B,C
    ENDIF
```

(e)
```
      CHARACTER *6 NUM,NAME1,NAME2,NAME3
      PRINT 10
    5 READ(5,20)NUM,NAME1,NAME2,NAME3
      IF(A .EQ. 0.)STOP
      PRINT(6,30)NUM,NAME1,NAME2,NAME3
      GO TO 5
   10 FORMAT(1H1,'STUDENT NO.    NAME')
   20 FORMAT(4A6)
   30 FORMAT(1X,A6,7X,3A6)
```

(f)
```
      PRINT*,'INPUT A POSITIVE INTEGER'
   10 READ*,N
      IF(N .LE. 0)THEN
         PRINT*,'YOUR NUMBER MUST BE POSITIVE. TRY AGAIN.'
         GO TO 10
      ELSE
   20    ISUM=ISUM+N
         N=N-1
         IF(N .GT. 0)GO TO 20
         PRINT 30,N,ISUM
      ENDIF
   30 FORMAT(1X,'1+2+...+',I4,'=',I6)
```

6. (Computer Science, Education, General)

Develop an algorithm and an interactive timesharing program that prints a geometric figure and asks the user to identify it. We suggest you use the following figures and corresponding codes: 1 = circle; 2 = square; 3 = rectangle; 4 = star; 5 = triangle. The program should output any one of these figures and then ask the user to enter the digit "1" if the user recognizes the figure as a circle, "2" if the figure is recognized as a square, and so on. If the user enters something other than 1–5, output a message asking the user to enter a code within the specified range.

If the user identifies the figure incorrectly, give another chance, up to three chances, then give the correct response. When a correct answer is given, the program should respond with an encouraging comment and ask if the user wants to keep on. If no, output the record of number of correct guesses on first try, second try, and third try, and the number wrong. If yes, randomly select another figure to output and proceed as described previously.

Your program should include the following:

(a) Output messages telling the user about the code for each figure.

(b) Output messages clearly describing the response expected from the user.

    (c) Output messages indicating correct or incorrect answer, and on which attempt it occurred.

    (d) Output information summarizing the user's score when he or she chooses to stop.

7. (General)

Develop an algorithm and a computer program to maintain a checking account balance. The input to the program follows:

    (a) Starting balance as XXXXXX.XX.

    (b) Starting date as DDMMYY where DD represents two digits for the day, MM two digits for the month, and YY represents two digits for the year.

    (c) For each account-transaction these data are required:

        Columns 1–4, date as DDMM.

        Column 5, transaction code, D for deposit, C for check.

        Columns 6–23, descriptive information.

        Columns 24–30, amount of transaction as XXXX.XX.

Output consists of two parts. Part I is a report of the transactions processed, and Part II is a summary report of the transactions and of the status of the account.

Part I output:

DATE	DESCRIPTION	CODE	AMOUNT	BALANCE
---	---	---	---	---
---	---	---	---	---
---	---	---	---	---

Part II output:

```
OPENING BALANCE       $
TOTAL DEPOSITS        $
TOTAL                 $
TOTAL OF CHECKS       $
SERVICE CHARGE        $
PENALTY               $
CLOSING BALANCE       $
TOTAL NUMBER OF DEPOSITS
TOTAL NUMBER OF CHECKS
TOTAL NUMBER OF NSF CHECKS
```

Rules for processing checks:

(a) A check is NSF (nonsufficient funds) if its amount exceeds the current balance. For each such check, output an asterisk immediately to the left of the description in the output for Part I. Also charge a penalty of $5.00, reducing the balance by $5.00 or to zero if the current balance is less than $5.00.

(b) For each check that causes the current balance to go below $100.00, charge a $0.25 service charge.

(c) If for any check the current balance goes below $100.00, charge a $3.00 service charge for the current statement period.

8. (Mathematics, Statistics, General)

Develop an algorithm and a program to input a value for N, followed by N pairs of numbers where for each pair the numbers are denoted X and Y. Compute all of the following:

$$\text{SUMX} = \sum_{i=1}^{N} X_i \qquad \text{SUMY} = \sum_{i=1}^{N} Y_i$$

$$\text{SUMXY} = \sum_{i=1}^{N} X_i Y_i \qquad \text{SUMX2} \sum_{i=1}^{N} X_i^2$$

$$\text{SUMY2} = \sum_{i=1}^{N} Y_i^2 \qquad \text{XMEAN} = \text{SUMX}/\text{N}$$

$$\text{YMEAN} = \text{SUMY} / \text{N}$$

$$\text{XVAR} = \text{SUMX2} / \text{N} - \text{XMEAN} * \text{XMEAN}$$

$$\text{YVAR} = \text{SUMY2} / \text{N} - \text{YMEAN} * \text{YMEAN}$$

$$\text{XSTD} = \sqrt{\text{XVAR}} \qquad \text{YSTD} = \sqrt{\text{YVAR}}$$

$$\text{B} = (\text{SUMXY} - \text{N} * \text{XMEAN} * \text{YMEAN}) / (\text{SUMX2} - \text{N} * \text{XMEAN} * \text{XMEAN})$$

$$\text{A} = \text{YMEAN} - \text{B} * \text{XMEAN}$$

Print the following equation where, instead of A and B as shown, the values for A and B computed above are printed.

```
Y = A * X + B
```

Use the following data that correspond to a random sample of fifteen males participating in a research project:

```
X = Age              = 45 43 46 49 50 37 34 30 31 26 22 58 60 52 27
Y = Cholesterol count = 30 52 45 38 62 55 25 30 40 17 28 44 61 58 45
```

9. (Mathematics, Education, General)
   Develop an algorithm and a program to compute the area of any of a selected number of geometric figures. The program is to be interactive in the sense that conversational kinds of comments and questions are presented to the user for his or her response. Also include output that describes for the user what choices of figures are available, and exactly what input is expected from the user once he or she has selected the figure of which area is to be computed. When processing for one figure is complete, have the program ask if the user wants to stop or continue. If the choice is to continue, present to the user the complete list of options each time.

10. (Mathematics, General)
    Develop an algorithm and a computer program that converts the common units of English measure into corresponding units of metric measure. The program is to be interactive in the sense that conversational kinds of comments and questions are presented to the user for response. Also include output that describes for the user what choices of measures are available, and exactly what input is expected from the user once the option to be used has been selected. The output resulting from the conversion from the English units to the metric might look something like this:

```
XX.X DEGREES FAHRENHEIT = XX.X DEGREES CENTIGRADE
XX.X MILES = XXX.X KILOMETERS
```

The program should produce output that asks the user whether or not to stop the program or continue processing other cases.

11. (Mathematics, General)
    Develop an algorithm and write a program to read a number and determine its value rounded to the Nth decimal place. We define the Nth decimal place as positive to the right of the decimal point and

negative to the left of the decimal point. For example, the number 6532.15 rounded to the −2 position would be 6500. The same number rounded to the 1st position is 6532.2. The output need only be the rounded value.

12. (General)

Develop an algorithm and write a program to read any letter of the alphabet and print the same letter in the size 2 inches high and about 1.2 inches wide. Letters like W and M may need to be wider; letters like I may look best if narrower. This will mean that you will have to print many characters arranged so that the result has the appearance of the letter input to the program. Therefore, your first step will probably be the drawing on squared paper of all twenty-six letters in the specified size so that you can know where characters must be printed to end up with a given letter. Each large letter should be made up of small letters of the same kind—for example, a large A should be made of small A's printed in the appropriate pattern.

13. (Computer Science, General)

Develop an algorithm and a program that accepts as input one of the expressions

A*B, A ** B, A / B, A − B, A + B

and produces as output the corresponding phrase like one of these:

```
A MULTIPLIED BY B
A RAISED TO THE POWER B
A DIVIDED BY B
B SUBTRACTED FROM A
B ADDED TO A
```

CHAPTER 7
Subprograms

Throughout this book thus far, it has been convenient to assume that each example program presented was constructed as one unit. There have, of course, been several subtasks to be completed within many of the programs, such as the input task, the computation task, and the output task; but we have made no concerted effort to suggest that it would be useful to group together within a program those statements that handle a specific function. As your problem-solving skills and programming skills develop to the point where you can deal with larger and more complex problems, you will find it very helpful to consider a problem in segments. It is more efficient for most people to develop solutions for smaller, more comprehensible problems than for massive, complex ones. Then by connecting together the solutions to the smaller problems, a satisfactory result can be obtained for the larger one. When computer programming is done in smaller sections such that each section is designed, written and tested separately, such sections are usually called *subprograms*. However, to qualify as a ForTran subprogram, a collection of statements must conform to one of several formats, each of which has specific programming rules to follow. It is those various rules that we discuss in this chapter along with examples to clarify each type of subprogram structure.

REASONS FOR SUBDIVIDING A PROGRAM

1. It is easier to comprehend the steps needed to solve a smaller, less complex segment of a problem than to grasp a large, complicated collection of tasks.
2. The logical steps in a *segment* of a computer program are more readily examined for errors than is a longer, more logically-complex program. Thus program segments are generally easier to debug.
3. For the reason given in 1 above, most people can structure a small computer program better than they can a large one.
4. When a large problem is segmented into several smaller units, it is often possible to simplify the logical steps in each segment so that when taken as a whole, the entire problem solution has simpler logic than if the solution had been developed from the viewpoint of the entire large problem.
5. By being able to simplify the logic of a computer program, it will generally take less time to code into ForTran, often require less computer memory; and usually can be run in less time on the computer system.
6. When subprograms are used appropriately, they can significantly reduce the amount of coding needed to accomplish a given task.

For these reasons it seems that it would be worthwhile to learn about subprogram structures in ForTran. We begin our introduction to subprograms by considering the function statement. This single statement is the simplest case of a subprogram.

FUNCTION STATEMENT

A function statement consists of a single line of ForTran code that has the appearance of an assignment statement. The role of a function statement in ForTran is similar to that of a formula in mathematics. A function is defined once and is used, with different parameters, as many times as needed.

Some sample function statements

1. `F1(X)=X**K`
2. `F2(X,Y,Z)=X*X+2 *Y*Y+3 *Z*Z`
3. `F3(X,Y)=SIN(X)+COS(Y)`
4. `F4(X,Y)=ALOG(X)+EXP(Y)`
5. `G1(X,Y)=ALOG10(X)*Y`
6. `G2(A,B)=(A+B)*FF(A)`
7. `I1(A,B)=A+B`

Analysis of sample function statements

In every function statement, what appears to the left of the equal sign is called the *function name*. It is not a subscripted variable but has all appearances of one. A function name must conform to the same rules as those for naming ForTran variables. The data type of a function may be declared by using an appropriate declaration statement or by making use of the default rules for data types, in which case the first letter of the function name determines if it is of type real or type integer. Whatever appears in the parentheses to the right of the function name is called the parameter of the function. It is the *parameter* of the function that is the key variable in the expression to the right of the equal sign. Every function statement must have at least one parameter. All parameters of a function statement, as well as other variables appearing in the expression on the right of a function statement, must have been defined, prior to the use of the function. A function statement can be used by including its name in an expression, an assignment statement, or as the parameter of another function statement or subprogram.

Let's consider the examples of function statements that have just been listed. In Example 1, F1(X) is defined as the Kth power of X. Thus, before F1(X) can appear elsewhere in the program, the parameter substituting for X as well as K must have been defined. For example, consider the following program in which a function statement like Example 1 appears:

```
PROGRAM   SAMPL1
REAL   A,F1
INTEGER   K
F1(X) = X**K
READ*, A, K
PRINT*, A, K, F1(A)
STOP
END
```

Suppose the values input for A and K are 1.5 and 3, respectively. Since F1(X) is defined in the function statement as the Kth power of X, then F1(A) when A = 1.5 is the third power of 1.5, or $1.5 \times 1.5 \times 1.5 = 3.375$. Therefore the output of the preceding example is

```
1.5    3    3.375
```

In the second example of function statements listed previously, F2 has three parameters, so all three parameters must either have been defined in the program preceding any appearance of F2 in an executable statement or the parameters must be constants. For example,

```
PROGRAM SAMPL2
INTEGER  X, Y, Z
REAL  R, F2
F2(X,Y,Z) = X*X + 2*Y*Y +3*Z*Z
R = 0.5 * F2(3,2,1)
PRINT *,R, F2(1,1,1)
STOP
END
```

would result in a value of 20 for F2(3,2,1), since the first parameter is multiplied times itself, the second parameter is multiplied times itself and that result multiplied by 2, and, finally, the third parameter is multiplied times itself and then multiplied by 3. Thus, R (in that program) has a value of 10, and its output is

```
10.0    6.0
```

As you can tell from the two examples discussed so far, the parameters in function statements are place-holders for actual variables or constants that appear as parameters of the function when it is used elsewhere in the program. Then, whatever operations are specified for the dummy parameters in the function statement are applied to the actual arguments that are provided at the time the function is used.

A look in the previous list of examples at 3, 4, 5, and 6 reveals that the defining expression for a function may include other functions. Note that in Example 3, the system library functions SIN and COS are used. Similarly, in Examples 4 and 5, the system library functions ALOG, EXP, and ALOG10 appear in the defining expression. Functions that have been defined in the program by the programmer may also be used as arguments in the defining expression of a function. A case in point appears in Example 6, where FF is such a function. Consider the following program:

```
PROGRAM  SAMPL3
REAL  FF,G2
INTEGER  A, B, X
FF(X) = 3*X*X - 2*X
G2(A,B) = (A+B) * FF(A)
PRINT*, G2(2,1)
STOP
END
```

What is the output of this program? We must determine what G2(2,1) is, so we look to the defining expression for G2. We learn that the two arguments, 2 and 1, must be added, and that the sum must be multiplied times FF(2). The value for FF(2) is obtained by multiplying 3 times 2 times 2, which gives 12, then subtracting the product of 2 and 2, which is 4. Thus FF(2) is 12 − 4 which is 8. Back now to evaluating G2(2,1), we note that G(2,1) = (2+1) * 8 = 24. Therefore, the output of the above program is the single number 24.

Example 7 is included to demonstrate a function statement in which the function name is of type integer. As a specific case, the value of I1(1.3,2.4) is the integer 3. Note that the defining expression for I1 specifies that the sum of the two parameters must be computed. For 1.3 and 2.4 this sum is 3.7. Then since the function name is I1 (an integer name under default rules), the truncated value of 3.7 is 3.

Rules concerning function statements

1. A function statement is a nonexecutable statement. Therefore, it should be defined among the nonexecutable statements at the beginning of the program or subprogram.
2. The parameters of a function statement must be simple variables. An array variable or function name is not a valid parameter for a function statement.
3. The data type of the name of a function statement should be declared by a data declaration statement or ForTran default rules for variable types will apply.
4. The value substituted for a parameter may be a constant, a variable, an expression, or another function name.
5. A function statement must have at least one parameter.

FUNCTION SUBPROGRAM

A function subprogram consists of three or more (usually more) program statements of which the last must be an END statement and the first must be a FUNCTION statement which has the following form:

Type FUNCTION NAME PAR1,PAR2,...,PARN

where "Type" is optional and is any of the ForTran type declarations, NAME is any ForTran variable name, and PAR1,PAR2,...,PARN represent the parameters (also called formal arguments or dummy arguments) of the function subprogram of which there must be at least one and no more than 63. Between the FUNCTION statement and the END statement appear as many program statements as needed. In particular, one of these program statements must assign a value to NAME. When a function subprogram has finished running it returns that value assigned to NAME back to the calling program. There may be more than one result produced by a function subprogram, in which case such results are passed back to the calling program by means of parameters. A function subprogram is called into action when its name is included, along with its actual arguments (the items whose values replace the parameters), in an expression, an assignment statement, or as an actual argument of another subprogram or function statement.

Because a computed value is associated with the name of a function, one must be aware of what *type* the function name is. Recall that if the data type of a variable is not declared explicitly by the use of statements for that purpose, a variable is of a given data type depending on the first letter of its name. These same rules apply to the data type of a function. Consider this example:

```
FUNCTION F1(A)
INTEGER F1
        .
        .
        .
FUNCTION F2(A)
CHARACTER F2
        .
        .
        .
FUNCTION NAME(F1,F2)
REAL NAME
```

In this example F1 is declared to be type integer, F2 is declared type character, and NAME is declared to be of type REAL.

Another method available for declaring the type of a function result is to comply with the general definition of a FUNCTION statement. For example, the above three functions could be assigned types by using the following FUNCTION statements instead of the form above:

```
INTEGER FUNCTION F1(A)
CHARACTER FUNCTION F2(A)
REAL FUNCTION NAME(F1,F2)
```

Note that the treatment of a function name with respect to its type must be consistent. Thus if a function is declared type integer through the use of a statement like the first of the above three, then it must be treated as being type integer in the main program or any other subprogram in which it appears. Consider this short program:

```
    PROGRAM MAIN
    A = 10.5
    PRINT (6,20) FUNC(A)
 20 FORMAT (A)
    STOP
    END
    CHARACTER FUNCTION FUNC(X)
    FUNC = 'A'
    RETURN
    END
```

In this example, FUNC is treated as type real in the main program (F-format specification is used) but the FUNCTION statement declares it to be of type character. This program would result in an error message.

Now consider a more-complex example.

Program

```
10    PROGRAM EXAMP1
20    INTEGER DIV,RES
30    REAL X,Y
40    READ*,X,Y
50    RES=DIV(X,Y)
60    IF(RES .EQ. 0)THEN
65        PRINT*,'NEITHER',X,'NOR',Y,'IS DIVISIBLE BY THE OTHER.'
70    ELSE IF(RES .EQ. 1)THEN
75        PRINT*,Y,'IS DIVISIBLE BY',X
76    ELSE IF(RES .EQ. 2)THEN
77        PRINT*,X,'IS DIVISIBLE BY',Y
78    END IF
80    END
100   INTEGER FUNCTION DIV(A,B)
105       INTEGER QUOT
108       REAL A, B
110       IF(A .GT. B)THEN
120          QUOT=A/B
130          IF(B*QUOT .EQ. A)THEN
140             DIV=1
150          ELSE
160             DIV=0
170          END IF
180       ELSE
190          QUOT=B/A
200          IF(A*QUOT .EQ. B) THEN
210             DIV=2
```

```
220      ELSE
230          DIV=0
240      END IF
250    END IF
255  RETURN
260  END
```

Analysis of previous example

In this example, the function subprogram extends from line 100 to line 260. The name of the function is DIV. It performs the task of determining whether or not two numbers are divisible one by the other. If the first number is divisible by the second, function DIV has the value 1. If the second is divisible by the first, DIV has the value 2. If neither number is divisible by the other, the value of function DIV is 0.

Therefore, in the main program (lines 10 through 80) we read two numbers, X and Y, then apply function subprogram DIV to the two numbers and test for a function value of 0 (line 60) or 1 (line 70). If the value is 0, line 65 is executed, which causes the printing of an appropriate message about the nondivisibility of the two numbers. If the value of DIV is 1, line 75 is executed, which causes the output to state that the first number is divisible by the second. If the value of DIV is 2, line 77 is executed causing the output to state that the second number is divisible by the first. Thus, we have considered all cases of two numbers being divisible by each other.

A new statement appearing in this program is the RETURN (line 255). Its purpose is to return program control to that point in the calling program immediately following the appearance of the function name. It may appear more than once in a subprogram. The RETURN statement is discussed more fully in the section that follows, but we note that only the first form described there may be used in a function subprogram.

Note that a function subprogram may contain a variety of ForTran statements as long as the first statement is the FUNCTION statement (discussed previously) and the last statement is END.

We shall consider more examples of function subprograms later in this chapter, but now we turn to the last subprogram structure available in ForTran, the SUBROUTINE subprogram.

SUBROUTINE SUBPROGRAM

This form of subprogram begins with a SUBROUTINE statement and concludes with an END statement. Between these two statements may appear any other ForTran statements discussed so far in this book, as well as one (the RETURN statement) not yet described.

The SUBROUTINE statement has this general form:

SUBROUTINE NAME PAR1,PAR2,...,PARN

where NAME is any acceptable ForTran variable name, and PAR1, PAR2, ...,PARN refer to parameters (or formal arguments). Parameters in a *subroutine* or *function* subprogram serve a somewhat different purpose than parameters in a function statement. In a function statement, parameters provide the means for passing information *to* the function while the result is available *from* the function to the main program only through the function name. On the other hand, in a subroutine or function *subprogram*, parameters are used as place-holders not only for information passed *to* the subprogram but also for results *from* the subprogram. A subroutine subprogram may have as many as sixty-three different parameters but may also have none, in which case the SUBROUTINE statement has this appearance:

SUBROUTINE NAME

The name assigned to a subroutine subprogram must comply with the rules for variable names in ForTran, but whether the name specifies a real or integer variable is immaterial. A subroutine subprogram's name is *not* a vehicle for bringing a result back to the calling program as in the case of a function subprogram. There may be more than one result from a subroutine subprogram and these are associated with one or more of its arguments. These characteristics will be illustrated when we present an example of a complete subroutine subprogram. The SUBROUTINE statement is a nonexecutable statement.

We consider next a statement of particular importance in the development of subroutine subprograms.

RETURN statement

There are two general forms of the RETURN statement:

1. RETURN
2. RETURN e

In the second form, e represents any acceptable expression in integer mode and is used to determine where program control transfers next. We postpone the explanation of this process to a later point in this book.

The purpose of either form of the RETURN statement is to return program control from the function or subroutine subprogram to the main program or other subprogram that called the subprogram into action. As we shall learn very soon, the calling of a subroutine subprogram into action is done by using a CALL statement. The first form of the RETURN statement always transfers control back to the statement immediately following the CALL statement that was last executed. A RETURN statement is an executable statement.

CALL statement

The general form of the CALL statement is:

CALL NAME ARG1,ARG2,...,ARGN

where NAME represents the subroutine subprogram name and ARG1,ARG2,...,ARGN the actual arguments of subprogram NAME, which are in one-to-one correspondence with PAR1, PAR2,..., PARN as discussed previously. If there are no parameters in the subroutine subprogram, the form of the CALL statement is

CALL NAME

In those cases where there are arguments, the number of arguments specified in the CALL statement should be exactly the same in number and type as the number of parameters specified in the SUBROUTINE statement of the subprogram. The values used for arguments are determined at the time of executing the CALL statement and are dependent on whatever are specified as actual arguments in the CALL statement.

When a CALL statement is executed, program control transfers to the subroutine subprogram whose name (as defined in the SUBROUTINE statement) is the same as the name specified in the CALL statement. Program control remains in the subroutine subprogram until either a RETURN statement or END statement is encountered, whereupon control returns to the calling program. The calling program may be either a main program or another subprogram (a function subprogram or a subroutine subprogram).

It is important to be aware that arguments of subroutine subprograms are the means whereby information is carried back and forth between the calling program and the subprogram. Typically one or more arguments carry information *from* the calling program to the subprogram, and at least one argument carries the results of the subprogram action back *to* the calling program. As you may recall from our discussion of parameters and arguments in connection with function statements, it is the *positions* of the actual arguments in the list of arguments of the calling statement and the positions of the formal arguments in the first statement in the subprogram that determine the information carried to and from the subprogram. In fact, the variables appearing in the same list-position of the calling statement and the FUNCTION or SUBROU-TINE statement actually occupy the same memory location, even though the names used for the variable may be different in the two argument lists.

There are other ways of making information available to both the calling program and the subprogram but we shall wait with any discussion of them until we have considered some examples of the use of subroutine subprograms. Here are four examples.

Example 1

```
PROGRAM EXAMP1
REAL   X,Y,Z
READ*,X,Y,Z
CALL SUB1(X+Y,Z)
PRINT*,' THE ABSOLUTE VALUE OF ',X,'+',Y,' IS ',Z
STOP
END
SUBROUTINE SUB1(A,B)
B = A
IF(A .GT. 0)RETURN
B = -A
RETURN
END
```

In the calling program values are input to variables X and Y. Then subprogram SUB1 is called, and the sum of X and Y is passed to the subprogram as the first argument, called A in the subprogram. In the subprogram, the second argument is assigned the value of the first argument if the first argument has a positive value. If the value of the first argument is not positive, the second argument is assigned a value that is the negative of the first argument. Upon return of program control to the calling program, the value of Z is the absolute value of the sum of X and Y, which value is then output with appropriate identifying information.

Example 2

```
PROGRAM EXAMP2
REAL A,X
READ *,A,X
X = FUNCT1(A,X)
PRINT*,A,X
STOP
END
FUNCTION FUNCT1(X)
REAL X
FUNCT1 = X + X **2 + X**3
END
```

In this example, an error message would result because when FUNCT1 is called into action in the calling program, it is done so with two arguments. Since only one parameter occurs in the FUNCTION line of the subprogram, a discrepancy results, which causes an error message.

Example 3

```
INTEGER A(10,3)
      .
      .
      .
CALL SUB2(A,A(10,1),B)  } Main program
      .
      .
      .
END
SUBROUTINE SUB2(X,Y,Z)
INTEGER X(10,3)
      .                     } Subprogram
      .
END
```

This example is given to demonstrate the handling of arrays in subprograms. Note that in the CALL statement the array name A appears (without subscripts) in the list of arguments. This means that the address of the first member of array A is passed to SUB2 so that all thirty computer words of array A are available to both the main program and the subprogram. In the subprogram the first argument is an array and should be properly defined as an array (as it is in this example). The form of array X in SUBROUTINE SUB2 is exactly the same as array A in the main program. Although doing so is not necessary, it is good programming practice to declare corresponding arrays in the main program and subprograms to be of identical form and size.

The second argument in the CALL statement is A(10,1) and in the SUBROUTINE statement it is Y. Thus, whatever operations involve Y in the subprogram will use the value stored in A(10,1) from the main program. Similarly, B in the main program will contain whatever the subprogram puts in its variable Z after completion of subprogram SUB2.

Example 4

Main Program

```
PROGRAM EXAMP4
REAL A(10)
INTEGER N
READ*,N
READ*,(A(I),I=1,N)
CALL SUB3(N,A)
IF(N .NE. 0)THEN
    PRINT*,N, ' IS THE ROUNDED SUM OF THE',
+               ' ELEMENTS OF A.'
ELSE
    PRINT*, ' THE SUM OF THE ELEMENTS OF A',
+               ' IS ZERO'
ENDIF
STOP
END
```

Sub-program

```
SUBROUTINE  SUB3(N,B)
REAL B(10), SUM
INTEGER  I,N
SUM = 0
DO 2 I = 1,N
    SUM = SUM + B(I)
2   CONTINUE
N = SUM + .5
RETURN
END
```

In Example 4 the values of the simple variable N and the array A are passed to SUB3 (see arguments in CALL statement). In the subprogram, N is equivalent to N and array B is equivalent to array A. The sum of the elements in the array is computed and the value of the simple variable N (specified as the first parameter of the subprogram) is returned to the main program (where it has the same name, N), the value of the sum rounded to the nearest integer. Here is an example of using an argument both for passing data *to* the subprogram and for passing a result back *from* the subprogram. Array A in the main program was used by the subprogram but its values were not affected by it.

Examples 3 and 4 are important because we want to make you aware of the care that must be taken when passing arrays as arguments between calling program and subprogram. The following guidelines are worth remembering:

1. When entire arrays are passed as arguments, only the array names (without subscripts) appear in the list of arguments.
2. When arrays or individual elements thereof are passed as arguments, it is easier to keep track of what is happening if the dimensions of arrays are declared identically in the calling program and in the subprogram. Although such identical declaration is not required by the ForTran compiler, the correspondence of storage locations between calling program and subprogram is more obvious and simpler to follow than when the declaration is different.
3. Any given argument may be used both for passing information *to* a subprogram and back *from* it. However, since the limitation on number of parameters is not very restrictive, we recommend for beginning programmers that certain parameters be used for passing data *to* the subprogram and different parameters be used to receive results *from* the subprogram.

COMMON STATEMENT

Near the beginning of our discussion of subprograms it was mentioned that there were some methods of making information available to both calling program and subprogram other than by using arguments. One of those methods will now be discussed—namely, the COMMON statement. The COMMON statement is a nonexecutable ForTran statement that sets aside specified portions of computer memory and makes those portions available to all subprograms as well as to the main program. It literally specifies portions of memory that are shared in common by the main program and by all subprograms in which appropriate COMMON statements appear. Here is its general form:

COMMON VAR1,VAR2,...,VARN

where VAR1,VAR2,...,VARN represent any acceptable ForTran variables or arrays. If any of the variables in the COMMON statement is an array, the COMMON statement may be used to also specify the dimension of the variable. If, however, you choose to place a subscripted variable in both a type declaration or DIMENSION statement and a COMMON statement, it must appear without subscript(s) in all but one of them.

Here are some examples of the use of the COMMON statement:

Example 1

```
INTEGER A(20),N
REAL B
CHARACTER X*6, FUNC2*6
COMMON A,B,N
READ*,N
   .
   .
   .
X=FUNC2(D)
   .
   .
   .
END
CHARACTER *6 FUNCTION FUNC2(D)
INTEGER X(20),M
REAL B
COMMON X ,B,M
IF(M .GT. 0)FUNC2='POSITV'
IF(M .EQ. 0)FUNC2='ZERO'
IF(M .LT. 0)FUNC2='NEGATV'
RETURN
END
```

When this program and function subprogram are executed, an array called A in the main program and X in the subprogram occupies the first twenty locations of common storage. A simple variable called B in both main program and subprogram occupies the twenty-first location in common storage, and a simple variable called N in the main program and M in the subprogram occupies the twenty-second location in common storage.

In this example the array and the first simple variable were not used by the subprogram, though they *were* available to it because of being specified in a COMMON statement. The only variable stored in common that is used is the variable called M in the subprogram and N in the main program. The value of this variable determines whether FUNC2 has the string value POSITV, ZERO, or NEGATV, and one of these strings is then assigned to variable X in the main program.

Though it causes no errors, it is not good programming practice to declare COMMON variables of different size and structure in a single program. The next two examples demonstrate difficulties arising from such actions. We urge readers to use identical variable array names with identical structures, if possible, when COMMON statements are used in a program. Assume default data type declaration based on first letter of variable name.

Example 2

```
COMMON A(2,3),B(5),N
      .
      .
      .
CALL SUB5(X)
      .
      .
      .
END
SUBROUTINE SUB5(Y)
COMMON X(10),A,NN
      .
      .
      .
END
SUBROUTINE SUB6
COMMON AA(10)
      .
      .
      .
END
```

This example is included to show how common memory is shared by the main program and subprograms. Table 7.1 shows the names by which the memory shared in common is known in each subprogram.

TABLE 7.1   Names assigned to common storage

Position of common storage	Name in main program	Name in SUB5	Name in SUB6
1	A(1,1)	X(1)	AA(1)
2	A(2,1)	X(2)	AA(2)
3	A(1,2)	X(3)	AA(3)
4	A(2,2)	X(4)	AA(4)
5	A(1,3)	X(5)	AA(5)
6	A(2,3)	X(6)	AA(6)
7	B(1)	X(7)	AA(7)
8	B(2)	X(8)	AA(8)
9	B(3)	X(9)	AA(9)
10	B(4)	X(10)	AA(10)
11	B(5)	A	
12	N	NN	

As one can tell from Table 7.1, the amount of storage set aside in common is whatever is specified by the *largest* number in any of the COMMON statements.

Here is another program in which COMMON statements, though correct, are not used very wisely in trying to achieve an understandable, well-structured program.

Example 3

```
10         PROGRAM  EXCOMM
20         REAL A(5),B,C,D,X
50         COMMON A, B(2,3),C(4),D
60         READ*,A,D
70         X = FUNC4(0)
80         PRINT*,B,X
90         CALL SUB7
100        PRINT*,A,D
110        PRINT20,B,C
120  20 FORMAT(2F10.4)
130        STOP
140        END
150C**************************************
160C*       FUNCTION SUBPROGRAM FUNC4      *
170C**************************************
180        REAL  FUNCTION FUNC4(A)
185        INTEGER  A
190        COMMON X(10),D
```

```
200       DO 10 I=1,5
210          X(11-I) = X(I)
220  10 CONTINUE
230       D = X(9) * 2.8
240       FUNC4 = X(10)
250       RETURN
260       END
270C**********************************************
280C*          SUBROUTINE SUBPROGRAM SUB7         *
290C**********************************************
300       SUBROUTINE  SUB7
305       REAL X,Y
310       COMMON X(3,4),Y(4)
320       X(3,3) = 15
330       X(3,4) = 20
340       DO 20 I =1,4
350          Y(I) = I*I
360  20 CONTINUE
370       RETURN
380       END
```

This example is included because we want to understand more clearly how the COMMON statement specifies memory locations when main program and subprograms are involved. Examine Table 7.2, which lists variable names in the common area of memory as interpreted by each program module.

TABLE 7.2   List of variables in common, Example 9

Position in common area	Name in main program	Name in FUNC4	Name in SUB7
1	A(1)	X(1)	X(1,1)
2	A(2)	X(2)	X(2,1)
3	A(3)	X(3)	X(3,1)
4	A(4)	X(4)	X(1,2)
5	A(5)	X(5)	X(2,2)
6	B(1,1)	X(6)	X(3,2)
7	B(2,1)	X(7)	X(1,3)
8	B(1,2)	X(8)	X(2,3)
9	B(2,2)	X(9)	X(3,3)
10	B(1,3)	X(10)	X(1,4)
11	B(2,3)	D	X(2,4)
12	C(1)		X(3,4)
13	C(2)		Y(1)
14	C(3)		Y(2)
15	C(4)		Y(3)
16	D		Y(4)

Line 60 causes the reading of six numbers because variable A has been dimensioned for five memory locations and appears without subscripts in the READ statement; hence, the entire array is to be read. The sixth number to be read is stored in variable D. Suppose that when this program is executed the input data provided are:

10, 20, 30, 40, 50, 60

After the READ statement has been executed, the first five computer words in the common area contain 10, 20, 30, 40, and 50, respectively. The sixth word, D, contains the value 60.

Line 70 calls subprogram FUNC4 into action. Because of the COMMON statement in FUNC4, the values for X(1) through X(5) are 10, 20, 30, 40, and 50, respectively. (A look at Table 7.2 will help you see why.) The DO loop at lines 200–220 assigns values to X(10) through X(6) that are equal, respectively, to the values at X(1) through X(5). Line 230 assigns to D (which is in the common area) the value equal to the product of X(9) and 2.8. Since X(9) contains the same value as X(2), which is 20, then D will contain 56. Table 7.3 shows the contents of the computer words in the common area and their names in the main program after line 70 has been executed. Note that up to this point, no values have been assigned to the twelfth through the fifteenth words of the common area. The sixteenth word in the common area contains the value 60 because of the reading of input data as mentioned earlier.

TABLE 7.3   Contents of the common area after line 70 is executed

Position in common area	Contents	Name in main program	Name in FUNC4
1	10.	A(1)	X(1)
2	20.	A(2)	X(2)
3	30.	A(3)	X(3)
4	40.	A(4)	X(4)
5	50.	A(5)	X(5)
6	50.	B(1,1)	X(6)
7	40.	B(2,1)	X(7)
8	30.	B(1,2)	X(8)
9	20.	B(2,2)	X(9)
10	10.	B(1,3)	X(10)
11	56.	B(2,3)	D
12		C(1)	
13		C(2)	
14		C(3)	
15		C(4)	
16	60.	D	

The value in variable X in the main program is FUNC4(0), which is X(10) in the subprogram and therefore equals 10.

At line 80, output occurs causing the printing of the entire array B and the single variable X. With no format specified, the output is five numbers per line resulting in the following:

```
50.0    40.0    30.0    20.0    10.0
56.0    10.0
```

Line 90 calls subprogram SUB7 into action, which sets X(3,3) in SUB7 (the ninth computer word in common) equal to 15. and X(3,4) in SUB7 (the twelfth word in common) to 20. Then Y(1) through Y(4) in SUB7 are assigned values equal to the squares of their subscripts. Table 7.4 shows the latest contents of the common area together with names associated with each memory location.

TABLE 7.4   Contents of the common area after execution of SUB7

Position in common area	Contents	Name in main program	Name in FUNC4	Name in SUB7
1	10.	A(1)	X(1)	X(1,1)
2	20.	A(2)	X(2)	X(2,1)
3	30.	A(3)	X(3)	X(3,1)
4	40.	A(4)	X(4)	X(1,2)
5	50.	A(5)	X(5)	X(2,2)
6	50.	B(1,1)	X(6)	X(3,2)
7	40.	B(2,1)	X(7)	X(1,3)
8	30.	B(1,2)	X(8)	X(2,3)
9	15.	B(2,2)	X(9)	X(3,3)
10	10.	B(1,3)	X(10)	X(1,4)
11	56.	B(2,3)	D	X(2,4)
12	20.	C(1)		X(3,4)
13	1.	C(2)		Y(1)
14	4.	C(3)		Y(2)
15	9.	C(4)		Y(3)
16	16.	D		Y(4)

After executing subprogram SUB7, control returns to line 100 of the main program, whereupon output occurs that causes the printing of entire array A and the simple variable D. The result is:

```
10.0    20.0    30.0    40.0    50.0
16.0
```

Next, line 110 causes the output of entire arrays B and C, resulting in the following:

```
50.0000    40.0000
30.0000    15.0000
10.0000    56.0000
20.0000     1.0000
 4.0000     9.0000
```

The reason for two numbers per line is the FORMAT statement at line 120, which specifies 2F10.4 format; but whenever there are more numbers to be printed than there are format specifications, the format returns to the rightmost right-facing parenthesis in the FORMAT specification (causing a new line) and repeats the format specifications already used. This repetition, with a new line each time, continues until all values in arrays B and C have been printed. In this example, the argument for FUNC4 has no effect. ForTran rules require FUNCTION subprograms to have at least one argument.

Before we leave the COMMON statement, let us emphasize again that any variables specified in common are available to all program modules in which the COMMON statement appears. However, the variable names associated with memory locations in common may be different in the main program and in each subprogram. It is the *position* of a variable in the list of variables in a COMMON statement that determines the name associated with it in each subprogram. This fact was clearly demonstrated in previous Examples 2 and 3. You must have noticed, however, especially in Example 3, how difficult it was to keep track of the contents of variables in the common area because COMMON statements were not identical in main program and subprograms. Only one COMMON statement of the type described here can appear in any given program module.

PROBLEMS WITH COMPLETE SOLUTIONS

Now we present two examples, complete with algorithms and programs, that demonstrate the use of subprograms.

Problem 1: Updating a file

Problem.   This problem is basically the same as Problem 2 toward the end of Chapter 6. We consider it again here in order to:

1. Make the solution more general by having it handle the creation of a data file if it does not already exist;

2. Demonstrate the concept of modular programming—that is, of separating the solution into several simple tasks, each task handled in its own subprogram.

Recall from Chapter 6 that the problem consists of two major tasks. If the data file MASTFL exists it is to be modified according to three codes, U, D, and A. A code of U means to *update* the specified customer's record with new information provided. A code of D means to *delete* the specified record from the data file. A code of A means to *add* the specified record to the file.

If data file MASTFL does not exist, create such a file from the provided input data, where it is understood that to be processed, data must have the code A (for add). Data with any other code are rejected.

Recall that the data in MASTFL are such that each record consists of two data elements: customer number and account balance. *Input data* records consist of the action code (U, D, or A), customer number, and the amount of the transaction to be added to the account balance if the code is U, or the account balance itself if the code is A. If the code is D only the customer number is present.

Problem solution.   We separate the problem into subdivisions where one main division handles some basic tasks and other sections (modules) are called into use as needed to perform specific tasks.

1. Main division.   This section performs the following tasks: (1) initializes data files needed, (2) reads one set of transaction data and one record from file MASTFL, and (3) calls appropriate subroutines to perform certain actions, depending on the value of input datum, CODE. When transaction data are read, the first task is to find in file MASTFL either the record to be updated or deleted (if CODE is "U" or "D"), or where to insert a record if CODE is "A" (for *add*). Subroutine LOCATE performs these tasks.

When CODE has a value other than "A", "D", or "U", the main program calls subroutine FINISH to terminate processing.

In addition to subroutines LOCATE and FINISH, the main program may call these subroutines:

ADD for adding a new record to MASTFL,
UPDATE for updating an existing record in file MASTFL,
DELETE for deleting an existing record from MASTFL.

2. Subdivision LOCATE.   This division reads record after record of file MASTFL until it locates the record to be used in processing the transaction data currently in hand, or until the end of MASTFL is reached, in which case a special variable EOF is set to "true."

3. Subdivision FINISH.   This division copies file TEMPF into file MASTFL and produces a printed copy of the new MASTFL.

4. Subdivision ADD.   This division receives a set of transaction data and, if the customer number in those data is *not* present in MASTFL, the new transaction data are added to MASTFL. If CODE is "A" and the customer number in the current transaction data *is* in MASTFL, this division prints an appropriate error message.

5. Subdivision UPDATE.   This division receives a set of transaction data and if the customer number in those data *is* in file MASTFL, this division updates the account balance on MASTFL. A positive transaction value is considered a purchase and a negative value a payment on the account. An appropriate message, based on the transaction value, is printed. If the account number is *not* in MASTFL, an appropriate message is printed.

6. Subdivision DELETE.   This division deletes customer account data from MASTFL. If the account has a balance, that amount is printed before the account is deleted. If the account balance is zero, a message indicating the deletion has taken place is printed.

Algorithms.   The following are algorithms for the main division and each of the subdivisions.

1. Main division

M.1 Initialize NUMPRV, EOF, MASTFL, and TEMPF.

M.2 Read one record from MASTFL and if the end of MASTFL is reached, set EOF to "true."

M.3 Read one set of transaction data and set the error flag for transaction data to "false."

M.4 If CODE is "A", "U", or "D" and if transaction data are in proper order, call subdivision LOCATE; otherwise, report an appropriate message saying transaction data are out of order. If CODE is not one of the above, copy the rest of MASTFL (if any remains) to TEMPF and call subdivision FINISH to terminate the program.

M.5 Call appropriate subdivisions depending on the value of CODE.

M.6 Repeat the process, beginning with step M.3.

2. Subdivision LOCATE

L.1 Read record after record of file MASTFL until the account number in the record read from MASTFL is greater than or equal to the account number in the current transaction data, or until the end of MASTFL is reached.

L.2 If the end of MASTFL is reached, set EOF to "true" and return to main division.

L.3 If the current transaction account number is less than or equal to the account number from MASTFL, return to the main division.

3. Subdivision FINISH

F.1 Initialize files MASTFL and TEMPF at their beginnings.

F.2 Copy TEMPF in its entirety to MASTFL and to an output area.

F.3 Close both files and return to main division.

4. Subdivision ADD

A.1 If the current transaction data intended for adding have an account number already in MASTFL, then report an appropriate message and return to the main division.

A.2 Else write the current transaction data to TEMPF and return to the main division.

5. Subdivision UPDATE

U.1 If the account number in current transaction data (intended for updating) does not exist in MASTFL, then print an appropriate message and return to the main division.

U.2 Compute the new account balance.

U.3 Write the transaction data with new balance into TEMPF.

U.4 Write an appropriate message based on the value of transaction amount.

U.5 Read the next record from MASTFL and when the end of MASTFL is reached, set EOF to "true."

U.6 Return to main division.

6. Subdivision DELETE

D.1 If the account number in current transaction data (intended for deleting) does not exist in MASTFL, then print an appropriate message and return to main division.

D.2 Write an appropriate message based on the account balance.

D.3 Read the next record from MASTFL and when the end of MASTFL is reached, set EOF to "true."

D.4 Return to main division.

Program

```
      PROGRAM  UPDAT3
C*********************************************************************
C*          Program definition                                     *
C*              For a complete description of the problem          *
C*              see 328.                                           *
C*                                                                 *
C*              Variable Definition                                *
C*                  NUMI= customer # read from transaction card*
C*                  NUMF= customer # read from file MASTFL         *
C*                  TRANS= Amount of transaction (purchase or      *
C*                          payment)                               *
C*                  BALANC= amount read from MASTFL as account     *
C*                          balance                                *
C*                  CODE= 'A' for adding new account               *
C*                        'U' for updating existing account        *
C*                        'D' for deleting an account              *
C*                  Anything else indicates the end of data.       *
C*                  NUMPRV  Previous transaction number.           *
C*                  EOF     To indicate the end of MASTFL.         *
C*                  ERROR   To indicate if there is any error      *
C*                          in transactions.                       *
C*********************************************************************
C*
      CHARACTER CODE
      INTEGER NUMI, NUMF, NUMPRV
      REAL BALANC,TRANS
      LOGICAL EOF, ERROR
C*
C*                  Initialization
C*
      NUMPRV = 0
      EOF = .FALSE.
      OPEN(1,FILE='MASTFL',STATUS='OLD')
      REWIND 1
      OPEN(2,FILE='TEMPF',STATUS='NEW')
      READ(1,10,END=5)NUMF,BALANC
C*********************************************************************
C*                  Read transactions one at a time and check   *
C*                  for appropriate action based on the content*
C*                  of the variable CODE.                       *
C*********************************************************************
```

```
C*
  3       READ 20, CODE, NUMI, TRANS
            ERROR = .FALSE.
            IF(CODE .EQ. 'A' .OR. CODE .EQ.'U' .OR. CODE .EQ. 'D')THE
              IF(NUMPRV .LT. NUMI .AND. .NOT. EOF)THEN
                CALL LOCATE(NUMF,BALANC,NUMI, EOF)
              ELSEIF(NUMPRV .GE. NUMI)THEN
                PRINT 109, NUMI, TRANS
                PRINT*, 'TRANSACTION NUMBER OUT OF SEQUENCE'
                ERROR = .TRUE.
              ENDIF
            ELSEIF(.NOT. EOF)THEN
 13           WRITE(2,10) NUMF,BALANC
                READ(1,10,END=30)NUMF,BALANC
              GOTO 13
            ELSE
 30           CALL FINISH
            ENDIF
C*
            IF(ERROR)GOTO3
C*
            NUMPRV = NUMI
          IF(EOF .AND. CODE .NE. 'A')THEN
            PRINT*, ' AT THIS POINT YOU CAN ONLY ADD TO MASTFL'
          ELSEIF(CODE .EQ. 'A')THEN
            CALL ADD(NUMF,NUMI,TRANS)
          ELSEIF(CODE .EQ. 'U' )THEN
            CALL UPDATE(NUMI,NUMF,TRANS,BALANC, EOF)
          ELSEIF(CODE .EQ. 'D' )THEN
            CALL DELETE(NUMI,NUMF,BALANC, EOF)
          ENDIF
        GOTO 3
  5     EOF = .TRUE.
        GOTO 3
C*
 10     FORMAT(I6,F8.2)
 20     FORMAT(A1,I6,F8.2)
109     FORMAT(' ERROR IN INPUT ',I6,1X,F8.2,' DATA IGNORED')
        END
C*
          SUBROUTINE  LOCATE(NUMF,BALANC, NUMI, EOF)
C*
C*****************************************************************
C*        This subroutine either finds record NUMI or the      *
C*        record just next to it. If the new transaction       *
C*        is an addition it is to be added befor the record*
C*        located by this subprogram.                          *
C*****************************************************************
```

```
C*
          LOGICAL EOF
  1       IF(NUMI .LE. NUMF)RETURN
            WRITE(2,10)NUMF,BALANC
            READ(1,10,END=30)NUMF,BALANC
          GOTO 1
  30      EOF = .TRUE.
          RETURN
  10      FORMAT(I6,F8.2)
          END
C*
          SUBROUTINE   FINISH
C*
C*******************************************************************
C*                    This subroutine is to copy the              *
C*                    content of MASTFL to printer and            *
C*                    terminate the program.                      *
C*                                                                *
C*                    The line "ENDFILE 2" puts an end-of-file    *
C*                    mark at the end of unit 2.                  *
C*******************************************************************
C*
  30      ENDFILE 2
          REWIND 1
          REWIND 2
          PRINT 18
  32      READ(2,10,END=50)NUMF,BALANC
          WRITE(1,10)NUMF,BALANC
          PRINT25,NUMF,BALANC
          GOTO 32
C*
C*******************************************************************
C*                    The following statement is to remove       *
C*                    unit 2 from the system. The CLOSE           *
C*                    statement has a parameter called STATUS.    *
C*                    If STATUS is omitted or is KEEP the unit    *
C*                    will be saved on the system. If STATUS      *
C*                    is DELETE the unit is  purged.              *
C*******************************************************************
C*
  50      CLOSE(2,STATUS='DELETE')
          CLOSE(1)
          STOP
  10      FORMAT(I6,F8.2)
  18       FORMAT('1',2X,'CUST #   BALANCE',///)
  25      FORMAT(2X,I6,F8.2)
          END
```

```
C*
          SUBROUTINE   ADD(NUMF,NUMI,TRANS)
C*
C*****************************************************************
C*          This subroutine adds a new record to MASTF1     *
C*          at location prior to NUMF. It also adds         *
C*          records to the end of MASTFL.                   *
C*****************************************************************
C*
          INTEGER   NUMI,NUMF
          REAL   TRANS
C*
          IF(NUMI .EQ. NUMF)THEN
            PRINT*, 'DUPLICATE REDORD ',NUMI,TRANS
          ELSE
            WRITE(2,10)NUMI,TRANS
            PRINT 110, NUMI, TRANS
          ENDIF
          RETURN
  10      FORMAT(I6,F8.2)
  110     FORMAT(' NEW ACCOUNT ADDED ',I6,1X,F8.2)
          END
C*
          SUBROUTINE   UPDATE(NUMI,NUMF,TRANS,BALANC,EOF)
C*
C*****************************************************************
C*          This subroutine  updates the current record.   *
C*          The number of the current record is NUMI.      *
C*****************************************************************
C*
          INTEGER   NUMI,NUMF
          REAL   TRANS, BALANC
          LOGICAL   EOF
C*
          IF(NUMI .EQ. NUMF)THEN
            BALANC = BALANC + TRANS
            WRITE(2,10)NUMI,BALANC
            IF(TRANS .LT. 0)THEN
              PRINT220, NUMI, -TRANS,BALANC
            ELSE
              PRINT210, NUMI,TRANS,BALANC
            ENDIF
            READ(1,10,END=5)NUMF, BALANC
          ELSE
            PRINT*, ' CUSTOMER # ',NUMI,' NOT IN MASTFL'
          ENDIF
          RETURN
```

```
 5       EOF = .TRUE.
         RETURN
10       FORMAT(I6,F8.2)
210      FORMAT(' CUSTOMER # ',I6,' NEW PURCHASE',F8.2,' BALANCE ',F8.2)
220      FORMAT(' CUSTOMER # ',I6,' PAYMENT ',F8.2,' BALANCE',F8.2)
         END
C*
         SUBROUTINE  DELETE(NUMI,NUMF,BALANC,EOF)
C*
C****************************************************************
C*       This subroutine deletes the current record from   *
C*       MASTFL. The number of the current record is NUMI.*
C****************************************************************
C*
         INTEGER  NUMI, NUMF
         REAL  BALANC
         LOGICAL  EOF
C*
         IF(NUMI .EQ. NUMF)THEN
           IF(BALANC .LE. .01)THEN
             PRINT 310, NUMI
           ELSE
             PRINT 320, NUMI,BALANC
             PRINT*, ' BUT IS DELETED ANYWAY '
           ENDIF
           READ(1,10,END=5)NUMF,BALANC
         ELSE
           PRINT*, ' CUSTOMER # ',NUMI,' IS NOT IN MASTFL'
         ENDIF
         RETURN
 5       EOF = .TRUE.
         RETURN
10       FORMAT(I6,F8.2)
310      FORMAT(' CUSTOMER # ',I6,' HAS BEEN DELETED')
320      FORMAT(' CUSTOMER # ',I6,' HAS A BALANCE OF ',F8.2)
         END
```

Problem 2: Matrix operations

Matrix operations apply to a great many situations in the field of science, and since ForTran was designed primarily for use in mathematics, statistics, and the natural sciences, we feel it is important for you to have some experience with programming applications to matrices. Some programming languages, like BASIC, provide matrix operations as a part of the language. In the case of ForTran, software packages can be purchased that, when made a part of the computer system on which one is working, allow the programmer to use programming techniques much like those associated with system library functions.

This example will demonstrate some of the capabilities of ForTran as related to matrices.

Before we present the problem to be solved, we shall state some definitions that apply to the problem.

Matrix definitions.

1. An arrangement of numbers in m rows and n columns is called a *matrix*. If we let the capital letter A represent the matrix and the small letter a with subscripts represent individual numbers in the matrix, the result is as follows:

$$A = \begin{pmatrix} a_{11} & a_{12} \cdots a_{1n} \\ a_{21} & a_{22} \cdots a_{2n} \\ \cdot \\ \cdot \\ \cdot \\ a_{m1} & a_{m2} \cdots a_{mn} \end{pmatrix}$$

where the brackets are used to enclose the members of the matrix. Note that the first subscript specifies the row in which a member appears and the second subscript the column.

2. A matrix is called a *square matrix* if the number of rows and columns are equal.

3. The transpose of a matrix A is the matrix B such that the rows of B are the columns of A and the columns of B are the rows of A, both in the same order as in matrix A. Note that if A is an *m*-by-*n* matrix, then B is an *n*-by-*m* matrix.

4. If A and B are both *m*-by-*n* matrices, then the matrix C, whose elements, c_{ij}, are determined by adding corresponding elements a_{ij} and b_{ij} of A and B, is called the *sum* of matrices A and B. We write C = A + B, where $c_{ij} = a_{ij} + b_{ij}$. Similarly, the matrix D obtained by subtracting corresponding elements of A and B is called the *difference* of A and B. We write D = A − B, where $d_{ij} = a_{ij} - b_{ij}$.

5. If A is a matrix whose elements a_{ij} are numbers, and k is a number like the elements of A, then the *scalar product* of k and A is the matrix P, whose elements, p_{ij} are the products of the number k and the elements of A. We write P = kA, where $p_{ij} = ka_{ij}$.

Now we are ready to state the problem.

Problem.   Write a program to do the following:

1. *Input* one or two matrices, A and B, whose dimensions are *m* by *n*, where *m* and *n* are both less than 20;
2. Find the *sum* of the matrices input if there are two of them;
3. Find the *difference* of the matrices input if there are two of them;
4. Find the *scalar product* of matrix A and number k.
5. Find the *transpose* of matrix A.

The program should be written so that the user of this program can accomplish one or more of these operations by specifying the desired operations as follows:

1. The word INPUT in columns 1 through 5 of a data card, and A or B in column 7 of the same card, followed by a second card having two numbers, *m* and *n*, specifying the number of rows and columns, would cause input into an array named A or B, as specified on the first card. The *m*-times-*n* members of the array are to be on as many cards as needed, separated by at least one blank space.
2. The word ADD in columns 1 through 3 of a data card would cause the sum of two arrays, A and B, to be computed and stored in array C, and also print the elements of C.
3. The word SUB in columns 1 through 3 of a data card would cause the difference of two arrays, A and B, to be computed and stored in array C, and also print the elements of C.
4. The word MULT in columns 1 through 4 of a data card and the letter A or B in column 7 of the same card, followed by a single number, k, in a second data card would cause the scalar product of k and A or B to be computed and stored in array C, and also print the elements of C.
5. The word TRANSP in columns 1 through 6, and A or B in column 7, would cause the transpose of the specified array to be stored in array C, and also print the elements of C.
6. The word TRANSF in columns 1 through 6, and A or B in column 7, would cause a copy of array C to be stored in the specified array.
7. The word FINISH in columns 1 through 6 would terminate all processing.

This example is included to give further experience with the use of subprograms and modular development in writing a computer program.

Algorithm for main program.

M1.   Input the process code, CODE, and matrix identifier, NAME, if any.
M2.   If CODE is FINISH, terminate all processing.
M3.   If CODE is INPUT, call subprogram INPUT.
M4.   If CODE is ADD, call subprogram ADDSUB.
M5.   If CODE is SUB, call subprogram ADDSUB.
M6.   If CODE is MULT, call subprogram MULT.
M7.   If CODE is TRANSP, call subprogram TRANSP.
M8.   If CODE is TRANSF, call subprogram TRANSF.
M9.   If CODE is none of the above, print message about incorrect process code.
M10. Repeat the process from step M1.

Algorithm for subprogram INPUT.   The argument for this subprogram, NAME, identifies the matrix name.

I1.   Read M and N, the number of rows and columns in the matrix to be read.
I2.   If either M or N is negative or is greater than 20, then return to the calling program.
I3.   If NAME is 'A', then read M rows of N values per row to matrix A and return the number of rows in A to variable MA and the number of columns in A to variable NA. Otherwise, read M rows of N values per row to matrix B and return the number of rows to variable MB and the number of columns to variable NB.
I4.   Return to the calling program.

Algorithm for subprogram ADDSUB.   The argument, I, of this subprogram is +1 for addition and −1 for subtraction.

A1.   If MA or NA is less than or equal to zero, then print appropriate message and return to the calling program.
A2.   If MB or NB is less than or equal to zero, then print appropriate message and return to the calling program.
A3.   If MA is not equal to MB or NA is not equal to NB, then print appropriate message and return to the calling program.
A4.   Compute C, the sum or difference matrix, using the formula $c_{ij} = a_{ij} \pm I^*b_{ij}$ for each of the elements in matrices A and B.
A5.   Return to the calling program.

Algorithm for subprogram MULT.   The argument, NAME, of this subprogram identifies the matrix name.

P1. Read the number, F, to be used as the multiplier.
P2. If NAME is 'A', then compute the product matrix C according to the formula $c_{ij} = F^*a_{ij}$ for each of the elements in matrix A.
P3. If NAME is not 'A', then compute the product matrix C according to the formula $c_{ij} = F^*b_{ij}$ for each of the elements in matrix B.
P4. Print matrix C.
P5. Return to the calling program.

Algorithm for subprogram TRANSP.   The argument, NAME, of this subprogram identifies the matrix name.

T1. If NAME is 'A', then
   T2. If MA or NA is less than or equal to zero, report the message that matrix A has not been defined; then return to calling program.
   T3. Compute transpose matrix C according to the equation $c_{ij} = a_{ij}$ for each of the elements in matrix A.
   T4. Set MC = MA and NC = NA and proceed to step T8.
   Else
   T5. If MB or NB is less than or equal to zero then report the message that matrix B has not been defined then return to the calling program.
   T6. Compute transpose matrix C according to the equation $c_{ij} = b_{ji}$ for each of the elements in matrix B.
   T7. Set MC = MB and NC = NB.
T8. Output matrix C.
T9. Return to calling program.

Algorithm for subprogram TRANSF.   The argument of this subprogram identifies the matrix name.

F1. If NAME is 'A', then perform steps F2 and F3. Otherwise perform steps F4 and F5.
F2. Transfer matrix C to matrix A by setting $a_{ij} = c_{ij}$ for each element c_{ij} in C.
F3. Set MA = MC and NA = NC and proceed to step F6.
F4. Transfer matrix C to matrix B by setting $b_{ij} = c_{ij}$ for each element c_{ij} in C.
F5. Set MB = MC and NB = NC and proceed to step F6.
F6. Output the resulting matrix, C.
F7. Return to calling program.

Program

```
      PROGRAM   MATRIX
C**************************************************************
C*          PROGRAM DEFINITION                               *
C*              THIS PROGRAM IS TO CARRY OUT ANY OF SIX       *
C*              OPERATIONS WITH ONE OR TWO MATRICES DEPENDING *
C*              ON THE INPUT DATA. THESE OPERATIONS ARE:      *
C*              1. INPUT A MATRIX INTO ARRAY A OR B.          *
C*              2. ADD MATRICES A AND B.                      *
C*              3. SUBTRACT MATRIX B FROM MATRIX A.           *
C*              4. MULTIPLY MATRIX A BY FACTOR F.             *
S*              5. TRANSPOSE MATRIX A.                        *
C*              6. TRANSFER MATRIX C INTO MATRIX A OR B.      *
C**************************************************************
C**************************************************************
C*          VARIABLE DEFINITION                              *
C*              A, B AND C ARE THREE ARRAYS TO HOLD THE TWO   *
C*                        POSSIBLE INPUT MATRICES AND THE     *
C*                        RESULT MATRIX.                      *
C*              MA, MB AND MC ARE NUMBER OF ROWS OF MATRICES  *
C*                        A, B, AND C, RESPECTIVELY.         *
C*              NA, NB, AND NC ARE NUMBER OF COLUMNS OF       *
C*                        MATRICES A, B, AND C, RESPECTIVELY.*
C*              F  IS A FACTOR TO MULTIPLY A MATRIX.          *
C*                 THIS IS DEFINED IN SUBROUTINE MULT.        *
C*              CODE  INDICATES ONE OF THE SIX OPERATIONS     *
C*                    TO BE PERFORMED.                        *
C*              NAME  IS THE NAME OF THE MATRIX, A OR B,      *
C*                    USED IN THE OPERATION.                  *
C**************************************************************
C*                                                           *
C*          LIST OF SUBPROGRAMS                              *
C*              INPUT: ENTERS ONE OR TWO MATRICES INTO ARRAYS *
C*                     A AND B.                               *
C*              ADDSUB: ADDS OR SUBTRACTS TWO MATRICES.       *
C*              MULT: FINDS THE SCALER PRODUCT OF A NUMBER    *
C*                    AND A MATRIX.                           *
C*              TRANSP: FINDS THE TRANSPOSE OF A MATRIX .     *
C*              TRANSF: PLACES A COPY OF ONE MATRIX INTO      *
C*                      ANOTHER.                              *
C**************************************************************
C*                                                           *
C*          START OF THE MAIN PROGRAM                        *
C*              THE MAIN PROGRAM READS THE CODE AND THE NAME  *
C*              OF THE MATIX AND CALLS APPROPRIATE SUBROUTINE *
C*              INTO ACTION.                                  *
C**************************************************************
```

```
C*
          REAL A(20,20), B(20,20), C(20,20)
          INTEGER MA, NA, MB, NB, MC, NC
          CHARACTER CODE*6, NAME
          COMMON /DATA/A,B,C,MA,NA,MB,NB,MC,NC
          DATA A,B,C,MA,NA,MB,NB,MC,NC/1206*0/
C*
  1       READ 10, CODE,NAME
            IF(CODE .EQ. 'FINISH')STOP
            IF(CODE .EQ. 'INPUT' )THEN
              CALL INPUT(NAME)
            ELSEIF(CODE .EQ. 'ADD')THEN
              CALL ADDSUB(1)
            ELSEIF(CODE .EQ. 'SUB')THEN
              CALL ADDSUB(-1)
            ELSEIF(CODE .EQ. 'MULT')THEN
              CALL MULT(NAME)
            ELSEIF(CODE .EQ. 'TRANSP')THEN
              CALL TRANSP(NAME)
            ELSEIF(CODE .EQ. 'TRANSF')THEN
              CALL TRANSF(NAME)
            ELSE
              PRINT*,' INVALID CODE:  ' ,CODE
            ENDIF
          GOTO 1
  10      FORMAT(A6,A1)
          END
C*
C*                SUBROUTINE TO READ A MATRIX
C*
          SUBROUTINE INPUT(NAME)
          REAL A(20,20),B(20,20),C(20,20)
          CHARACTER NAME
          COMMON /DATA/ A,B,C,MA,NA,MB,NB,MC,NC
C*
          PRINT* ,' ENTER MATRIX ROWS AND COLUMNS'
          READ*,M,N
          IF(M.LE.0 .OR. M.GT.20 .OR.N.LE.0 .OR. N.GT.20)THEN
            PRINT*,' ERROR IN MATRIX SIZE: ',M,N
            RETURN
          ENDIF
          PRINT* ,' ENTER YOUR MATRIX ROW BY ROW'
```

```
C*
          IF(NAME .EQ. 'A')THEN
            READ*,((A(I,J),J=1,N),I=1,M)
            MA = M
            NA = N
          ELSE
            READ*,((B(I,J),J=1,N),I=1,M)
            MB = M
            NB = N
          ENDIF
          RETURN
          END
C*
C*        SUBPROGRAM ADDSUB
C*            THE ARGUMENT OF THIS SUBPROGRAM IS
C*            1 FOR ADDITION AND IS -1 FOR
C*            SUBTRACTION.
C*
          SUBROUTINE ADDSUB(I)
          REAL A(20,20),B(20,20),C(20,20)
          COMMON /DATA/A,B,C,MA,NA,MB,NB,MC,NC
C*
          IF(MA .LE. 0 .OR. NA .LE. 0)THEN
            PRINT*,' ARRAY A IS NOT DEFINED'
            RETURN
          ELSEIF(MB .LE. 0 .OR. NB . LE. 0)THEN
            PRINT*,' ARRAY B IS NOT DEFINED'
            RETURN
          ELSEIF(MA .NE. MB .OR. NA .NE. NB)THEN
            PRINT*,' A AND B ARE NOT THE SAME SIZE'
            RETURN
          ELSE
            DO 40 J = 1, MA
              DO 30 K= 1,NA
                C(J,K) = A(J,K) + I*B(J,K)
30            CONTINUE
            PRINT*,' ROW ',J
            PRINT*,(C(J,K),K=1,NA)
40          CONTINUE
          ENDIF
          MC = MA
          NC = NA
          RETURN
          END
```

```
C*
C*          SUBPROGRAM   MULT
C*
       SUBROUTINE MULT(NAME)
       CHARACTER  NAME
       REAL A(20,20),B(20,20),C(20,20)
       COMMON /DATA/A,B,C,MA,NA,MB,NB,MC,NC
C*
       PRINT*, 'ENTER FACTOR'
       READ*,F
       IF(NAME .EQ. 'A')THEN
         DO 60 I=1,MA
           DO 70 J=1,NA
             C(I,J) = F * A(I,J)
70         CONTINUE
60       CONTINUE
         MC = MA
         NC = NA
       ELSE
         DO 90 I=1,MB
           DO 80 J=1,NB
             C(I,J) = F * B(I,J)
80         CONTINUE
90       CONTINUE
         MC = MB
         NC = NB
       ENDIF
C*
       DO 100 I=1,MC
         PRINT*,'ROW ',I
         PRINT*,(C(I,J),J=1,NC)
100      CONTINUE
       RETURN
       END
C*
C*          SUBPROGRAM TRANSP
C*
       SUBROUTINE TRANSP(NAME)
       CHARACTER NAME
       REAL A(20,20),B(20,20),C(20,20)
       COMMON /DATA/ A,B,C,MA,NA,MB,NB,MC,NC
```

```
C*
                IF(NAME .EQ. 'A')THEN
                  IF(MA .LE.O .OR. NA .LE. 0)THEN
                    PRINT*,' ARRAY A IS NOT DEFINED'
                    RETURN
                  ELSE
                    DO 20 I=1,MA
                      DO 15 J=1,NA
                        C(J,I) = A(I,J)
   15               CONTINUE
   20             CONTINUE
                  NC = MA
                  MC = NA
                  ENDIF
                ELSE
                  IF(MB .LE.O .OR. NB .LE. 0)THEN
                    PRINT*,' ARRAY B IS NOT DEFINED'
                    RETURN
                  ELSE
                    DO 30 I=1,MB
                      DO 25 J=1,NB
                        C(J,I) = B(I,J)
   25               CONTINUE
   30             CONTINUE
                  NC = MB
                  MC = NB
                  ENDIF
                ENDIF
C*
                DO 40 I = 1,MC
                  PRINT*,' ROW',I
                  PRINT*,(C(I,J),J=1,NC)
   40           CONTINUE
                RETURN
                END
C*
C*              SUBPROGRAM   TRANSF
C*
                SUBROUTINE TRANSF(NAME)
                CHARACTER NAME
                REAL A(20,20),B(20,20),C(20,20)
                COMMON /DATA/A,B,C,MA,NA,MB,NB,MC,NC
```

```
C*
            IF(NAME .EQ. 'A')THEN
              DO 300 I=1,MC
                DO 310 J=1,NC
                  A(I,J)=C(I,J)
310             CONTINUE
300           CONTINUE
              MA = MC
              NA = NC
            ELSE
              DO 330 I=1,MC
                DO 320 J=1,NC
                  B(I,J) = C(I,J)
320             CONTINUE
330           CONTINUE
              MB = MC
              NB = NC
            ENDIF
C*
            DO 350 I=1,MC
              PRINT*,'ROW ',I
              PRINT*,(C(I,J),J=1,NC)
350         CONTINUE
            RETURN
            END
```

Comments about previous program. This program does not edit all input data as well as it might. In particular, the name of an array to be processed is always assumed to be A or B. We check only to see if the name is A and if it is not, we assume it is B. If the name were *not* B, this program would not detect such an error.

Another weakness of this program is that it always produces output showing the results of the matrix operation being done. There may be times when such output could cause confusion, depending on the sequence of matrix operations requested. Here is an example.

Suppose matrix A is given by

$$A = \begin{vmatrix} 3 & 5 & 10 \\ 8 & -2 & 6 \\ 2 & 7 & 1 \end{vmatrix}$$

and A is to be added to the transpose of A by means of the previous program. Following are the input cards needed to accomplish these tasks, including, at the left, the order of reading of each card:

1. `INPUT A`
2. `3,3`
3. `3,5,10`
4. `8,-2,6`
5. `2,7,1`
6. `TRANSPA`
7. `TRANSFB`
8. `ADD`

The first five cards of input will result in array A being properly stored. Card 6 will cause the transpose of A to be computed and stored in array C. Also the transpose array is printed. This may cause confusion because the user only needs the final result—namely, the sum of array A and its transpose.

Data card 7 causes the transpose array (in array C) to be transferred to array B. Again, output is produced—maybe causing some confusion. Finally, with the reading of card 8, the sum of array A and its transpose (in array B) is computed and stored in array C as well as being printed.

As you can tell from the previous discussion, the program we have presented could be improved. We leave it as an exercise to the interested reader to make such improvements. Some may want to include even more changes to include interactive communication with the user if an interactive computer system is available.

Those readers with adequate background in linear algebra may find it interesting to enlarge the program to include matrix multiplication. Labeled common statements were used in this program. For a discussion, see page 437.

SORT AND SEARCH TECHNIQUES

In a great many computer applications the need arises for the capability of sorting information into some order, or searching for certain information in a list or in an array. In fact, earlier in one of the examples in this chapter, we employed a sorting procedure. The method demonstrated is easy to understand but is very wasteful of computer time if the arrays to be sorted are large. It is called the *linear sort* and will be included in the present section.

There are many other sorting techniques as well as a variety of search techniques. The topic of sorting and searching is a very rich one in computer science, with entire books being devoted to it. We shall discuss just a few techniques, none of which requires any extensive knowledge of data structures or significant depth in mathematical background. The procedure we shall follow will be to discuss the sorting or searching technique, present an algorithm for accomplishing it, and finally give a ForTran subprogram that could be called by any ForTran program. In the case of all sorting procedures, we shall assume the desired result is a list of numbers sorted in *ascending* order. The reader should be able to make the necessary changes to sort in *descending* order if desired.

Linear sort

Suppose there are N numbers in the list to be sorted. We compare the first number with all other N−1 numbers in the list. If after any comparison it is determined that a number other than the first one is smaller, we compare this new smaller one to the rest of the numbers and remember the position where it was found. When all N−1 comparisons have been made, the smallest number in the list will be exchanged with the number in the first position. Then we move to the second number in the list and repeat the whole process, comparing that number with all N−2 numbers in positions beyond the second. When that process is completed, the second smallest number in the list will be exchanged with the number in the second position. If this procedure is continued to the number in position N−1, the list will be sorted in ascending order. Table 7.5 shows the numbers in a list at all stages of applying the above sorting technique.

TABLE 7.5 Example of linear sorting

Original list	End of stage 1	End of stage 2	End of stage 3	End of stage 4
121	67	67	67	67
217	217	81	81	81
67	121	121	121	121
81	81	217	217	200
200	200	200	200	217

Note in Table 7.5 that there is no exchange during the steps of stage 3 because the third number at the end of stage 3 was already the third smallest number.

Algorithm for linear sort.   Here, then, is an algorithm for executing a linear sort. We assume array A already contains N unsorted numbers. We also prevent more than one interchange by remembering the position of the smallest element in each stage and only doing one interchange at the end of that stage.

1. Initialize index i to 1, where i is the position index that will take us through all N positions of array A.
2. Set $k = i$ where k is the position number of the smallest number. (We assume to begin with that the first number in array A is smallest.)
3. Set a second index, j, equal to $i+1$ so we can compare the ith number in array A with each of the numbers from position $i+1$ and on.
4. If A(k) is larger than A(j), then set $k = j$.
5. Increase j by 1. If j is less than or equal to N, repeat the process from step 4.
6. Interchange A(k) and A(i).
7. Increase i by 1. If i is less than or equal to N-1, repeat the process from step 2.
8. Return to calling program.

Subprogram for linear sort

Algorithm   *ForTran*
steps   *statements*

```
      C**********
      C*
      C*    ****SUBPROGRAM SORT1****
      C*
      C*    THE ARGUMENTS OF THIS SUBPROGRAM
      C*    ARE THE ARRAY, A, TO BE SORTED AND N,
      C*    THE NUMBER OF ELEMENTS IN THE ARRAY.
      C*

            SUBROUTINE   SORT1(A,N)
            REAL   A(N)
1.          DO 20 I = 1,N-1
2.             K = I
3.             DO 10 J = I+1, N
4.                IF(A(K) .GT. A(J))K = J
5. 10          CONTINUE
6.             TEMP = A(I)
               A(I) = A(K)
               A(K) = TEMP
7. 20       CONTINUE
            RETURN
            END
```

Bubble sort

This procedure involves the comparison of two adjacent members of the list, beginning with the first two. If the one farther on in the list is smaller, then interchange the two adjacent elements. If not, no interchange is needed. Continue through the list, comparing adjacent numbers until one of the two adjacent numbers is the last number in the list. The entire list is in ascending order when a pass can be made through the list in the above-described manner without making a single interchange of numbers.

To help clarify the procedure we shall work through an example by hand. Suppose the original array with position numbers is as follows:

(1) 121   (2) 217   (3) 67   (4) 81   (5) 200

When the first and second numbers are compared, no interchange is necessary since they are already in ascending order. When the second and third numbers are compared, an interchange is necessary, so the list becomes

(1) 121   (2) 67   (3) 217   (4) 81   (5) 200

Now the third and fourth numbers are compared and an interchange is necessary, resulting in

(1) 121   (2) 67   (3) 81   (4) 217   (5) 200

Finally, the fourth and fifth numbers are compared; another interchange is required, which gives these results:

(1) 121   (2) 67   (3) 81   (4) 200   (5) 217

Now we must start again at the beginning to see if a complete pass through the array can be made without any interchanges. When elements one and two are compared, an interchange is required and gives

(1) 67   (2) 121   (3) 81   (4) 200   (5) 217

Next, elements two and three are compared, resulting in their interchange, and yielding

(1) 67   (2) 81   (3) 121   (4) 200   (5) 217

When elements three and four are compared, no interchange is required, as is the case when elements four and five are compared.

Again we must start at the beginning to see if one complete pass can be made through the list without a single interchange. A quick check by you will assure you that such a pass can be made; the list has been sorted.

Algorithm for bubble sort.

1. Initialize index i to 1 and variable FLAG to zero.
2. If A(i) is greater than A(i+1) then interchange the two elements and set FLAG = 1.
3. Increase i by 1. If i is less than or equal to N−1 repeat the process from step 2.
4. If FLAG=1 repeat the process from step 1.
5. Return to calling program.

Subprogram for bubble sort

Algorithm steps	ForTran statements

```
      C**********
      C*
      C*   ****SUBPROGRAM SORT2****
      C*   THE ARGUMENTS OF THIS SUBPROGRAM ARE
      C*   THE ARRAY, A, TO BE SORTED AND N,
      C*   THE NUMBER OF ELEMENTS IN THE ARRAY.
      C*
            SUBROUTINE  SORT2(A,N)
            REAL  A(N), TEMP
            INTEGER FLAG
1. 1        FLAG = 0
            DO 10 I = 1, N-1
2.            IF(A(I) .GT. A(I+1))THEN
                TEMP = A(I)
                A(I) = A(I+1)
                A(I+1) = TEMP
                FLAG = 1
              ENDIF
3. 10       CONTINUE
4.          IF(FLAG .NE. 0)GOTO 1
5.          RETURN
            END
```

Computer scientists have shown that the bubble sort algorithm is somewhat less wasteful of computer time than is the linear sort.

Shell sort

The last sorting technique we shall consider is called the *shell sort*. This method is very much like the bubble sort except that, instead of comparing *adjacent* numbers in the list, we compare two numbers whose positions in the list are d positions apart. Except for that, however, the shell sort proceeds exactly like the bubble sort. After a pass through the array to compare elements d positions apart results in no interchanges, then a new d is computed and the process repeated from the beginning again with the new d. A commonly used starting value for d is $(N+1)/2$. New values of d are computed according to the formula:

new $d = (\text{old } d + 1)/2$

The list will be sorted if, when $d=1$, a complete pass can be made through the list without making any interchanges, just as with the bubble sort.

Algorithm for shell sort.

1. Initialize $d=N$.
2. Set $d=(d+1)/2$.
3. Set FLAG=0 and index $i=1$.
4. If $A(i)$ is greater than $A(i+d)$, then interchange the two numbers and set FLAG to 1.
5. Increase i by 1 and, if i is less than or equal to $N-d$, repeat the process from step 4.
6. If FLAG=1 or d is greater than 1, repeat the process from step 2.
7. Return to calling program.

Subprogram for shell sort

Algorithm   *ForTran*
steps   *statements*

```
C**********
C*
C*    ****SUBPROGRAM SORT3****
C*
C*    THE ARGUMENTS FOR THIS SUBPROGRAM ARE
C*    THE ARRAY, A, TO BE SORTED AND N,
C*    THE NUMBER OF ELEMENTS IN THE ARRAY.
C*
```

```
          SUBROUTINE  SORT3(A,N)
          REAL A(N), TEMP
          INTEGER FLAG, D
1.        D = N
2.  2     D = (D+1)/2
3.  3     FLAG = 0
          DO 20 I=1, N-D
4.          IF(A(I) .GT. A(I+D))THEN
              TEMP = A(I)
              A(I) = A(I+D)
              A(I+D) = TEMP
              FLAG = 1
            ENDIF
5.  20    CONTINUE
6.        IF(FLAG .EQ. 1 .OR. D .GT. 1)GOTO 2
7.        RETURN
          END
```

Searching techniques

Data are usually stored in computer memory or mass storage devices so that they may be examined later for any of a variety of reasons. For example, records of airlines reservations are stored so that reservations can be confirmed at a later date. Lists of credit card holders are stored so that charge purchases may be approved. Lists of students properly admitted to a university are stored for a number of reasons—among them, to determine eligibility for enrolling. Lists of items sold in stores together with their prices are stored so that when purchases are made at the point-of-sale (POS) terminal, the correct prices appear on the sales receipt. The applications that could be cited are numerous.

One way of searching for a given item in a list is to compare the item sought to each item in the list from beginning to end of the list until its match is found or it is clear that it is not in the list. This procedure is very time-consuming. If, for example, a stored list contains N items and M items are to be verified for presence or absence in the list, it can be shown that it would, on the average, require $N*(M/2)$ comparisons. In particular, if a store handles 1000 items and if, at any given time, there are ten cashiers using the store's POS terminals, using the searching technique just mentioned would require a wait by any given cashier of from two to thirty seconds, depending on the computer being used. Maybe a customer would tolerate the shorter waits, but if every customer had to wait the length of time required if each item purchased took thirty seconds to be processed at the check-out counter, the store would soon be without customers.

Therefore, there is good reason to try to discover faster searching techniques, just as it was worthwhile to consider faster sorting techniques. We will examine only one searching technique here, but we call your attention to the fact that computer science literature abounds with discussions of a variety of search techniques. The one presented here is relatively efficient yet is easy to understand. It is called the *binary search.*

Binary search.   We assume that we are searching a list sorted in ascending order containing N items. Using a normal linear search, one would begin at the top of the list and compare all items, item by item downward, moving toward the bottom of the list until a special item is found or we determine it is not there. In the binary search, we begin at the middle of the list and move in either direction depending on the results of the first comparison. If the item being sought is less than the middle item, only the half with smaller items need be searched further. Next, that half is searched by beginning at the middle, and similar results are obtained. Each time a comparison has been made, the number of items left to search is half as many as before. Therefore, one comparison reduces the size of the list yet to be searched to $N/2$ items. Two comparisons reduces it to $N/4$; three comparisons to $N/8$; and k comparisons to $N/2^k$ items. This rapid reduction in the size of the list remaining to be searched is the major reason the binary search technique is so efficient.

Now we present a general algorithm for the binary search method.

Algorithm for binary search.

1. Set the upper, or HIGH, and lower, or LOW, limits for the range of positions in the array A remaining to be searched. Initially set HIGH=N and LOW=1.
2. Determine the position in array A at which the next comparison is to be made. We call this position POINT and set POINT=(HIGH +LOW)/2.
3. IF HIGH is less than or equal to (LOW−1), then the item being sought is not in the list and a value of 0 is returned to the calling program as the position of ITEM (the item being sought) to indicate ITEM has not been found. Note that when HIGH is less than or equal to LOW−1, there are no more members of array A to compare with ITEM.
4. If ITEM is equal to the element in position POINT of array A, then return the value of POINT to the calling program.
5. If ITEM, the item being sought, is less than the element in position POINT of array A, then examine the lower half (the smaller numbers) of array A by setting HIGH=POINT−1.

6. If ITEM is greater than the element in position POINT then examine the upper half (the larger numbers) of array A by setting LOW =POINT+1.
7. Repeat the process from step 2.

Program for binary search

Algorithm steps		ForTran statements
		`INTEGER  FUNCTION SEARCH(A,N,ITEM)`
		`REAL  A(N)`
		`INTEGER  HIGH, LOW, POINT`
1.		`HIGH = N`
		`LOW = 1`
		`SEARCH = 0`
2.	2	`POINT = (HIGH + LOW)/2`
3.		`IF(HIGH LE. LOW-1) RETURN`
4.		`IF(ITEM .EQ. A(POINT))THEN`
		`    SEARCH = POINT`
		`    RETURN`
5.		`ELSEIF(ITEM .LT. A(POINT))THEN`
		`    HIGH = POINT -1`
6.		`ELSE`
		`    LOW = POINT + 1`
		`ENDIF`
7.		`GOTO 2`
		`END`

Upon return from this subprogram, the calling program must check the value of function SEARCH to determine the results of the search. If SEARCH=0, no item was found equal to X in array A. If SEARCH is not 0, then the value of SEARCH is the location in array A where ITEM resides.

SUMMARY

In this chapter we have discussed concepts related to writing program modules. Following are programming structures that were presented:

1. Function statement
General form:

NAME (PAR1, PAR2, ..., PARn)=expression

where "NAME" represents the function name which must comply with the usual rules for variable names, and "PAR1, PAR2, ..., PARn" represent parameters of which there must be at least one, and "expression" represents any arithmetic, logical, or string expression.

2. Function subprogram
General form:

type FUNCTION NAME (ARG1, ARG2, ..., ARGn)
  Statement 1
  Statement 2

   .
   .
   .

  Statement n
RETURN
END

where "type" refers to one of the variable data available in ForTran, "NAME" refers to the function name which must comply with rules for variable names, and "ARG1, ARG2, ..., ARGn" refers to formal arguments of which there must be at least one and no more than sixty-three. Note that "type" may be omitted, in which case "name" specifies the type as integer or real according to what the first letter is; "statement 1, statement 2, ..., statement n" represent acceptable ForTran statements and RETURN may appear 0, one, or more times; and END must be the last statement in the function.

3. Subroutine subprogram
General form:

SUBROUTINE NAME (ARG1, ARG2, ..., ARGn)
  Statement 1
  Statement 2

   .
   .
   .

  Statement n
RETURN
END

In the above general form, "NAME" represents the subprogram name which must comply with the usual rules for ForTran variables; "ARG1, ARG2, ..., ARGn" represent formal arguments; "statement 1, statement 2,

..., statement n" represent any acceptable ForTran statements; and RE-TURN must appear at least once in the subroutine but may appear more than once. END must be the last statement in the subroutine.

4. CALL statement
General form:

CALL NAME (ARG1, ARG2, ..., ARGn)

where "NAME" represents the name of a subroutine subprogram and "ARG1, ARG2, ..., ARGn" represent the actual arguments of the subroutine subprogram. The arguments and parentheses enclosing them may be omitted if the subroutine being called has no formal arguments.

We also discussed the COMMON statement as a means of making information available to more than one program module in a given program. Its general form is:

COMMON VAR1, VAR2, ..., VARn

where "VAR1, VAR2, ..., VARn" represent variable names of simple or array variables.

Some sort and search techniques were presented, including these: (a) linear sort; (b) bubble sort; (c) shell sort; and (d) binary search.

EXERCISES

1. Indicate which of the following statements are true and which are false. (Note that a statement must be considered false if any part of it is false.) Be prepared to make changes in those that are false so that they will be true.
   (a) A COMMON statement must appear in a program after any type declaration statements.
   (b) A DATA statement serves the purpose of assigning values to the elements of the array.
   (c) When COMMON statements appear in the main program and in subprograms, any array appearing in such COMMON statements must have identical dimensions wherever it appears.
   (d) The statement

   ```
   DATA A(10)/10*0/
   ```

   will set A(1) through A(10) to zero.

(e) The following statement is syntactically correct:

```
DATA A/5.2/, B/6.7/
```

(f) The main reason for using a COMMON statement is to reduce the number of different variable names used in a program.

(g) The DATA statement is used to initialize the contents of variables and may appear anywhere in the program.

(h) Any INTEGER statements in a program must follow all DATA statements.

(i) A FUNCTION statement may appear only in the main program and cannot be in any subprogram.

(j) SUBROUTINE subprograms must have at least one argument.

(k) A variable appearing in a COMMON statement of the main program may be an argument of a function subprogram as long as there is no COMMON statement in the subprogram.

(l) Any given program may have no more than one COMMON block.

(m) Any variable that appears in a COMMON statement of a subprogram must also appear in a COMMON statement in the main program.

(n) A function defined by a function statement must have no arguments.

(o) A FORMAT statement in the main program may be referenced by an input or output statement in a subprogram.

(p) All complete ForTran programs must have exactly one main program but may have as many subprograms as desired.

(q) A function statement must not appear in a function subprogram.

(r) If a FUNCTION statement is to appear in a program it must precede the first executable statement in that program.

(s) A function defined in a function statement may appear as an argument of a subprogram.

(t) One difference between a subroutine subprogram and a function subprogram is that the function subprogram returns a value to a memory location identified by the function's name and, therefore, the function name may be included in an arithmetic expression.

(u) A variable name may appear in more than one COMMON statement of a given program.

(v) If, when using an interactive computer system, you want to save your program in mass storage, you must give it a name in the first line of the program.

(w) Suppose the following statement is the first line of a subroutine subprogram:

```
SUBROUTINE SUB(N,X)
```

Then the statement

```
CALL SUB(1,SIN(X))
```

would be a valid statement for calling the subprogram into action.

(x) Suppose the following statement is the first line of a subroutine subprogram:

```
SUBROUTINE SUB1(A,B,I,C,J)
```

Then the statement

```
CALL SUB1(X,1.,K,A,4)
```

would be a valid statement for calling the subprogram.

(y) If the statement

```
FUNC(X,Y) = X * Y
```

appears in a program, then the statement

```
A=FUNC(B,B)
```

is a valid statement as long as it does not precede the first statement above.

(z) A function subprogram can only return one value to the calling program and that only through the subprogram name.

(aa) A DATA statement cannot be used to initialize an array variable.

(bb) A DATA statement must precede an INTEGER statement if both appear in a program.

(cc) If an array A is to be included in a COMMON statement, it must appear in a DIMENSION statement previous to its appearance in the COMMON statement.

2. Following are ForTran statements, some of them correct, others incorrect. Assume that all variables that appear in any statement have

been properly defined in the program before the appearance of the given statement. Identify those that are incorrect and suggest changes that would make them valid.

(a) READ*,I,J,K
(b) DATA A,B(10)/2.1,3.9/
(c) DIMENSION DATA(10),B(10,2)
(d) DATA A/5/,B/6.0/,C,D/2*10.5/
(e) COMMON A,B(10),I OR J
(f) COMMON A,A(100),B
(g) DATA A,(B(I),I=1,10)/10*2.5)
(h) DATA A,B,C/'AB','BC','A'/
(i) DATA A/5./,B/6./,C/10./
(j) GO TO N
(k) DATA (A(I,J),J=1,5)/25*10./
(l) DATA (B(I,1),I=1,10)/10*0.,1/
(m) DATA A,B,C/2+3,7,8/
(n) COMMON DATA,OR,DIM
(o) F3(X,Y) = X * SIN(X) + Y * SIN(Y)
(p) FUNCTION F1(A,B,X)
(q) SUBROUTINE NAME = (3.*G)/X
(r) SUBPROGRAM S2(A,B,C)
(s) CALL S2(2.5,R,S)
(t) CALL SUB1(X+Y,5)
(u) X = FUNCT1(5,B)
(v) COMMON A,B(5),X(20)

3. Identify all syntax errors in each of the following program segments. Assume default data type based on the first letter of a variable name unless otherwise declared.

```
(a)     INTEGER A(2,10)
        DATA A/1,2,3,4,5,6,7,8,9,10*0,10/
        REAL B(20)
        DO 10 I=1,10
    10 B(I) = A(I,1)
        DO 20 I=11,20
    20 B(I) = B(A(2,I-10))
(b)     INTEGER X,Y
        DIMENSION X(10),Y(10)
        DATA X/10*1/,Y/10*0/
        DO 10 Y = 1,10
    10 X(Y) = Y*Y
```

(c)
```
        DATA I(5)/5/
        GO TO I(5)
```
(d)
```
        DATA A/-5/,B/10-A/
        READ*,A,B
```
(e)
```
        DIMENSION ARY(10)
        DATA ARY/10*0/
        F(X) = ARY(1)*X
        ARY(I) = I*10
        G(X) = F(X)+ARY(X)
        READ*,X,I
```
(f)
```
        N=10
        GO TO N
```
(g)
```
        READ* N
        GO TO N
```
(h)
```
        READ* N
     10 GO TO(10,20,30)N
```
(i)
```
        DATA(A(I),I=1,10)/10*2.0/
        DIMENSION A(20)
```
(j)
```
        F(X) = X*X+1
        DATA FF/10./
        READ *,X
        CALL F(X)
```
(k)
```
        F(X) = SIN(X)+COS(X)
        G(X) = SIN(X)-COS(X)
        READ*,X
        PRINT*,X,F(X),G(X),SIN(X),COS(X)
```
(l)
```
        COMMON A,B
        DIMENSION A(10)
          .

          .

          .
        SUBROUTINE XX(I)
        COMMON A,B,C(9)
        DATA A,B/2.3,7.1/
        DATA C/9*0/
```
(m)
```
        CALL SUB(S)N(X))
          .

          .

          .
        SUBROUTINE SUB(X)
        IF(X .GT. 0)PRINT *,'RIGHT'
        IF(X .LE. 0)PRINT*,'LEFT'
        X=X+1
        CALL SUB(X)
```

```
(n)    CALL SUB(F(X))
                .

                .

                .

       SUBROUTINE SUB(X)
       F(X) = X*X+2*X+1
(o)    READ*,X,Y
       F(X) = X*X+Y
       CALL SUB(F(X),Y)
                .

                .

                .

       SUBROUTINE SUB(G(X))
(p)    COMMON A,B
       DATA A,B/10,-10/
       CALL SUB(A,B,A+B)
                .

                .

                .

       SUBROUTINE SUB(A,B,J)
     5 K = A+1
```

4. (Computer Science, Education, General)
   Refer to Exercise 6 of Chapter 6. Write a program to solve that problem; this time use a separate subprogram for each geometric figure.

5. (General)
   Refer to Exercise 7 of Chapter 6. Write a program to maintain a checking-account balance as in that problem, but include these modifications:
   (a) All deposits are to be processed by function subprogram DEPS.
   (b) All withdrawals are to be processed by function subprogram WITH.
   (c) A subroutine subprogram, OUTPUT, having one argument, I, is to produce either or both parts of the output depending on the value of I.
   (d) The main program should handle all other functions as well as call the appropriate subprograms when needed.

6. (Mathematics, Education, General)
   Refer to Exercise 9 of Chapter 6. Rewrite the program to include the following modifications:
   (a) A main program to describe what can be accomplished by the program and to give instructions regarding its use. All input should be processed by the main program.

(b) As many subprograms as needed, one for each geometric figure, to compute the area and produce the output.

7. (Mathematics, General)

Refer to Exercise 10 of Chapter 6. Rewrite the program so that a separate subprogram handles each different conversion—for example, one that handles the conversion of miles to kilometers. Another one that converts pounds to kilograms, and so on.

8. (Mathematics, General)

Refer to Exercise 11 of Chapter 5. Rewrite the program for that exercise so that the main program handles the input for the two matrices, one subprogram computes the sum matrix and produces associated output, and a second subprogram computes the difference matrix and produces associated output.

9. (Mathematics, Engineering, Computer Science)

In calculus, one learns how to compute the value of the definite integral of a function—say $f(x)$—from some point, A, to another point, B. What this means geometrically is that one determines the area of the plane bounded by the curves $y = f(x)$, $X = A$, $X = B$, and the X axis, as indicated in this diagram.

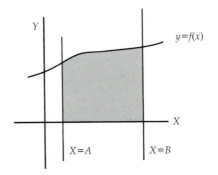

In order to compute the value of definite integrals for functions not readily integrable, mathematicians have developed techniques that make it possible to approximate the true value of such integrals by performing computations according to certain formulas. Two such formulas are the trapezoidal rule and Simpson's rule. We shall not discuss the mathematical background for these except to say the following:

A. In the case of the trapezoidal rule, the distance from A to B (as shown in the above diagram) is divided into equal segments of length $h = (B - A)/n$ for arbitrarily chosen n. Then, by drawing vertical lines at the ends of the horizontal segments of length h and connecting the points where those vertical lines and the lines $X = A$ and $X = B$ intersect with the curve $y = f(x)$, the following diagram is obtained; it shows the results if $n = 4$:

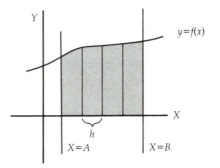

The trapezoidal rule approximates the exact area shaded by computing the sum of the areas of the four trapezoids shown in the shaded area of the preceding diagram. Further mathematical analysis will show that the sum of the areas of the trapezoids is given by the general formula

$$I = (h/2)[f(A)+2f(A+h)+2f(A+2h)+ \cdots +2f(A+(n-1)h)+f(A+nh)]$$

Thus I is an approximation to the definite integral of $f(x)$ from A to B.

B. In the case of Simpson's rule, the required area is separated into strips as with the trapezoidal rule, but this time the upper boundary of each strip is a portion of a parabola instead of a straight line. The mathematical analysis is somewhat more difficult and the interested reader is referred to almost any book on numerical methods for details. The result, however, is the formula

$$I=(h/3)[f(A)+4f(A+h)+2f(A+2h)+4f(A+3h)+2f(A+4h)+ \cdots + 2f(A+(n-2)h)+4f(A+(n-1)h)+f(A+nh)]$$

which is called Simpson's rule. Note that for this rule, n must be an even number.

Having presented the preceding background on approximating the value of definite integrals, we are now ready to propose an exercise.

Write a main program that reads values for A, B, and n (as in the preceding discussion), that calls two subroutine subprograms as defined below, and that outputs the results of both subroutines so that the results can be compared.

The first subroutine should include a function statement for defining a function, $f(x)$, as in the preceding background discussion, then apply the trapezoidal rule to compute the definite integral of $f(x)$ from A to B.

The second subroutine should include the same function statement as the first one; then apply Simpson's rule to compute the definite integral.

10. (Mathematics, General)

Write a program to make an ordinary printer (terminal or line printer) act as a plotter. The output will be a graph that looks something like this:

```
1.0  *********************************************************
     *     *     *     *     *     *     *     *     *     *     *
     *     *     *     *     *     *     *     *     *     *     *
0.9  *********************************************************
     *     *     *     *     *     *     *     *     *     *     *
     *     *     *     *     *     *     *     *     *     *     *
0.8  **********************************************0**********
Y    *     *     *     *     *     *     *     *     *     *     *
     *     *     *     *     *     *     *  0  *     *     *     *
T 0.7 *********************************************************
i    *     *     *     *     *     *     *     *     *     *     *
m    *     *     *     *     *     *0 *     *     *     *     *     *
e 0.6 *********************************0*************************
     *     *     *     *     *     *     *     *     *     *     *
     *     *     *     *     *     *     *     *     *     *     *
0.5  *********************************************************
     *     *     *  0  *     *     *     *     *     *     *     *
     *     *     *     *     *     *     *     *     *     *     *
0.4  *********************************************************
     *     *     *     *     *     *     *     *     *     *     *
     *     *  0  *     *     *     *     *     *     *     *     *
0.3  *********************************************************
     *     *     *     *     *     *     *     *     *     *     *
     *   * 0   *     *     *     *     *     *     *     *     *
0.2  *********************************************************
     *     *     *     *     *     *     *     *     *     *     *
     *     *     *     *     *     *     *     *     *     *     *
0.1  *********************************************************
     *     *     *     *     *     *     *     *     *     *     *
     *     *     *     *     *     *     *     *     *     *     *
0.0  *********************************************************

     0.0   0.1   0.2   0.3   0.4   0.5   0.6   0.7   0.8   0.9   1.0

                        X  Distance
```

Note that it is a 10-by-10 grid that the *Y* axis runs vertically along the left edge of the grid and the *X* axis runs horizontally along the bottom edge of the grid. Note also the labels printed along these axes. The 0's on the grid indicate points on the function being graphed.

The main program is to read an integer, *N* (not to exceed 100), followed by *N* real values to be stored in an array, *X*, and *N* real values to be stored in an array *Y*. Also to be read (using A format) are labels (not to exceed sixteen characters each) to be printed along the *X* and *Y* axes as shown in the previous example. Assume that all values for *X* and *Y* are not less than 0 nor greater than 1.

The main program calls a subroutine subprogram, GRAPH, which is to produce an output array whose elements contain the exact characters to be printed as the graph. For example, in the sample graph shown at the beginning of this problem, there are sixty-eight spaces from left to right starting at the column that contains the label for the *Y* axis, and thirty-five spaces from top to bottom counting from the top row of asterisks to the bottom row that contains the label for the *X* axis. That is a total of 68 times 35, or 2380 spaces in the output array. Thus the characters stored in the first row of the output array—say OUTPUT—for the sample graph shown would be as follows:

OUTPUT(1,1) and OUTPUT(1,2) contain blank spaces.
OUTPUT(1,3) contains 1.
OUTPUT(1,4) contains a period.
OUTPUT(1,5) contains 0.
OUTPUT(1,6) contains blank space.
OUTPUT(1,7) through OUTPUT(1,68) contain asterisks.

Similarly, the contents of row 2 in array OUTPUT would be blank spaces and asterisks appropriately placed. Subsequent rows of array OUTPUT would contain the characters shown in corresponding rows of the sample graph.

Subprogram GRAPH calls a subprogram GRID, which is to store all necessary characters in appropriate cells of array OUTPUT so as to later print out the asterisks of the grid and the labels for the axis.

After subprogram GRID has been called, then subprogram POINTS will be called by GRAPH. The function of POINTS is to store 0's in the appropriate cells of array OUTPUT, which correspond to the points whose coordinates are given by the values stored in arrays *X* and *Y*. Remember that everything stored in array OUTPUT must be stored as a character.

The last task for subprogram GRAPH is to call subprogram PRPLOT, which is to print the contents of array OUTPUT. If all cells in array OUTPUT have been correctly filled with the right characters by subprograms GRID and POINTS, the job of PRPLOT can be accomplished by using A1 format to print all of the elements (cells) of array OUTPUT.

We suggest that you use COMMON statements to make information available to all subprograms.

11. (General)

Write a program to score a multiple-choice test and output certain associated reports to be described later.

A. Limitations of the program:
1. At most there should be 150 items in the test.
2. At most 200 students will take the test.
3. The correct response to each test item will be a single character—a letter or a digit.
4. Each item will have a point-value consisting of an integer from 1 to 9.

B. Program input.
1. Instructor-related data:
   (a) The number, N, of students taking the test.
   (b) The correct responses in order from item 1 and on. (Note that if there are eighty items or less, only one card (line) of correct responses is needed. Otherwise two cards (lines) will be needed.)
   (c) The point-value of all test items in the same order as the correct responses. (As with the correct responses, one or two cards (lines) are needed for the point-values depending on the number of test items.)
2. Student-related data (one set for each student):
   (a) Student identification number in the first six columns (spaces).
   (b) Responses to each test item, in the same order as the instructor-provided correct responses. The student's response to test item 1 should be entered in column 7 (immediately following identification number). The response to item 2, in column 8; the response to item 3, in column 9; and so on. (As with the instructor's correct answers, one or two student-response cards will be needed depending on the number of test items.)

C. Program output.
  1. Mean score for all students, and standard deviation.
  2. A table summarizing the results of the test as follows:

ITEM NUMBER	CORRECT RESPONSES	INCORRECT RESPONSES	% CORRECT RESPONSES

  3. A table showing student identification number and score sorted in descending order of score.
  4. A table showing student identification number and score sorted in ascending order of student identification number.
  5. For each student, a report showing the following information:

STUDENT NUMBER XXXXXX   SCORE XXXX
STUDENT RANK ON TEST XXX   GRADE X

(For grade, assign A if score is in the top 10% of the class. Assign B if score is in the next 20%. Assign C if score is in the next 40%. Assign D if score is in the next 20% and F if score is in the bottom 10%.)
  6. A table showing the following:

ITEM NUMBER	STUDENT RESPONSES	CORRECT RESPONSES	POINTS PER ITEM

CHAPTER 8
Further Options Available in ForTran

The preceding chapters of this book have discussed the features of For-Tran needed by an effective programmer in that language. All of the commonly used properties of ForTran have been described and most readers of this book will neither need nor desire further capabilities.

Although we discuss the topics of this chapter in sufficient detail to clarify what is being described, the reader will notice that in this chapter fewer examples are presented and narrative is less detailed than in previous ones. We believe the reader will appreciate such an approach at this point in the process of learning ForTran because a rather extensive background in that language is now available to fill in any gaps. We assume also that those who read this chapter have a mathematics background beyond that of algebra and trigonometry.

DECLARATION STATEMENTS

These statements are all nonexecutable but have the function of designating the structure of data.

More on declaration statements

You have already been introduced to most of the statements that belong to the category of type-declaration statements. In this section we discuss some of these in more detail. You will recall that these statements specify the type of data stored in a given variable.

The DOUBLEPRECISION statement.   This statement is included in full ForTran 77 and causes the data stored in a given variable to have twice as many significant digits as usual—hence the name "double precision." Such increased number of digits is made possible by using *two* storage locations rather than one. That is, when a variable has been declared DOUBLEPRECISION, any mention of the variable name in a program causes the digits in *two* storage locations to be affected rather than only one location. Double precision variables are always real variables with twice as many significant digits as usual for any given computer.

Here is the form of the DOUBLE PRECISION statement:

```
        DOUBLEPRECISION V1,V2,....,VN
or
        DOUBLE PRECISION V1,V2,....,VN
```

where V1, V2,...,VN represent variable, array, or function subprogram names. In a mixed-mode expression, double precision type is the dominating type.

The format specification to be used for input and output of double precision values is the D specification. It has the same general properties as the E specification and its form is D$w.d$ where w represents the total number of spaces in the width of the field being input or output, and d represents the number of decimal (fractional) positions in the input or output value, similar to the E specification.

To designate a double precision *constant* one uses the exponential form, but with the letter D replacing the letter E of single precision constants. Here is an example:

```
        DOUBLE PRECISION X
        X = 27.691D9
        WRITE(6,10)X
     10 FORMAT(D10.5)
```

Note the D in the assignment statement, replacing what would be E in a single precision constant of exponential form. The D-format specification functions exactly like the E specification as far as the form of the output produced. It differs, however, in that it signals the computer to retrieve the value for X from two adjacent memory locations rather than one.

More on the COMPLEX statement.   The COMPLEX statement is included in full ForTran 77 and has this form:

```
        COMPLEX V1,V2,....,VN
```

where V1,V2,...,VN are ForTran variables names, including arrays, or function subprogram names. This statement declares that any variable or function in the list of names is to be treated as a complex value. This means that each such complex value is stored in two adjacent memory locations, each location containing a real number. One of the real numbers is the so-called *real* part of a complex number and the other is the real coefficient of i from the *imaginary* part of the complex number, where $i = \sqrt{-1}$. In particular, if the complex number is Z, where $Z = X + iY$, then X and Y are the two real numbers stored in the two adjacent memory locations set aside for the complex variable Z.

Complex-value constants are defined by using a pair of parentheses in which are enclosed two real numbers, the first one being the real part of the complex number and the second one being the imaginary part. For example, consider the following program segment:

```
COMPLEX Z,W
Z=(2 ,3 )
W=(0 ,3.5)
```

In this case Z is the complex constant $2 + i*3$ and W is the complex constant $0 + i*3.5$ or just $i*3.5$.

Whenever complex values are processed, one must remember that there are two real numbers making up each complex number. Thus, in input or output operations, two fields must be reserved for each complex value. Consider this program example:

```
    PROGRAM COMPLX
    COMPLEX W,Z,T
    READ 10,W,Z
10  FORMAT(2F10.2,2F10.3)
    T = W + Z
    PRINT 20,W,Z,T
20  FORMAT(F10.2,2X,F10.2)
    STOP
    END
```

In the previous example, all three of W, Z, and T are declared COMPLEX, so it makes sense to perform the sum of W and Z, the result of which is also a complex number stored in two adjacent memory locations. If $W = X1 + i*Y1$ and $Z = X2 + i*Y2$, then $T = W + Z = (X1+X2) + i*(Y1+Y2)$ where, of course, the sums X1+X2 and Y1+Y2 are again real numbers, each stored in one of two adjacent memory locations. Note that in statement 20, two F formats are provided in order to output the two parts of a complex number.

Recall from the study of complex numbers that if $Z = X + i*Y$, then the absolute value of Z, denoted by $|Z|$, is given by $\sqrt{X*X+Y*Y}$. Recall, also, that if

$$Z1 = X1 + i*Y1 \text{ and } Z2 = X2 + i*Y2,$$

then the product,

$$Z1*Z2 = (X1*X2-Y1*Y2) + i\,(X1*Y2+X2*Y1)$$

All of these arithmetic operations are handled automatically by the ForTran compiler when complex data types have been declared.

More on the LOGICAL statement.   Recall from Chapter 3 our discussion of logical operations and that the result of such operations is either TRUE or FALSE. In all previously discussed situations it has not been necessary to save the results of a logical expression for later use. The truth or falsity of the expression was tested immediately and caused either of two actions, depending on the result of the logical expression. There do arise cases where it would be useful to have available the result of a logical expression that appeared previously in the program. This is made possible in ForTran by declaring a variable to be a logical variable. Here is the form of the LOGICAL statement:

```
LOGICAL V1,V2,...,VN
```

where V1,V2,...,VN are names of variables (including arrays) or function subprograms. The appearance of a variable in a LOGICAL statement makes it permissible to store only logical values (TRUE or FALSE) in such a variable. Similarly, if a function name appears in a LOGICAL statement, the value produced by that function may be only TRUE or FALSE.

Variables that have been declared LOGICAL may appear in output statements. When they do, a special format specification is required—namely, the L specification. Its form is the same as the I specification in that an integer is written to the right of the L indicating how many spaces to allow for the output. The output is always T or F corresponding to TRUE or FALSE.

Here is a program example using the LOGICAL statement:

```
   PROGRAM LOGEX
   LOGICAL A,B,C
   A = .TRUE.
   B = .FALSE.
   C = A .OR. B
   PRINT 10,A,B,C
10 FORMAT(3(2X,L1))
   STOP
   END
```

In this example, A, B, and C are all declared logical variables so that logical values can be assigned to them. Furthermore, the line

```
   C = A .OR. B
```

causes a logical value to be assigned to C. The output of this program is as follows:

```
 T  F  T
```

Note that the L specification in the FORMAT (see statement 10 above) is used for writing logical variables.

Consider now a problem that lends itself nicely to the use of logical variables in the program solution.

Problem: Input a number, N, and compute N factorial if N is greater than 0 and less than or equal to 15. If N = 0, the output should be "0!=1." If N is greater than 15, output the sentence "N IS TOO LARGE." If N is negative, output the sentence "N IS NEGATIVE."

Solution: To solve this problem, we use four logical variables. One of them is set to FALSE if N is greater than 15. The next one is set to FALSE if N is less than zero. The third variable is set to FALSE if N=0. For values of N other than those specified, these three logical variables have the value TRUE. The fourth variable is assigned the value TRUE only if all three of the other variables have the value TRUE. Thus, this fourth variable is used to determine when N factorial is to be computed. Following is a program that complies with this proposed solution. The program has been executed for these values of N: 20,15,7,0,−12, and 6.

Program

```
      PROGRAM LOGIC
C***************************************************************
C*              THIS IS A THIRD ANSWER TO THE FACTORIAL   *
C*              PROBLEM OF CHAPTER 3.                      *
C***************************************************************
      LOGICAL A,B,C,D
      DATA A,B,C/3*.TRUE./
C*
      READ*,N
      IF(N.GT.15)A=.FALSE.
      IF(N.LT.0) B=.FALSE.
      IF(N.EQ.0) C=.FALSE.
      D = A .AND. B .AND. C
C***************************************************************
C*              D HAS A VALUE OF TRUE ONLY IF N-FACTORIAL*
C*              IS TO BE COMPUTED.                        *
C***************************************************************
      IF(D)THEN
        K=N
        DO 50 I=1,N-1
          K=K*(N-I)
50      CONTINUE
        PRINT*,N,' FACTORIAL IS ',K
      ELSEIF(.NOT.A)THEN
        PRINT*,N,' IS TOO LARGE'
      ELSEIF(.NOT.B)THEN
        PRINT*,N,' IS NEGATIVE'
      ELSE
        PRINT*,' 0 FACTORIAL IS 1'
      ENDIF
      END


      20    IS TOO LARGE
      15    FACTORIAL IS  2004310016
       7    FACTORIAL IS  5040
       0 FACTORIAL IS 1
     -12    IS NEGATIVE
       6    FACTORIAL IS  720
```

The previous example shows how logical variables can be conveniently used in decision-making processes. Sometimes the understandability as well as the efficiency of a program is improved by using logical variables.

Other declaration statements

Situations arise when it would be very helpful to conserve computer memory by assigning two or more names to a given memory location or group of memory locations. For example, suppose that in a given program the following DIMENSION statement appears:

```
DIMENSION STATP(50,2)
```

Suppose that array STATP is used in the first part of the program to store data related to the names and populations of the fifty states in the United States. After all processing is complete relative to the population data, suppose the program processes data related to the governors of the states and the state budgets. In this latter portion of the program it would be convenient to denote the data related to governors by the array GOVN(50) and the state budgets by the array BUDGT(50). Thus the same 100 memory locations would be used first for state names and populations, then for governors and budget figures. Figure 8.1 shows the memory locations and the two sets of names for them. The rectangular boxes represent the memory locations, the names above the boxes show the designation during the first part of the program, and the names below the boxes show the designations during the latter part of the program. Note that each box has two names associated with it. The appearance of either of the two names in a program statement will cause a reference to the same computer word. To maintain consistency of data type, all variables in this example must be declared type character.

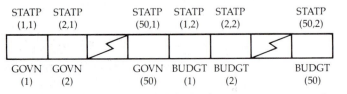

Figure 8.1 Diagram of a block of computer memory where each word has two names

ForTran provides for the kind of multiple names for memory locations just described, through the use of the EQUIVALENCE statement.

The EQUIVALENCE statement. The general form of this statement is as follows:

```
EQUIVALENCE(V1,V2,...,VN)
```

where V1,V2,...,VN are simple or array variable names all of which will use the same memory location and therefore conserve memory. Conserving memory is the main purpose for using the EQUIVALENCE statement. In order to help you understand how the EQUIVALENCE statement works, we consider several examples of program segments using it.

Examples of applications of the EQUIVALENCE statement

Example 1

```
REAL A(5),B(10),C,D,E
EQUIVALENCE (B(3),A(2)),(C,D,E)
```

Figure 8.2 shows the effect on memory locations as far as their names are concerned.

B(1)	B(2)	B(3)	B(4)	B(5)	B(6)	B(7)	B(8)	B(9)	B(10)	C
	A(1)	A(2)	A(3)	A(4)	A(5)					D
										E

Figure 8.2   Names associated with certain memory locations because of the statements of Example 1

Note that the EQUIVALENCE statement in Example 1 causes B(3) and A(2) to be two names for the same memory location. However, since A is an array of 5 locations and B an array of 10 locations, the names B(3) and A(2) associated with the same memory location forces B(2) and A(1) to be associated with the same memory location, and similarly for B(4) and A(3), B(5) and A(4), and B(6) and A(5). (See Figure 8.2.) The appearance in the EQUIVALENCE statement of C, D, and E in the same pair of parentheses causes all three names to be associated with the same memory location.

Example 2

```
CHARACTER A*4,B*4,C(2)*3
EQUIVALENCE (A,C(1)),(B,C(2))
```

The CHARACTER statement sets aside four words for A, four words for B, and three words for each of C(1) and C(2). The EQUIVALENCE statement forces the first character of C(1) and the first character of A to be stored in the same memory location. It also causes the first character of B and the first character of C(2) to occupy the same location. Figure 8.3 shows this situation.

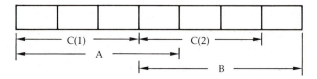

Figure 8.3   Names associated with certain portions of memory because of the statements of Example 2

Example 3

```
INTEGER A(2,3),B(6),C(10)
EQUIVALENCE (A(1,1),C(1)),(B(1),C(5))
```

In this example, A and B are arrays of six words each and C is an array of ten words. The appearance within the same parentheses of A(1,1) and C(1) in the EQUIVALENCE statement causes the six words of array A to coincide with the first six words of array C. The appearance of B(1) and C(5) in the same parentheses of the EQUIVALENCE statement causes the six words of array B to coincide with C(5), C(6), C(7), C(8), C(9), and C(10). Figure 8.4 shows the results.

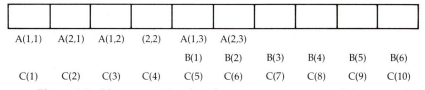

Figure 8.4   Names associated with certain computer words because of the statements of Example 3

Example 4

```
INTEGER A(3) B
EQUIVALENCE (A(1),B),(A(2),B)
```

This is an invalid situation because it is impossible for both variables A(1) and A(2) to be identified with the same memory location as would be required by the appearance of B with both A(1) and A(2) in the EQUIVA-LENCE statement.

Example 5

```
REAL A(2)
DOUBLEPRECISION D
EQUIVALENCE (A(1),D)
```

Since the double precision variable D occupies two computer words, and the real array A occupies two computer words, the EQUIVALENCE statement in Example 5 causes A(1) and the most significant half of D to be stored in one computer word and A(2) and the least significant half of D to be stored in the adjoining word. Figure 8.5 shows the situation:

Figure 8.5 Names associated with certain computer words as a result of the statements in Example 5

Example 6

```
REAL A(3)
DOUBLEPRECISION D(3)
EQUIVALENCE (A(1),D(1)),(A(2),D(2))
```

In this example array A occupies three adjacent memory locations because of the REAL statement. Array D, consisting of three double precision elements, occupies six adjacent memory locations. This EQUIVALENCE statement *is invalid* because the first element of array A cannot be associated with the same computer word as the most significant part of D(1) at the same time that A(2) is associated with the most significant part of D(2). Figure 8.6 shows the assignment of names and memory locations if only A(1) and D(1) are "equivalenced."

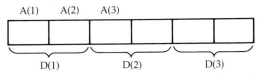

Figure 8.6 Correct assignment of variable names if the last part of the EQUIVALENCE statement in Example 6 is omitted

Note from Figure 8.6 that it would be impossible to assign A(2) and D(2) to the same memory location.

The labelled COMMON statement.   In Chapter 7 we discussed a type of COMMON statement which is often called the unlabelled or blank COMMON statement to distinguish it from the one we are about to discuss. The form of the labelled COMMON statement is as follows:

```
COMMON /NAME/V1,V2,...,VN
```

where NAME is any acceptable ForTran variable name, and V1,V2,...,VN are the names of single or array variables. NAME is called the common block name. Similarly, one may refer to a blank common block, or unlabelled common block in connection with the COMMON statements of Chapter 7.

The *size* of a common block is the number of memory locations specified by the list of variables to the right of the common block name that is enclosed by slashes, plus any additional locations that may become specified because of EQUIVALENCE associations.

Before we proceed to discuss further the labelled COMMON statement, let us examine reasons for using this type rather than the blank COMMON. The major advantage of this new statement is that it allows the programmer to identify certain portions of common block by using the block name in whatever subprograms make use of data in the common block. Recall from Chapter 7 that if a subprogram made use of only *some* of the data in the common block, dummy variable names were required in the COMMON statement so as to exactly match the variable name with specific locations from the common block. By using labelled COMMON statements, portions of the common block are identified without such artificial procedures. This advantage will be clarified later through examples. A second reason for using a labelled COMMON statement is that you may use more than one such statement per program or subprogram. Recall that a COMMON statement of the type discussed in Chapter 7 may occur only once in any given program.

Now we consider some examples to help clarify some of the concepts we have mentioned.

Examples of applications of labelled COMMON statements

Example 1

```
PROGRAM EXAMP1
COMMON /A/ X(10),B,N
   .

   .

   .
END
SUBROUTINE SUB1
COMMON /B/ Y(5),X
   .

   .

   .
END
```

```
FUNCTION FUNC1(D)
COMMON /A/ Y(11),I, /B/ X(5),C
   .
   .
   .
END
```

In Example 1 there are two named common blocks, one called A and the other B. Note that block A appears in the main program, EXAMP1, and in the function subprogram, FUNC1. In both programs block A has size 12 where the first eleven words are real and the last word is an integer. Block B appears in subroutine SUB1 and in function FUNC1, both times having size 6, all real. We assume default data types.

Note that though the block name for a given common block must be the same in each subprogram or main program, the variables making up a common block may be called by different names in different subprograms. Corresponding portions of a common block must be of the same data type, however. It is good programming style to keep the structure and the name of the variables in separate common blocks identical. Note also from Example 1 that, as in function FUNC1, more than one named common block may be declared in the same COMMON statement.

Example 2

```
PROGRAM EXAMP2
COMMON /AA/ X(5),Y
REAL B(7)
EQUIVALENCE (B(1),X(1))
   .
   .
   .
END
```

In this example, the size of common block AA is 6, as specified in the COMMON statement. However, B is an array of seven words, and B(1) is associated with X(1) in the EQUIVALENCE statement. Therefore, since array X is in common block AA, array B is also in block AA. But because the REAL statement dimensions B as having seven words, common block AA is of size 7.

Because Example 2 made use of an EQUIVALENCE statement as well as COMMON, this is an appropriate time to caution you about the use of these two statements in the same program:

1. Variables that appear in different-named common blocks must not be associated in an EQUIVALENCE statement.

To illustrate this restriction consider the following pair of statements:

```
COMMON /A/ X,Y,Z, /B/ R,S
EQUIVALENCE (X,R)
```

Since X is a part of common block A and R is a part of common block B, these two variables must not be associated in an EQUIVALENCE statement.

Related to the above restriction is the naming of a given variable in two different named common blocks. For example:

```
COMMON /A/ X,Y,Z, /B/ X,R,S
```

The variable X may *not* appear in two different common blocks as here shown.

2. EQUIVALENCE statements must not cause a named common block to be extended in memory to locations *preceding* the ones specified in the COMMON statement.

In Example 2, above, you will recall that the combination of the REAL and EQUIVALENCE statements caused the common block to be extended to one memory location *further on* in memory than what was specified in the COMMON statement. Such extensions are permitted. Consider this next program segment as one that illustrates an action that is not permitted:

```
COMMON /A/ X(3)
REAL N(4),X
EQUIVALENCE (N(2),X(1))
```

Figure 8.7 shows the resulting allocation of memory.

common block A

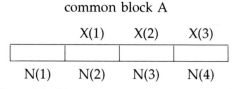

Figure 8.7 Names incorrectly associated with certain memory locations as a result of a specific combination of COMMON, REAL, and EQUIVALENCE statements

Notice from Figure 8.7 that, because of the equivalence association of N(2) and X(1), N(1) is forced to precede the specified common block. Such forced extensions of a common block toward the beginning of memory are not permitted.

As our last example in this section we demonstrate some forbidden uses of COMMON and EQUIVALENCE statements.

Example 3

```
1    PROGRAM EXAMP3
2    COMMON A(10)
3    DATA A /10*1/
4    COMMON /B/ X,Y,Z
5    REAL N(5)A,X,Y,Z
6    EQUIVALENCE (N(4),Z)
  .    .
  .    .
  .    .
99   END
100  SUBROUTINE SUB3
101  COMMON /B/ X(3),Y
102  COMMON Z(7)
103  EQUIVALENCE (Z(1),X(1))
104  REAL X,Y,Z
  .    .
  .    .
  .    .
150  END
```

The first violation in Example 3 occurs in line 3, where a DATA statement is used to initialize an unlabelled (blank) common block. As you know from previous comments, only *named* common blocks may be so initialized.

A second violation results from lines 4 and 101 when common block B has a size of three words in the main program and a size of four words in subprogram SUB3. Named common blocks must have the same size in all places where they are specified.

Statements 4, 5, and 6 result in a forbidden extension of the common block, to memory locations preceding the specified common block. That is, N(4) and Z occupy the same computer word, N(3) and Y the same word, and N(2) and X occupy the same word. This forces N(1) to be associated with a computer word *preceding* common block B, a situation that is not permitted.

The last violation in Example 3 occurs as a result of lines 101, 102, and 103. Line 103 forces portions of a blank common block to occupy the same computer memory locations as the named common block B. It is never possible for variables in two different common blocks, whether labelled or unlabelled, to be associated in an EQUIVALENCE statement.

COMPUTER FILE PROCESSING

In previous chapters we have used data files in a few of our examples. At those times, we discussed only the facts about data files needed for an understanding of what was happening in the context of those examples. Now, however, we shall consider some special ForTran statements designed to make it easy to process data files.

The ENDFILE statement

This statement causes the insertion of a special character called the end-of-file (EOF) symbol, at the end of all records contained in a given data file. The form of this statement is

```
ENDFILE u
```

where u is the logical unit number (an integer) associated with the particular data file.

A single ENDFILE statement may be used to place an EOF symbol on more than one file. For example:

```
ENDFILE (u1,u2,...,un)
```

would cause an EOF symbol to be placed at the end of each of the files whose logical unit numbers are u1,u2,...,un.

The BACKSPACE statement

This statement causes the internal pointer that designates the position from which the next reading from a file will occur to be moved backwards one record. This statement is essential if a record that has already been read from a file is to be read again. Here is its form:

```
BACKSPACE u
```

where u is the logical unit number associated with the data file we want to process.

It is possible to backspace one record on more than one data file with only a single BACKSPACE statement. Here is an example:

```
BACKSPACE (u1,u2,....,un)
```

This statement would cause a backspace of one record on each of the files whose logical unit numbers are u1,u2,...,un.

The REWIND statement

This statement causes the repositioning of the internal pointer to the initial point in the file—that is, to the beginning of the first record in the file. Where the BACKSPACE statement causes repositioning of the internal pointer back one record, the REWIND statement moves the pointer backwards past all records to the beginning of the first one.

There are two forms to this statement:

```
REWIND u
```

or

```
REWIND (UNIT=u,ERR=n)
```

where u is the logical unit number associated with the file to be processed, and n is the reference number of the next statement to be executed if an error occurs during the rewind process.

Here are two examples:

```
REWIND 4
```

This causes the repositioning of the internal pointer to the beginning of the data file associated with logical unit 4.

```
REWIND(UNIT=4, ERR=50)
```

This statement does exactly what the first example does and in addition specifies the statement whose reference number is 50 to be the next statement executed if an error occurs during the process of rewinding.

The OPEN statement

This statement is used to connect a specific file to a program or another file. In other words, before a data file can be processed, it must have been "opened" by an OPEN statement. The form of the OPEN statement is as follows:

```
OPEN (P1,P2,...,P9)
```

where P1,P2,...,P9 refer to nine items (parameters), the first of which *must* be specified but the others of which are optional. The list of nine parameters are these:

P1. UNIT = logical unit number of the file. Note: the characters UNIT= may be omitted.
P2. IOSTAT = I, where I is an integer variable set to zero by the system if the file is opened without error, and set to a system-dependent positive value if an error is encountered in opening the file.
P3. FILE = filename where filename must comply with rules for naming ForTran variables.
P4. ERR = reference number of the next statement to be executed if an error occurs in processing the file.
P5. STATUS = XXX, where XXX is one of these: OLD, NEW, SCRATCH, or UNKNOWN.
P6. ACCESS = XXX, where XXX is one of these: SEQUENTIAL or DIRECT.
P7. FORM = XXX, where XXX is one of these: FORMATTED or UNFORMATTED.
P8. RECL = record length.
P9. BLANK = XXX, where XXX is one of these: NULL or ZERO.

Every compiler has default values for all except the first of these parameters. It is required that the programmer specify the logical unit number of the file being opened. You should find out what the default values are for the other eight parameters rather than assume they will be what you expect them to be.

Here is an example of an OPEN statement:

```
OPEN(2, FILE='STUDNT', STATUS='OLD')
```

Notice that some of the parameters are omitted. The first parameter is always the logical unit number; in this example it is 2. The execution of this statement would cause a previously existing (OLD) file called STUDNT on logical unit 2 to be attached to a program and thus be made available for input or output.

The CLOSE statement

This statement causes the placing of an EOF symbol at the end of the specified file and causes the file to be disconnected from the program.

The general form is

```
CLOSE(P1,P2,P3,P4)
```

where P1,P2,P3, and P4 are as follows:

P1. UNIT = logical unit number of file. *Note:* the characters UNIT= may be omitted.
P2. IOSTAT = same as for OPEN.
P3. ERR = same as for OPEN.
P4. STATUS = XXX, where XXX is KEEP if this is an existing file that is to continue existing; or XXX is DELETE if this is an existing file that is to cease existing. The default value for XXX is KEEP unless the file was a SCRATCH file previous to the execution of the CLOSE statement, in which case the default value is DELETE.

Here are two examples of CLOSE statements:

```
CLOSE (2)
CLOSE (3, IOSTAT=I, STATUS='DELETE')
```

The first of these causes an EOF symbol to be placed at the end of the data file currently identified as logical unit 2. Since no other parameters are specified, the system default conditions apply. In the second example, an EOF symbol is placed at the end of the data file identified as logical unit 3. Variable I will contain zero if the file was closed without error, otherwise it will contain a positive value. After the CLOSE has been executed, the file no longer exists because of the DELETE parameter.

The INQUIRE statement

This statement is used to determine certain information about a data file. The form of the statement is

```
INQUIRE (P1,P2,...,P16)
```

where P1,P2,...,P16 are parameters defined as follows:

P1. Logical unit number of file.

P2. `IOSTAT=V2`

where V2 is a variable to which the system assigns the value zero if no error occurred during the INQUIRE, and assigns a positive value if an error occurred. If no error occurred but an EOF condition was encountered, V2 is assigned a negative value.

P3. `ERR=N`

where N is the reference number of the next statement to be executed if an error occurs during the INQUIRE.

P4. `EXIST=V4`

where V4 is a logical variable to which the system assigns the value TRUE if the file exists, FALSE if it does not exist.

P5. `OPENED=V5`

where V5 is a logical variable to which the system assigns the value TRUE if the file is currently connected to an input/output unit, FALSE if it is not.

P6. `NUMBER=V6`

where V6 is an integer variable to which the system assigns the logical unit number of the input/output unit to which the file is connected. If no unit is currently connected to the file, V6 is undefined.

P7. `NAMED=V7`

where V7 is a logical variable to which the system assigns the value TRUE if the file has a name, and FALSE if it has not been assigned a name.

P8. `NAME=V8`

where V8 is a character variable to which the system assigns the name of the file, if it has one. If the file has no name, V8 remains undefined.

P9. `ACCESS=V9`

where V9 is a character variable to which the system assigns the value SEQUENTIAL if the file is connected for sequential access. The value DIRECT is assigned to V9 if the file is connected for direct access. If the file is not currently connected, V9 remains undefined.

P10. `DIRECT=V10`

where V10 is a character variable to which the system assigns the value YES if the file is a direct file, NO if it is not, and UNKNOWN if the system cannot determine.

P11. `FORM=V11`

where V11 is a character variable to which the system assigns the value FORMATTED if the file is connected for formatted input/output, or the value UNFORMATTED if the file is connected for unformatted input/output. If the file is not connected, V11 remains undefined.

P12. `FORMATTED=V12`

where V12 is a character variable to which the system assigns the value YES if formatted input/output is permitted, or the value NO if formatted input/output is not permitted. If the system cannot determine, the value UNKNOWN is assigned to V12.

P13. `UNFORMATTED=V13`

Same as for P12 with unformatted replacing formatted.

P14. `RECL=V14`

where V14 is an integer variable to which the system assigns the value of the record length if the file is connected for direct access. If the file is not connected or if the connection is not for direct access, V14 remains undefined.

P15. `NEXTREC=V15`

where V15 is an integer variable to which the system assigns the record number of the next record to be read or written. If the file is not connected for direct access or if the record position cannot be determined because of a previous error condition, V15 remains undefined.

P16. `BLANK=V16`

where V16 is a character variable to which the system assigns the value NULL if null blank control is in effect for the file, or the value ZERO if zero blank control is in effect provided the file is connected for formatted input/output. Otherwise, V16 remains undefined.

An alternate form of the INQUIRE statement is

INQUIRE (filename,P2,...,P16)

where "filename" represents the file name and P2,...,P16 are as discussed in the other form.

Here are three examples of the INQUIRE statement: the first is

```
INQUIRE(UNIT=5, FORM=F, RECL=L)
```

After the execution of this statement, the file associated with logical unit 5 is examined and the variable F will contain FORMATTED, UNFOR-MATTED, or UNKNOWN, and the variable L will contain the value of the record length in the file.

The second example is

```
INQUIRE(2, NAME=N)
```

After the execution of this statement, the file associated with logical unit 2 will be examined and variable N will contain the file name if it has been assigned a name.

The third example is

```
INQUIRE('EXAMPL', OPENED=S, NUMBER=M)
```

After the execution of this statement, the file named EXAMPL is examined and if it has been opened, variable S will contain TRUE; otherwise, S will contain FALSE. Variable M will contain the logical unit number to which the file is currently connected if it is connected. Otherwise M is undefined.

ADDITIONAL FORMAT SPECIFICATIONS

In Chapter 6 we discussed the use of the FORMAT statement and in connection with it the use of the apostrophe, slash, parentheses, and format specifications I, F, X, A, and H. Earlier in this chapter we discussed the use of the D-format specification for double precision values and L format for logical variables. At this time we present information about format specifications T, TL, and TR.

T-, TL-, and TR-format specifications

There are sometimes situations in the tasks of input and output (especially output) when it would be helpful to be able to move quickly to the right or left before performing printing or reading. This capability is similar to that of tabulation on an ordinary typewriter. The three format specifications we are about to discuss take their names from tabulating. You may think of the T specification as meaning "tabulate," the TL specification as meaning "tabulate left," and the TR specification as meaning "tabulate right."

The form of these three specifications is the similar:

Tn TLn TRn

where in each case n represents an integer specifying the number of positions to be moved. The specification Tn means "tabulate n positions from the leftmost position." TLn means "tabulate n positions to the left of the current location." TRn means "tabulate n positions to the right of the current location."

Here is an example:

```
      PRINT 10
   10 FORMAT (T20,'ABC',TL1,'XYZ',T18,'RST',TR3,'OPQ')
```

Let's analyze the output resulting from these two program lines.

1. Specification T20 positions the output device to begin output in position 20. Therefore, the letters ABC w ll be output in positions 20, 21, and 22.
2. Specification TL1 causes movement to the left one position from the current location (position 23), thus positioning the output device so as to begin its output in position 22. Therefore, the letters XYZ will be output in positions 22, 23, and 24. Note that this causes X to be printed over C.
3. Specification T18 causes the output device to be moved to position 18 so that the letters RST are output in positions 18, 19, and 20. This causes T to be printed over A.
4. Specification TR3 causes the output device to be moved three positions to the right, resulting in OPQ being output in positions 24, 25, and 26. Since Z has already been output in position 24, this latest action causes O to be printed over Z.

Let's review what the final result of the two statements is:

Position	18	contains	R
"	19	"	S
"	20	"	A (T over A)
"	21	"	B
"	22	"	C (X over C)
"	23	"	Y
"	24	"	Z (O over Z)
"	25	"	P
"	26	"	Q

SOME ADDITIONAL FEATURES

We conclude this chapter by discussing some features available in full ForTran 77 that enhance the methods for accomplishing certain tasks. Since most of these features are not available in many of the ForTran compilers currently in use, we recommend that you determine the capabilities of your local situation before proceeding to use these features. If the ForTran compiler available to you is full ForTran 77, then not only the features we are about to consider are available to you but also all of those previously mentioned in this book.

Although some of the system library functions for string processing were mentioned in Chapter 2, in the following section we describe these functions in considerably more detail.

System library functions for string processing

Several library functions are available in ForTran 77 that help a great deal when processing character strings. We discuss four of those here.

Function LEN. This function determines the length (number of characters) of the character variable or constant specified as its argument. For example,

```
CHAR='JOHN A. PETERSON'
L=LEN(CHAR)
```

would result in the integer "16" being stored in L.

Function INDEX. This function has two arguments that are character variables or constants. The value of this function is an integer indicating the position in the character string specified by argument one where the first character of the substring specified by argument two can first be found. For example,

```
C1='ABCXRSCXRNCX'
C2='CXR'
N=INDEX(C1,C2)
```

would result in the integer "3" being stored in variable N because the C of CXR is in position 3 of character string C1.

Function CHAR. This function identifies the character associated with a given integer in a coding sequence. For example, if we assume a coding sequence of 00 through 25 for the letters of the alphabet where A is coded as 00, B as 01, and so on, then if the integer 10 were the argument to the function CHAR, the function value would be K because K is the character whose code is 10. The argument of CHAR may be an integer constant, or variable or arithmetic expression. If we assume the coding sequence for the letters of the alphabet as just described, here is an example using function CHAR:

```
I=4
J=7
C=CHAR(I+J)
PRINT*,C
```

This program segment would produce as output the letter L because L is the character whose code is 11 (the sum of I and J).

Function ICHAR.   This function is the inverse of function CHAR in that its argument is a character variable or constant of length 1 and its value is an integer equal to the coding sequence of the character specified by the argument. If we assume the coding sequence for letters of the alphabet as discussed in the previous section, then in this example

```
I=ICHAR('H')
PRINT*,I
```

the output would be 7 because the code for H is 07, which would be output without the leading zero.

System library functions for making string comparisons

Four library functions are available that make it possible to compare two character strings lexically. To compare lexically is to compare the internal code for two character strings, and on that basis the library function (which is a *logical* function) is made the value TRUE or FALSE.

Each of these functions has two arguments, specifying the two character strings being compared. The first character string is compared to the second in determining the function value of TRUE or FALSE.

Lexically greater than.   The system name for this function is LGT, and has the form

LGT(ARG1,ARG2)

The value of this function is TRUE if the internal code for the contents of ARG1 is greater than the internal code for the contents of ARG2. Otherwise the function value is FALSE.

Lexically greater than or equal to. The system name for this function is LGE, and has the form

LGE(ARG1,ARG2)

The value of this function is TRUE if the internal code for the contents of ARG1 is greater than or equal to the internal code for the contents of ARG2. Otherwise the function value is FALSE.

Lexically less than. The system name for this function is LLT, and has the form

LLT(ARG1,ARG2)

The value of this function is TRUE if the internal code for the contents of ARG1 is less than the internal code for the contents of ARG2. Otherwise the function value is FALSE.

Lexically less than or equal to. The system name for this function is LLE, and has the form

LLE(ARG1,ARG2)

The value of this function is TRUE if the internal code for the contents of ARG1 is less than or equal to the internal code for the contents of ARG2. Otherwise the function value is FALSE.

Now let's assume that the coding sequence (also called *collating sequence*) for the letters of the alphabet is 01 for A, 02 for B,...,26 for Z. Assume also that the coding sequence for a blank is 00. Then consider this program segment:

```
STATE1='MINNESOTA'
STATE2='TEXAS'
IF (LLE(STATE1,STATE2))GO TO 200
```

The next statement executed would be the one whose reference number is 200 because the code for MINNESOTA is 1309141440519152001 and for TEXAS is 2005240119. When making this comparison, the function LLE fills blank spaces on the right into the code for the shorter string in order to make the strings of equal length. The numeric code for blank is inserted in the numeric code for the corresponding string before the specified comparison takes place.

Note that all of these functions compare the string specified by the first argument with the string specified by the second argument and it is that order that determines greater than, greater than or equal, less than, and less than or equal.

PARAMETER statement

Sometimes it is useful to be able to refer to a constant by a symbolic name—for example, the constant pi of mathematics or the current withholding tax rate for employees of given characteristics. A statement is available in ForTran 77 to accomplish this. Here is its general form:

PARAMETER (NAME1=exp1,NAME2=exp2,...)

where NAME1, NAME2, and so on refer to acceptable ForTran symbolic names and exp1, exp2, and so on, refer to a constant or an arithmetic expression containing only constants. The type of the symbolic name must correspond to the type of the constant expression. Following is a specific example:

```
PARAMETER(PI=3.14159,WHRATE=12.7,HEAD='REPORT A')
```

In any subsequent program statements, whenever the symbol PI is used, the value 3.14159 replaces it. Similarly for WHRATE and HEAD.

Note that when a symbolic name appears in a PARAMETER statement its value must not be modified anywhere in that program.

Subroutines with multiple entry and return points

In any previous discussion of subroutines, when the CALL statement was executed, the subroutine was entered at its first executable statement. Similarly, return to the calling program was always to the first executable statement after the CALL statement. Situations do arise where these restrictions seem unnecessary and ForTran 77 makes provision for avoiding them. Two new statements must be discussed.

The RETURN exp statement.   This statement causes return to the calling program at various points depending on the value of exp. Here is the form:

RETURN exp

where exp represents any integer constant or expression. When this form of the RETURN statement is used, the CALL statement must contain in the parentheses following the subroutine name as many statement reference numbers as there are to be return values to the calling program. Each of these statement numbers must be preceded by an asterisk to distinguish them from arguments. Whatever the value is of exp in the RETURN statement, it specifies the *position* of the reference number in the CALL statement that controls where return to the calling program is to occur.

Consider this example:

```
        .
        .
        .
    CALL SUB(N,*100,*150,*200)
        .
        .
        .
100 READ*,A
        .
        .
        .
150 READ*,B
        .
        .
        .
200 READ*,C
        .
        .
        .
    END
    SUBROUTINE SUB(I,*,*,*)
        .
        .
        .
    IF(I.EQ.1)RETURN 1
        .
        .
        .
    IF(I.EQ.2)RETURN 2
        .
        .
        .
    RETURN 3
        .
        .
        .
    END
```

Notice in this example that the CALL statement contains three reference numbers each preceded by an asterisk. In the subroutine, you will see that the argument, I, provides a means for the subroutine to select which of the three returns is to apply. If I=1, then RETURN 1 is selected and the return to the calling program is to line 100. If I=2, the return to the main program is to statement 150 since 150 is the second asterisked reference number in the CALL statement. The third return is to statement 200.

It should not be difficult to conceive of situations where this multiple return feature could be very useful. In the example just presented, you can see that a completely different variable is input at each of the returns. Similarly, different actions could be taken following each of those READ statements. This feature makes it possible to write a program having considerably greater flexibility while using the same subroutine.

The ENTRY statement.   This statement makes it possible to enter a subprogram at more than one point in the subprogram. Such additional entry points must not be in the range of a DO loop nor in the range of an IF-THEN-ELSE structure. The general form of this statement follows:

ENTRY entry name (arguments)

where "entry name" refers to the name to be used for calling the subroutine so as to enter it at the alternate point, and "arguments" refers, as usual, to any subroutine arguments.

Here is a program example to illustrate its use:

```
10          PROGRAM  MAIN
15          REAL  A,X
20          READ*, A
30          CALL SUB(A,*10,*20)
40          A = FUNC(A)
50          GOTO 15
60    10    CALL SUB1(A)            Main
70    15    PRINT*, A               program
80          STOP
90    20    X = FUNC1(A)
100         PRINT*,X
110         STOP
120         END
```

```
130        SUBROUTINE  SUB(X,*,*)  ⎫
140        X = X*X                 ⎪
150        IF(X .LE. 100) RETURN 1 ⎪
160        IF(X .GT. 900) RETURN 2 ⎬ Subroutine
170        ENTRY  SUB1(X)          ⎪ SUB
180        X = 3*X                 ⎪
190        RETURN                  ⎪
200        END                     ⎭
210        REAL FUNCTION  FUNC(A)  ⎫
220        FUNC = SQRT(A)          ⎪
230        RETURN                  ⎪
240        ENTRY   FUNC1(A)        ⎬ Function
250        A = SQRT(A)             ⎪ FUNC
260        FUNC1 = SQRT(A)         ⎪
270        RETURN                  ⎪
280        END                     ⎭
```

We urge you to run this program with the input data given below and determine whether or not the corresponding output shown is what the program produces and whether or not the sets of output are correct.

Input:	10	40	100	-34	0	7	30
Output:	300	6.32456	10	5.83095	0	147	51.9615

Now we analyze the possible paths that could be taken through this program. As you can tell from the braces drawn at the right in the preceding program, the main program includes lines 10 through 120. Subroutine SUB includes lines 130 through 200, and function FUNC includes lines 210 through 280.

At line 20 a value is input to A, following which subroutine SUB is entered at line 130. There the value of A (which was brought into the subroutine as argument X) is squared; then the squared value is compared with 100 and possibly, depending on its value, with 900 (at lines 150 and 160). If the square of A is less than 100, then return to the main program takes place at the statement whose reference number is 10 (line 60). There subroutine SUB is called again (this time under its new entry name, SUB1) and entry to SUB is at line 170. When line 180 is executed, the value brought into subroutine SUB by the argument X is multiplied by 3 and control returns to the main program at line 70. There the current value of A (which is now three times the original value read in) is output and execution stops.

If, at line 150, the square of A (carried into the subroutine as argument X) is greater than 100, then line 160 is executed where the square of A is compared with 900. If it is greater than 900, return to the main program occurs at line 90 (reference number 20). There function FUNC is called by its alternate-entry name of FUNC1 and entry to the function occurs at the alternate entry point at line 240. Then at line 250 the square root of A (which at this point is really A squared because of previous action at line 140) is computed; this is followed at line 260 by a second square-root computation. Therefore, when return to the main program occurs at line 100, the value that is output will be the square root of the original value of A that was input at line 20.

If at line 160 the square of A (brought into the subroutine as argument X) is not greater than 900, lines 170 and 180 are executed and cause the square of A (argument X) to be multiplied by 3; this is followed by a return to the main program at line 40. There, function FUNC is called and the argument A (actually its value now is three times the square of A as originally input) is processed by the square root library function at line 220. Then return to the main program is at line 50, which transfers control to line 70 (reference number 15), where output takes place and execution stops. The value output at line 70 this time is the square root of three times the square of the original value input for A.

Computed GO TO statement

We now leave our special consideration of subprograms and the problem of making information available among the main program and its subprograms. The next ForTran statement we shall discuss is the computed GO TO. Here is its general form:

GO TO $(n_1,n_2,...,n_k)$,INDEX

where $n_1,n_2,...,n_k$ are positive integers and are reference numbers of statements in the program. The comma before INDEX is optional in ForTran 77 but required by some other ForTran compilers. INDEX is a positive integer variable or arithmetic expression such that INDEX is not less than 1 nor greater than k. INDEX has the function of specifying which of statements $n_1,n_2,...,n_k$ is the next one to be executed after the computed GO TO; it does so by indicating the *position* of the integer inside parentheses that is the reference number of the statement next to be executed. Since INDEX may be an arithmetic expression, the GO TO occurs in that case only after a computation has been done, hence the name "computed GO TO." Let's look at a program that lends itself to the use of this structure.

Example

Problem: We need a program to retrieve and update eight different data files stored on magnetic tape or disk. These data files contain information related to six categories of students plus one each for staff and faculty. Assume the following codes:

Category of person	Code
Freshman	1
Sophomore	2
Junior	3
Senior	4
Master student	5
Doctor student	6
Staff	7
Faculty	8

Suppose there are eight subprograms already written such that each one of the subprograms is used to process the data for one of the above-mentioned files. The program we now want to show is to read a code and from the value of the code select the appropriate subprogram for processing one of the data files. Let us assume that if the code is 9, that means we are to stop the program. Here is one way to accomplish the desired task:

```
      PROGRAM   MAIN
      INTEGER   CODE
10    READ*, CODE
      IF(CODE .EQ. 1)GOTO 100
      IF(CODE .EQ. 2)GOTO 200
      IF(CODE .EQ. 3)GOTO 300
      IF(CODE .EQ. 4)GOTO 400
      IF(CODE .EQ. 5)GOTO 500
      IF(CODE .EQ. 6)GOTO 600
      IF(CODE .EQ. 7)GOTO 700
      IF(CODE .EQ. 8)GOTO 800
      IF(CODE .EQ. 9)GOTO 900
      PRINT*,' ERROR IN CODE ',CODE
      GOTO 10
100   OPEN(1,FILE='FRESH',STATUS='OLD')
      REWIND 1
      CALL SUB1
      GOTO 10
```

```
200      OPEN(2,FILE='SOPHM',STATUS='OLD')
         REWIND 2
         CALL SUB2
         GOTO 10
300      OPEN(3,FILE='JUNIR',STATUS='OLD')
         REWIND 3
         CALL SUB3
         GOTO 10
400      OPEN(4,FILE='SENIR',STATUS='OLD')
         REWIND 4
         CALL SUB4
         GOTO 10
500      OPEN(5,FILE='MASTR',STATUS='OLD')
         REWIND 5
         CALL SUB5
         GOTO 10
600      OPEN(6,FILE='DOCTR',STATUS='OLD')
         REWIND 6
         CALL SUB6
         GOTO 10
700      OPEN(7,FILE='STAFF',STATUS='OLD')
         REWIND 7
         CALL SUB7
         GOTO 10
800      OPEN(8,FILE='FACUL',STATUS='OLD')
         REWIND 8
         CALL SUB8
         GOTO 10
900      STOP
         END
```

The subprograms SUB1,SUB2,...SUB8 are user-written for the purpose of updating each of the eight data files mentioned in the problem statement of this example.

However, these characteristics of the program are not of major concern to us right now. We are interested in learning how to simplify the program through the use of a computed GO TO statement. Here are three lines of program code to replace the *first* eleven lines of code in the preceding program:

```
     PROGRAM MAIN
     INTEGER CODE
10   READ*,CODE
     GO TO(100,200,300,400,500,600,700,800,900)CODE
```

By making this exchange of three lines for eleven, and leaving the rest of the program unchanged, everything is accomplished as described earlier. An integer value is input to variable CODE at line 10. Then if CODE=1, the computed GO TO statement causes program control to transfer to line 100. If CODE=2, program control goes to line 200, and so on, for values of CODE up to and including 9. It is important that CODE not have values outside of the range of integers 1 through 9 since there are nine reference numbers listed inside the parentheses of the computed GO TO. Should CODE contain a value outside that range, the compiler will disregard the computed GO TO statement and control will be transferred as if it were a CONTINUE statement.

Note also that not only is the number of program statements reduced by using the computed GO TO, but the resulting program flow is easier to follow.

Assigned GO TO statement

There is one more type of transfer statement available in ForTran that provides for a multiple-branch situation as the computed GO TO does. This statement is called the *assigned GO TO*, and gets its name from the fact that the key variable in it is *assigned* a value (using a special ASSIGN statement) previous to the execution of the assigned GO TO in which the key variable appears. There are two forms of the assigned GO TO:

1. GO TO NVAR
2. GO TO NVAR,$(n_1,n_2,n_3,...,n_k)$

In the case of the first form, variable NVAR is an integer variable that has previously been assigned a value through the use of the ASSIGN statement and whose assigned value is the reference number (also called statement number) of some statement in the program. When it is executed, program control transfers to the statement whose reference number is equal to the contents of NVAR.

The second form of the assigned GO TO also requires that NVAR be assigned a value previous to the execution of the computed GO TO. But this time NVAR must be assigned a value equal to one of the integer constants $n_1,n_2,n_3,...,n_k$. Then when the statement is executed, program control transfers to the program statement whose reference number is the same as the contents of NVAR. Here are two program segments using these two forms of the computed GO TO:

```
1. ASSIGN 50 TO N
   GO TO N
```

In this program segment, N is assigned the value 50 that must be the reference number of a statement in the program. Then when GO TO N is executed, program control transfers to the statement whose reference number is 50.

```
2.        ASSIGN 30 TO IN
          GO TO IN,(10,20,30,40,50)
    10    READ 30,X
    30    FORMAT(A2)
          GO TO 50
    20    READ 100,X
   100    FORMAT(A6)
          GO TO 50
    40    READ*,X
    50    PRINT*,X
```

This example has an error in it because statement 30 is a FORMAT statement and is, therefore, nonexecutable. It is not possible to transfer control to a nonexecutable statement. Depending on what the programmer intended, this error could be corrected by replacing 30 in the statement numbers listed in the computed GO TO by the reference number of any *executable* statement. The value assigned to IN could have been any of the other reference numbers appearing in the assigned GO TO statement.

The ASSIGN statement may also be used to assign a format statement number in an input or output statement. This makes it possible to select *one* FORMAT statement from among several that appear in the program. Suppose, for example, that there are four FORMAT statements in the program having reference numbers 60, 70, 80, and 90. The following pair of statements could be effective in accomplishing a specific output, namely the output specified in the FORMAT whose reference number is 80.

```
          ASSIGN 80 TO FRMAT
          WRITE (6,FRMAT) NAME,RATE,WAGE
```

ORDER OF FORTRAN STATEMENTS

Throughout this book we have discussed two major categories of statements: executable and nonexecutable. The order of executable statements follows the logic of the algorithm designed to solve the problem.

The order of nonexecutable statements is dependent on the ForTran compiler, but for ForTran 77 is as follows:

1. PROGRAM, FUNCTION, or SUBROUTINE statement.
2. PARAMETER statement(s).
3. IMPLICIT statement(s).
4. INTEGER, REAL, COMPLEX, DOUBLE PRECISION, LOGICAL, or CHARACTER statements.
5. DIMENSION, COMMON, or EQUIVALENCE statements.
6. DATA statement(s).
7. Any function statement definitions.
8. Executable statements.

The PROGRAM statement (for the main program) is optional in ForTran 77, though some compilers may require it in order to specify input/output devices. The above order is valid for the main program and for all subprograms.

SUMMARY

In this chapter we have presented some features of ForTran 77 that the person first learning the ForTran language will likely not use. However, for those readers familiar with the more commonly used capabilities of ForTran 77, the features discussed here will make ForTran 77 an even more powerful programming language for you. Because this chapter discusses topics in somewhat condensed form, our summary here will simply show general forms of the ForTran 77 statements and programming concepts presented in this chapter. We discussed three type-declaration statements:

1. DOUBLE PRECISION statement.
   General form:

   ```
   DOUBLE PRECISION V1, V2, ..., VN
   ```

   where V1, V2, ..., VN represent variable names or function names.
2. COMPLEX statement.
   General form:

   ```
   COMPLEX V1, V2, ..., VN
   ```

   where V1, V2, ..., VN represent variable names or function names.

3. LOGICAL statement.
   General form:

   `LOGICAL V1, V2, ..., VN`

   where V1, V2, ..., VN represent variable names or function names.

   The EQUIVALENCE statement was presented as a means of conserving memory by allowing a single memory location to be identified by two or more names. Its general form is

   `EQUIVALENCE (V1, V2, ..., VN)`

   where V1, V2, ..., VN represent simple or array variable names.
   The labelled COMMON statement was introduced as a method for making blocks of memory available to two or more program modules. Its form is:

   `COMMON /NAME/ V1, V2, ..., VN`

   where NAME is the block name and must comply with the usual rules for naming variables, and V1, V2, ..., VN represent variables.
   Other statements discussed were the following:

1. PARAMETER statement.
   General form:

   `PARAMETER (NAME1=expl, NAME2=exp2,....,NAMEn=expn)`

   where NAME1, NAME2, and so on, are acceptable ForTran names and exp1, exp2, and so on, are constants or expressions involving only constants.
2. ENDFILE statement.
   General form:

   `ENDFILE u`

   where u is an integer identifying the logical unit number on which the end-of-file mark is to be placed.

3. BACKSPACE statement.
   General form:

   ```
   BACKSPACE u
   ```

   where u is an integer identifying the logical unit number of the file to be backspaced one record.
4. REWIND statement.
   General forms:

   ```
   REWIND u
   REWIND (UNIT=u, ERR=u)
   ```

   where u is an integer specifying the logical unit number of the file to be repositioned at its beginning record, and u is an integer specifying the reference number of the next statement to be executed if the rewinding process runs into an error situation.
5. OPEN statement.
   General form:

   ```
   OPEN (P1, P2, ..., P9)
   ```

   where P1, P2, ..., P9 are certain parameters.
6. CLOSE statement.
   General form:

   ```
   CLOSE (P1, P2, P3, P4)
   ```

   where P1, P2, P3, and P4 are certain parameters.
7. INQUIRE statement.
   General form:

   ```
   INQUIRE (P1, P2, ..., P16)
   ```

   where P1, P2, ..., P16 are certain parameters.

Three additional format specifications were presented: T, TL, and TR. All three have to do with positioning the output device in preparation for producing output. These are similar to tabulation on a typewriter.

Some features related to the processing of character strings were discussed. These include the following:

1. Finding the number of characters in a string.
   This is done by using the LEN system function.
2. Locating the position of occurrence of one character string within a second one.
   This is accomplished by using the system function INDEX.
3. Identifying the character associated with a given code number.
   The system function CHAR does this task.
4. Identifying the numeric code for a specified character.
   The system function ICHAR accomplishes this job.
5. Functions for comparing the numeric codes of two characters.
   The four functions LGT, LGE, LLT, and LLE make such comparisons possible.

The capability of being able to refer to a constant by its usual name (for example, using PI to refer to 3.14159) can be accomplished through the use of the PARAMETER statement, whose general form is:

PARAMETER (NAME1=exp1, NAME2=exp2, ..., NAME3=exp3)

The final topic of the chapter was that of subroutines with more than one entry point or return point. The means for accomplishing these resides in the ENTRY statement and a special RETURN statement.

1. ENTRY statement.
   General form:

   ENTRY name (arg1, arg2, ..., arg*n*)

   where "name" represents the name to be used for calling the subroutine so as to enter it at a point other than its beginning, and "arg1, arg2, ..., arg*n*" represent the usual subprogram arguments.

2. RETURN exp statement.
   General form:

   RETURN exp

   where "exp" represents any integer constant or integer arithmetic expression and is used to identify the position in the CALL statement of the reference number to which return is to occur. In the CALL statement such reference numbers are preceded by asterisks.

There are some features of ForTran 77 omitted from any consideration in this book. These are all rather sophisticated capabilities and are most often used by professional programmers with extensive experience in ForTran programming. For the reader interested in a complete statement of the features of ForTran 77, you are referred to the publication by the American National Standards Institute (ANSI) called *American National Standard Programming Language FORTRAN*, available as publication ANSI X3.9-1978 from American National Standards Institute, Inc., 1430 Broadway, New York, New York 10018.

EXERCISES

1. Determine which of the following statements are true and which are false.
    (a) When using subprograms it is essential that COMMON statements be used so that all program units needing access to the same data will be able to have that access.
    (b) It is not permissible to use both the DATA statement and COMMON statements in the same program.
    (c) A variable appearing in a blank COMMON statement cannot be initialized by a DATA statement.
    (d) Labelled common blocks having the same name must be of the same size.
    (e) The size of a common block may be legitimately changed by an EQUIVALENCE statement.
    (f) The IMPLICIT statement will not change the type of a variable whose type is specified by another statement.
    (g) In order to use a labelled common block in a subprogram it must first have been defined in the main program.
    (h) There must be no more than one blank common block in a program.
    (i) To define a blank common block, you simply omit the block name in the COMMON statement.
    (j) The PARAMETER statement makes it possible to pass information between subprograms and main program.
    (k) Before you can use any variable in a parameter statement you must assign a value to it.

(l) If there is a PARAMETER statement in a program it must precede any DIMENSION statements if there are any.

(m) An IMPLICIT INTEGER statement cannot be used in the same program as an INTEGER statement.

(n) No more than one PARAMETER statement may appear in a program.

(o) Variables included in a labelled common block must not be passed as arguments in a subprogram.

(p) There is no means of providing for more than one entry point to a subprogram.

(q) When a subprogram has been called and has completed its task, control always returns in the calling program to the statement immediately following the call statement.

(r) Variables specified in COMPLEX statements must also be defined as REAL or INTEGER.

(s) The statement

```
X=(2.5,3.0)
```

is valid only when X has been specified as a double precision variable.

(t) All variables in any given arithmetic expression must be of the same type or else the compiler will say there is an error.

(u) If an arithmetic expression includes one complex variable then all variables must be complex in that expression.

(v) If an arithmetic expression includes one double precision variable then all variables in that expression must be either double precision or complex.

(w) The ENTRY statement makes it possible to enter a subprogram at some point other than the beginning.

(x) A statement of the form

```
RETURN N
```

appearing in a subprogram will cause control to return to the statement in the calling program whose reference number is the current value stored in N.

(y) The REWIND statement is used to position the internal input/output pointer to the beginning of a data file.

(z) The BACKSPACE statement makes it possible to read a record from a data file more than once.

2. Check the following statements or program segments for possible errors in syntax. If a statement is incorrect, make changes in it so it becomes valid.

(a) SUBPROGRAM SUB(X,Y)

(b) ENTRY X,Y

(c) ENTRY X(Y)

(d) FUNCTION F(X)=X*X+1

(e) SUBROUTINE SUB(X,*,*)

(f) PARAMETER P1,P2

(g) ENTRY X(Y,*,*)

(h) RETURN 2

(i) CALL SUB(X,10,20)

(j) COMMON A,B
    DATA A,B/2*0/

(k) DIMENSION A(10)
    COMMON /A/B(10)

(l) SUBROUTINE SUB
```
        .
        .

        .
    ENTRY
        .
        .
        .
```

(m) COMMON A(10),B(20)
    EQUIVALENCE (A(1),B(10))

(n)   COMPLEX X
     READ 10,X,A
    10 FORMAT(F10.2,A6)

(o) DOUBLE PRECISION X,Y
    READ*,X,Y
    Z=X+Y

(p)   DIMENSION Z(10),A(10)
     COMPLEX Z
     READ 10,Z,A
    10 FORMAT(10(F10.2,A1))

(q) CALL SUB(X,10)
```
        .
        .
        .
```

```
        SUBROUTINE SUB(Y,*)
           .

           .

           .
        RETURN 10
 (r)  CALL SUB1(X)
           .

           .

           .
        SUBROUTINE SUB(X,Y)
           .

           .

           .
        ENTRY SUB1(Z)
           .

           .

           .
```

3. Describe the output of each of the following programs. Assume the programs are correct and will run without error and assume default data types where necessary.

    (a)
```
        PROGRAM ONE
        IMPLICIT INTEGER (A-Y)
        READ*,VAR1,VAR2
        Z=(VAR1+VAR2)/2.0
        PRINT VAR1,VAR2,Z
        END
```
Assume input data of 2 and 7.

    (b)
```
        PROGRAM TWO
        COMMON /A/ B(10)
        DO 10 I=1,10
 10  B(I)=I*I
        CALL SUB
        PRINT*,B
        CALL SUB1
        PRINT*,B
        END
        SUBROUTINE SUB
        IMPLICIT INTEGER (A-Z)
        COMMON /A/ X(5),Y(5)
        DO 10 I=1,5
 10  X(I)=Y(I)/X(I)
        ENTRY SUB1
        DO 20 I=1,5
```

```
         20 Y(I)=Y(I)/X(I)
            RETURN
            END
(c)         PROGRAM SORTIT
            LOGICAL FLAG
            COMMON A(100)
            FLAG= .TRUE.
            READ*,N
            IF(N .LE. 0 .OR. N .GT.100)THEN
               PRINT*,'ERROR IN NUMBER OF NUMBERS.'
               FLAG= .FALSE.
            ELSE
               READ*,(A(I),I=1,N)
            ENDIF
            CALL SORT(FLAG,*30,*20)
         20 PRINT*,(A(I),I=1,N)
         30 STOP
            END
            SUBROUTINE SORT(F,*,*)
            LOGICAL F
            COMMON A(100)
            IF(.NOT. F)RETURN 1
            DO 10 I=1,N-1
              K=I
              DO 20 J=I+1,N
                IF(A(K) .GT. A(J)) K = J
         20   CONTINUE
              T=A(I)
              A(I)=A(K)
              A(K)=T
         10 CONTINUE
            RETURN 2
            END
```

Assume input for the preceding program are the following data in the order given:

8, 3, 5, 10, 2, 12, 9, 6, 5

```
(d)         PROGRAM MAT
            INTEGER MATR(20,20)
            COMMON /ONE/ MATR
            DO 20 I=1,20
            CALL ROW(I)
```

```
   20 CONTINUE
      PRINT 30,MATR
   30 FORMAT(20A2,/)
      END
      SUBROUTINE ROW(N)
      COMMON /ONE/ MATR(20,20)
      DO 10 J=1,20
   10 MATR(N,J)=N
      RETURN
      END
```

(e)
```
      COMPLEX X,Y,Z
      X=(2.5,3)
      Y=X+X
      Z=X+Y
      X=X-Y
      Y=X+Y
      PRINT 10,X,Y,Z
   10 FORMAT (2F10.2)
      D=CABS(X+Y)
      PRINT*,D
      END
```

You may want to refer to Appendix D to check the results of the library function CABS.

(f)
```
      LOGICAL A,B,C
      A= .TRUE.
      B= .FALSE.
      C= .NOT. B
      B=A .OR. B .AND. C
      A= .NOT. (A .OR. B) .AND. (B .OR. .NOT. C)
      C=(A .OR. B .AND. C) .OR. (A .AND. B .OR. C)
      PRINT 10,A,B,C
   10 FORMAT(3L2)
      END
```

4. Identify any syntax errors in the following programs.

(a)
```
      PROGRAM MAIN
      ENTRY SUB1
      READ*,A
      CALL SUB(A)
      IF(A .LE. 0)CALL SUB1
      PRINT*,A
      END
```

```
              SUBROUTINE SUB(B)
              IF(B .GT. 30 .OR. B .LE. 0)B=0
              IF(B .EQ. 0)RETURN
              DO 10 I=1,100
                 B=B/2
                 IF(B .GT. 1)GO TO 10
                 B=I
                 RETURN
           10 CONTINUE
              PRINT*,'BAD INPUT'
              STOP
              END
    (b)       PROGRAM MAIN
              COMMON /A/ B(3),AB(5)
              DATA B,AB/3*1.,5*2./
              CALL SUB
              PRINT *,B,AB
              END
              SUBROUTINE SUB
              COMMON /A/ X(5),Y(3),Z
              DIMENSION B(5)
              EQUIVALENCE (Y(1),B(1))
              DATA B/2.,1.,2.5,3.5,6.1/
              DO 10 I=1,5
           10 X(I)=B(6-I)
              RETURN
              END
    (c)       PROGRAM MAIN
              X=FUNC1(1)
              Y=FUNC2(1)
              PRINT*,X,Y
              END
              FUNCTION FUNC(X)
              INTEGER X
              ENTRY FUNC1(I)
              FUNC1 = X*I
              RETURN
              ENTRY FUNC2(I)
              FUNC2 = I*I
              RETURN
              END
```

(d)
```
      PROGRAM MAIN
      COMPLEX X,Y,Z
      COMMON /A/ X,Y
      READ 10,X,Y
   10 FORMAT(2F10.2)
      CALL SUB
      PRINT 10,X,Y
      Z=FUNC(X,Y)
      PRINT 10,Z
      END
      SUBROUTINE SUB
      COMMON /A/ VAR1,VAR2
      DOUBLE PRECISION VAR1,VAR2
      VAR1=VAR1+VAR2
      VAR2=VAR1-VAR2
      RETURN
      END
      FUNCTION FUNC(X,Y)
      COMPLEX FUNC
      FUNC=CONJ(X+Y)
      RETURN
      END
```

You may want to refer to Appendix D to check the purpose of library function CONJ.

For exercises in programming, we suggest that you select exercises from earlier chapters and try to write better programs by using the additional ForTran features described in this chapter.

APPENDIX A

Number Systems and Internal Representation of Data

In order to better understand techniques employed by computer designers in the electronic representation of information, we begin by presenting some basic facts about number systems.

NUMBER SYSTEMS

Decimal numbers

First we review certain concepts about the *decimal* number system. Consider the decimal number 235. Everyone reading this book knows that

$$235 = 2 \times 100 + 3 \times 10 + 5 \times 1$$

or

$$235 = 2 \times 10^2 + 3 \times 10^1 + 5 \times 10^0$$

recalling from mathematics the definition $b^0 = 1$ for any real number, b. The righthand side of the preceding equation is called the *polynomial form* of the number 235. Though the digit 2 is the smallest digit in the previous example, 2 represents the greatest portion of 235 because of its position in the number. The position value of a digit is called its *rank*. Thus in the example 235,

2 has rank 10^2
3 has rank 10^1
5 has rank 10^0

In general, the rank of a decimal number n positions to the left of the decimal point is $10^{(n-1)}$.

The number 10 in the preceding discussion is called the *base* of the number system. The symbols used for digits in the decimal system are, of course, 0, 1, 2, 3, 4, 5, 6, 7, 8, and 9.

As you know, decimal numbers exist that have fractional parts, for example, 235.718. In this number, the rank of 7 is 1/10, of 1 it is 1/100, and of 8 it is 1/1000. Another way of writing this is to say that the rank of 7 is $(1/10)^1$, of 1 it is $(1/10)^2$, and of 8 it is $(1/10)^3$. Note that each position move to the right results in a decrease in the rank by a factor of 1/10. We summarize our review of decimal numbers as follows:

1. The base of the decimal number system is the number 10.
2. The ranks of digits in a decimal number are either a power of 10 or of 1/10.
3. The rank increases by a power of 10 for each position move to the left of the decimal point.
4. The rank decreases by a power of 1/10 for each position move to the right of the decimal point.

Binary numbers

The theory of number systems makes it possible to design number systems with any base. Since we are so indoctrinated with the decimal system, when we discuss other systems we shall refer to their base numbers in terms of decimal numbers. Also, our symbols for the digits in other number systems will be the symbols for decimal digits insofar as that is possible. For example, the number system whose only digit symbols are 0 and 1 has as its base number the decimal number 2 and is called the binary system. Consider the number 1011.101 from the binary system:

$$1011.101 = 1 \times 2^3 + 0 \times 2^2 + 1 \times 2^1 + 1 \times 2^0 + 1 \times (\tfrac{1}{2})^1 + 0 \times (\tfrac{1}{2})^2 + 1 \times (\tfrac{1}{2})^3$$

As you study the righthand side of the preceding equation you will recognize that the ranks of digits in the binary system are either powers of 2 or of ½.

We can summarize this discussion of the binary system with statements similar to those we used for the decimal system:

1. The base of the binary number system is the decimal number 2.
2. The ranks of digits in a binary number are either powers of 2 or of ½.

3. The rank increases by a power of 2 for each position move to the left of the binary point.
4. The rank decreases by a power of ½ for each position move to the right of the binary point.

There are two more number systems important to computer science, the octal system and the hexadecimal system.

Octal numbers

The octal number system has as its base the decimal number 8. The digits in this system are 0, 1, 2, 3, 4, 5, 6, and 7. As an example of an octal number take 156.24. We write

$$156.24 = 1 \times 8^2 + 5 \times 8^1 + 6 \times 8^0 + 2 \times (\tfrac{1}{8})^1 + 4 \times (\tfrac{1}{8})^2$$

Note that the base number, 8, appears in connection with the value associated with each digit. As we did for the decimal and binary systems, we now state for the octal system:

1. The base of the octal system is the decimal number 8.
2. The ranks of digits in an octal number are either powers of 8 or of ⅛.
3. The rank increases by a power of 8 for each position move to the left of the octal point.
4. The rank decreases by a power of ⅛ for each position move to the right of the octal point.

Hexadecimal numbers

The hexadecimal number system (also referred to as the hex system) has as its base the decimal number 16. This number system requires more digit symbols than the decimal system since there are as many digits as the base number, namely, 16. The symbols most commonly used are 0, 1, 2, 3, 4, 5, 6, 7, 8, 9, A, B, C, D, E, and F. Thus, the first six letters of the alphabet are used as symbols for digits. A is used to represent the hexadecimal equivalent of the decimal number 10. Similarly, B in hex equals 11 in decimal; C in hex equals 12 in decimal; and so on through F.

Let's consider the hexadecimal number 5A7.B2.

$$5A7.B2 = 5 \times 16^2 + 10 \times 16^1 + 7 \times 16^0 + 11 \times (\tfrac{1}{16}) + 2 \times (\tfrac{1}{16})^2$$

Note that in the preceding equation, the hexadecimal number is on the left side while the decimal equivalent is on the right side.

The following statements are true for the hexadecimal system:

1. The base of the hexadecimal system is the decimal number 16.
2. The ranks of digits in a hexadecimal number are either powers of 16 or of $\frac{1}{16}$.
3. The rank increases by a power of 16 for each position move to the left of the hexadecimal point.
4. The rank decreases by a power of $\frac{1}{16}$ for each position move to the right of the hexadecimal point.

Converting numbers from one system to another

You should be aware that when you read a number in any system other than the decimal system, you cannot use the names for positions such as fifteen, twenty, hundred, thousand, and so on, because those names are associated with the position for which they are used. Therefore, the binary number 1011.1 must be read, "one, zero, one, one, point, one". The same is true for numbers in other systems except the decimal system.

For anyone planning to do much work with computers, it is important to know how to convert numbers from one system to another. This is so because the number system commonly used by most people is the decimal system, while that of computers is the binary system. As bridges between these two systems we often use the octal system or the hexadecimal system. We now discuss some algorithms for converting numbers.

Converting an integer from decimal to any other system.   A decimal integer, I, may be converted to an integer in the number system whose base is the decimal integer b, as follows:

1. Set $M = I$.
2. Divide M by the base, b, which gives a quotient, Q, and a remainder, R. R is a digit of the equivalent integer in the new system.
3. If $Q = 0$, the conversion is complete and the equivalent integer is the number obtained by writing the remainders, R, obtained in step 2 in the order computed from right to left. If Q is zero, stop processing.
4. Set $M = Q$ and go to step 2.

Here are some examples applying the above algorithm.

Example A.1.   Convert the decimal integer 2610 to its octal equivalent.

$$2610/8 = 326, \text{ remainder } \underline{2}$$
$$326/8 \ = \ 40, \text{ remainder } \underline{6}$$
$$40/8 \ \ = \ \ \ 5, \text{ remainder } \underline{0}$$
$$5/8 \ \ \ = \ \ \ 0, \text{ remainder } \underline{5}$$

Therefore the octal equivalent is 5062.

Example A.2.   Convert the decimal integer 6825 to its hexadecimal equivalent.

$$6825/16 = 426, \text{ remainder } \underline{9}$$
$$426/16 \ = \ 26, \text{ remainder } \underline{10}$$
$$26/16 \ \ = \ \ 1, \text{ remainder } \underline{10}$$
$$1/16 \ \ \ = \ \ 0, \text{ remainder } \underline{1}$$

Therefore the hexadecimal equivalent is 1AA9. (Recall that in the hex system, the decimal number 10 is written A.)

Converting a decimal fraction to any other system.   A decimal fraction, F, may be converted to its equivalent in a number system whose base is b as follows:

1. Multiply F by b and call the result Y. If $Y=0$, go to step 4. If not, go to step 2.
2. Note the integer part of Y (it may be zero) since it is a digit in the equivalent number we are producing in the system whose base is b.
3. Set F equal to the fractional part of Y (maybe all of Y) and go to step 1.
4. The conversion is complete and the equivalent number is obtained by writing the integer parts of Y from step 2, in the order computed from left to right following a period.

Note that the above algorithm may never produce $Y=0$ at step 1. In this case, stop the algorithm when as many digits as desired have been computed.

Example A.3.   Convert the decimal fraction .2 to its binary equivalent.

$$.2 \times 2 = \underline{0}.4$$
$$.4 \times 2 = \underline{0}.8$$
$$.8 \times 2 = \underline{1}.6$$
$$.6 \times 2 = \underline{1}.2$$
$$.2 \times 2 = \underline{0}.4$$
$$.4 \times 2 = \underline{0}.8$$

Although the product, Y, is never zero, we stop at this point to write the result .001100 as the binary fraction equivalent to the decimal fraction, .2.

Example A.4.   Convert the decimal number 126.25 to its equivalent octal number.

We convert the integer portion, 126, by applying the previous algorithm for integers, and the fractional portion by applying the previous algorithm for fractions.

$$126/8 = 15, \text{ remainder } \underline{6}$$
$$15/8 \ = \ 1, \text{ remainder } \underline{7}$$
$$1/8 \ \ = \ 0, \text{ remainder } \underline{1}$$

The octal equivalent of 126_{10} is 176_8. (We often write subscripts to indicate the base of the number system to which a given number belongs.) Now we convert the fractional part.

$$.25 \times 8 = \underline{2}.0$$
$$.0 \ \ \times 8 = \underline{0}.0$$

The octal equivalent of $.25_{10}$ is $.2_8$. Therefore, $126.25_{10} = 176.2_8$, and the conversion is complete.

Converting a base b number to a decimal number.   Perhaps the easiest method of converting a number from a base b number to a decimal number is to express the number in its polynomial form, then compute the decimal equivalent of the polynomial form.

Example A.5. Convert the binary number 101.01011 into its decimal equivalent. (The right side of the equation below is polynomial form.)

$$101.01011 = 1 \times 2^2 + 0 \times 2^1 + 1 \times 2^0 + 0 \times (\tfrac{1}{2}) + 1 \times (\tfrac{1}{2})^2 + 0 \times (\tfrac{1}{2})^3$$
$$+ 1 \times (\tfrac{1}{2})^4 + 1 \times (\tfrac{1}{2})^5$$
$$= 4 + 0 + 1 + 0 + .25 + .0625 + .03125$$
$$= 5.34375$$

Conversions among binary, octal, and hexadecimal numbers. Because the base numbers of these three systems are related to each other as powers of 2, conversions among these systems are relatively easy to perform. Since $8 = 2^3$ and $16 = 2^4$, we can conclude that each position in an octal number has the value of three positions in the binary system. Similarly, each hexadecimal digit has the value of four binary digits. Therefore,

$$10101.01101_2 = \underline{010}\ \underline{101}.\underline{011}\ \underline{010}_2 = 25.32_8$$

Notice that beginning at the binary point, we have underlined groups of three binary digits. If the leftmost group does not have three digits, one or two leading zeros are added to make it three. Similarly, if the rightmost fractional group does not have three digits, one or two trailing zeros are added to make it three. Then each group of three binary digits is converted into its octal equivalent, keeping the corresponding position in the octal number that the triplet had in the binary number. Let's now consider each of the foregoing underlined triplets as it occurs from left to right:

$$010_2 = 2_8$$
$$101_2 = 5_8$$
$$011_2 = 3_8$$
$$010_2 = 2_8$$

Therefore, $10101.01101_2 = 25.32_8$.

If we want to convert an octal number into a binary number, we reverse the procedure just described. That is, we expand each octal digit into its equivalent in terms of three binary digits. Suppose we convert 25.32_8 into binary equivalent:

$$\begin{array}{cccc} 2 & 5 & . & 3 & 2 \\ \hline 010 & 101 & . & 011 & 010 \end{array}$$

Below each octal digit we have written its binary equivalent. By placing the binary point directly below the octal point, we have retained the correct positioning of integer and fractional digits in the binary number.

We turn now to converting from binary numbers to hexadecimal numbers. Recall that we pointed out previously that since $2^4 = 16$, we conclude that every group of four binary digits converts to one hexadecimal digit. Here is an example. Convert 110101.11_2 to its hexadecimal equivalent.

$$\underline{0011} \ \underline{0101}.\underline{1100}_2$$
$$3 \qquad 5 \ . \ C_{16}$$

As we did with binary-to-octal conversions, we add leading 0's and trailing 0's (if necessary) in order to create quadruplets of binary digits. Then each quadruplet is converted into its hexadecimal equivalent and the hexadecimal point keeps the same relative position as the binary point. Recall that the letters A, B, C, D, E, and F are digits (equivalent to decimal 10, 11, 12, 13, 14, and 15, respectively) in the hexadecimal system.

Converting from hexadecimal to binary means that we reverse the above process. That is, every hexadecimal digit gives rise to four binary digits. Study this example and the procedure should be clear:

$$3 \qquad D \ . \ 9 \qquad F_{16}$$
$$\underline{0011} \ \underline{1101}.\underline{1001} \ \underline{1111}_2$$

REPRESENTING NUMBERS INSIDE A COMPUTER

Having discussed procedures for converting numbers from one numeration system to another, we are ready to consider the *internal representation* of information in a digital computer. We begin with integers.

Internal representation of integers

Any information stored in a computer is represented internally in a form logically equivalent to a collection of binary digits. (See Chapter 1 for a discussion of some of the actual hardware to accomplish this.) Furthermore, most computers are equipped to store a fixed number of binary digits in each storage unit, called a *word*. (See Chapter 1 for more information about computer words.) Computer word size varies from twelve bits to sixty-four bits, with thirty-two bits being a very common word size. Let's assume for this discussion that our computer has a word size of thirty-two bits.

When a decimal integer is stored in a computer it is converted to its binary-system equivalent, then stored right-adjusted in the computer word. One bit of the word is usually reserved to specify the *sign* of the number that is stored, usually as the leftmost bit. Typically, 0 means positive and 1 means negative. For example, the decimal number +123 would be stored like this in a 32-bit word:

$$\boxed{0\,0\,0\,0\,0\,0\,0\,0\,0\,0\,0\,0\,0\,0\,0\,0\,0\,0\,0\,0\,0\,0\,0\,0\,1\,1\,1\,1\,0\,1\,1}$$

The leftmost bit is zero because the number stored is positive. The number itself is right-adjusted in the word, so if you analyze the binary number shown you will see that

$$1111011_2 = 123_{10}$$

The technique just described is a common method for storing integers in computers.

Complements of binary numbers. There is another commonly used method of dealing with negative numbers in computers. It involves the concept of the complement of a number. In the binary system, the complement of a single digit is the opposite digit. Thus, the complement of 0 is 1, and of 1 the complement is 0. Given a binary number, X, of more than one digit, its complement, denoted by $\overline{X}$, is the number made up of digits such that each one is the complement of its counterpart in X. For example, if we consider the 16-bit word

$$X = 0000000000101100,$$

then

$$\overline{X} = 1111111111010011$$

This form of the complement is called the 1's *complement*.

The computer operation of complementing a number is electronically easy to accomplish, so if use can be made of it to cause arithmetic operations to take place, that would be an intelligent application of complementing. The arithmetic operation of subtraction can be accomplished by adding the 2's complement of the subtrahend. The 2's *complement* of a binary number is the result of adding 1 to its 1's complement. That is, given a binary number, X, then the 2's complement is $\overline{X} + 1$.

$$X = 0000000000100101$$
$$\overline{X} = 1111111111011010$$
$$\overline{X}+1 = 1111111111011011$$
$$X+(\overline{X}+1) = 0000000000000000$$

Since the sum of any number (in any numeration system) and its negative is 0, it appears from the preceding example that $\overline{X}+1$ is the negative of X. Thus, the 2's complement of a binary number, X, is one way of representing the negative of X—that is, $-X$.

If we use the 2's complement form of representing negative integers, then the range of values for integers that may be stored is -2^{m-1} to $(2^{m-1} - 1)$ where m is the number of bits in a word.

Internal representation of real numbers

The internal representation of real numbers differs among the various available computers. There are some common features that we shall mention, however, after which we will describe the details of one method employed in some computers.

When representing real numbers, most computers employ the method of designating certain bits as specifying the coefficient and other bits as specifying the exponent of the real number. By "coefficient" and "exponent" we mean something very much like .625 and 2, respectively, when 62.5 is represented as

$$.625 \times 10^2$$

The difference is that in a computer, all numbers are represented as bits. Not only do certain bits specify the coefficient and the exponent, but it is necessary to have one bit specify the sign of the coefficient. Normally, a sign-bit of 0 indicates positive while a sign-bit of 1 denotes negative.

In order to show how 62.5 would be stored internally in a word of sixteen bits, we make some assumptions. Suppose the sixteen bits are identified as bits 0 through 15 from right to left. Then assume the coefficient is stored in bits 0 through 7, the exponent in bits 8 through 13, the sign of the exponent in bit 14, and the sign of the coefficient in bit 15. Figure A.1 shows these assumptions in diagram form:

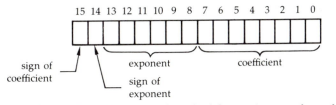

Figure A.1   Diagram of method for storing a real number

Since all numbers are represented internally as binary numbers, if we are to represent 62.5_{10} we must convert it to its binary equivalent. First we convert 62:

$62/2 = 31$, remainder $\underline{0}$
$31/2 = 15$, remainder $\underline{1}$
$15/2 = 7$, remainder $\underline{1}$
$7/2 = 3$, remainder $\underline{1}$
$3/2 = 1$, remainder $\underline{1}$
$1/2 = 0$, remainder $\underline{1}$

Therefore, $62_{10} = 111110_2$. Next we convert $.5_{10}$ as follows:

$.5 \times 2 = \underline{1}.0$

Therefore, $.5_{10} = .1_2$. Now we can write the entire binary equivalent of the decimal number 62.5 as 111110.1, which is equal to $.1111101_2 \times 2^6$, or $.1111101_2 \times 2^{110}{}_2$. This last form has both the coefficient and the exponent in binary form.

Now we insert the coefficient, .1111101, in bits 0 through 7, and the exponent, 110, in bits 8 through 13. Bit 15 is 0 because the coefficient is positive, as is bit 14 because the exponent is positive. All other bits are 0's. Figure A.2 shows the result:

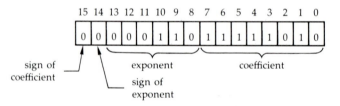

Figure A.2 Diagram of storing the decimal number 62.5

Since the writing of even sixteen bits (as in Figure A.2) is a tedious chore, internal representations are sometimes written in octal form or in hexadecimal form. If we write the internal binary representation from Figure A.2 as a hexadecimal number we would get 0 6 F A, obtained as follows:

$$\underbrace{0000}_{0} \quad \underbrace{0110}_{6} \quad \underbrace{1111}_{F} \quad \underbrace{1010}_{A}$$

Each underlined quadruplet of binary digits is converted to its hexadecimal equivalent.

In the preceding example, we have assumed that the normal form for the coefficient stored internally is such that the binary point is understood to be at the extreme left end of the coefficient field of the computer word. In most computers the situation is such that the binary point is understood to be at the extreme *right* of the coefficient field. This fact does not change the validity of our previous discussion, but it does have the effect of producing more negative exponents in the internal representations of real numbers. That is, if the binary point is assumed to be to the right of where it really would be in the number being represented, then it takes a negative exponent to indicate that the binary point really belongs to the left of where it is assumed to be.

REPRESENTING CHARACTER DATA INTERNALLY

Nonnumeric (or character) data are represented internally as bits arranged according to some coding scheme. Table A.1 shows a standard character set of sixty-three characters and the code for each character in each of the following coding schemes: Hollerith punch code (zone and digit punches), External BCD code (octal), ASCII punch code (zone and digit punches), and ASCII code (hexadecimal).

By referring to Table A.1, you can verify that the internal representation of the word DOGS in a 32-bit word using ASCII code would be:

$$0100 \quad 0100 \quad 0100 \quad 1111 \quad 0100 \quad 0111 \quad 0101 \quad 0011$$
$$\underline{4} \quad \underline{4} \quad \underline{4} \quad \underline{F} \quad \underline{4} \quad \underline{7} \quad \underline{5} \quad \underline{3}$$

ASCII is the acronym for American Standard Code for Information Interchange and was designed to be the standard 8-bit code for the computer industry. Some computers like the Cyber series of Control Data Corporation have words of sixty bits in size. For these computers it is more conservative of computer storage space to use External BCD code since that 6-bit code allows for the storage of ten characters per word. Thus, in such a machine, the word COMPUTERS would appear as follows:

$$110\,011\,100\,110\,100\,100\,100\,111\,010\,100\,010\,011\,110\,101\,101\,001\,010\,010$$
$$\underline{6}\,\underline{3}\quad\underline{4}\,\underline{6}\quad\underline{4}\,\underline{4}\quad\underline{4}\,\underline{7}\quad\underline{2}\,\underline{4}\quad\underline{2}\,\underline{3}\quad\underline{6}\,\underline{5}\quad\underline{5}\,\underline{1}\quad\underline{2}\,\underline{2}$$

You can verify the preceding by referring to Table A.1.

SUMMARY

Surely more could be said about number systems, coding schemes, and internal representation. However, our purpose has been to introduce you to decimal, binary, octal, and hexadecimal number systems. We have also demonstrated common methods for storing integers and real numbers. Finally, we considered the representation of character data in computer storage employing either of two coding schemes.

TABLE A.1　Standard character sets

ASCII graphic subset	Hollerith punch (026)	External BCD code	ASCII punch (029)	ASCII code	ASCII graphic subset	Hollerith punch (026)	External BCD code	ASCII punch (029)	ASCII code
:	8-2	00	8-2	3A	6	6	06	6	36
A	12-1	61	12-1	41	7	7	07	7	37
B	12-2	62	12-2	42	8	8	10	8	38
C	12-3	63	12-3	43	9	9	11	9	39
D	12-4	64	12-4	44	+	12	60	12-8-6	2B
E	12-5	65	12-5	45	−	11	40	11	2D
F	12-6	66	12-6	46	*	11-8-4	54	11-8-4	2A
G	12-7	67	12-7	47	/	0-1	21	0-1	2F
H	12-8	70	12-8	48	(	0-8-4	34	12-8-5	28
I	12-9	71	12-9	49	)	12-8-4	74	11-8-5	29
J	11-1	41	11-1	4A	$	11-8-3	53	11-8-3	24
K	11-2	42	11-2	4B	=	8-3	13	8-6	3D
L	11-3	43	11-3	4C	blank	no punch	20	no punch	20
M	11-4	44	11-4	4D	, (comma)	0-8-3	33	0-8-3	2C
N	11-5	45	11-5	4E	. (period)	12-8-3	73	12-8-3	2E
O	11-6	46	11-6	4F	#	0-8-6	36	8-3	23
P	11-7	47	11-7	50	[	8-7	17	12-8-2	5B
Q	11-8	50	11-8	51	]	0-8-2	32	11-8-2	5D
R	11-9	51	11-9	52	%	8-6	16	0-8-4	25
S	0-2	22	0-2	53	" (quote)	8-4	14	8-7	22
T	0-3	23	0-3	54	_ (underline)	0-8-5	35	0-8-5	5F
U	0-4	24	0-4	55	!	11-0 or 11-8-2	52	12-8-7 or 11-0	21
V	0-5	25	0-5	56					
W	0-6	26	0-6	57	&	0-8-7	37	12	26
X	0-7	27	0-7	58	' (apostrophe)	11-8-5	55	8-5	27
Y	0-8	30	0-8	59	?	11-8-6	56	0-8-7	3F
Z	0-9	31	0-9	5A	<	12-0 or 12-8-2	72	12-8-4 or 12-0	3C
0	0	12	0	30					
1	1	01	1	31	>	11-8-7	57	0-8-6	3E
2	2	02	2	32	@	8-5	15	8-4	40
3	3	03	3	33	\	12-8-5	75	0-8-2	5C
4	4	04	4	34	∧ (circumflex)	12-8-6	76	11-8-7	5E
5	5	05	5	35	; (semicolon)	12-8-7	77	11-8-6	3B

APPENDIX B
Summary of Programming Style

This appendix is essentially a list of do's and don't's to help you develop a programming style that makes your programs understandable, efficient, and easier to debug. We suggest that for more complete discussions of these topics you refer to *The Elements of Programming Style* by Brian W. Kernighan and P. J. Plauger (published by McGraw-Hill, 1974) and to *Program Style, Design, Efficiency, Debugging, and Testing* by D. VanTassel (published by Prentice-Hall, 1974).

1. Writing understandable programs.
    (a) Don't try to be too clever. Rather write clearly and in a straightforward manner.
    (b) Avoid the use of temporary variables unless there is no other way to do what must be done.
    (c) Choose variable names that are meaningful to *anyone* reading your program even if such names are longer and therefore more tiresome to use.
    (d) Use parentheses to help avoid any sense of ambiguity.
    (e) Avoid unnecessary branches.
    (f) If a logical expression is hard to understand, transform it into a simpler expression using such tools as De Morgan's Laws.
    (g) Choose carefully the data structure you use. It can greatly simplify the program.
    (h) Use GO TO's very sparingly and then only to implement fundamental program structures.

(i) Make use of system functions whenever possible. Not only will this increase the understandability of your program but will usually increase its efficiency.

(j) Make sure comments and code agree.

(k) Use comments freely but not just to repeat a code. Make every comment count.

2. Writing efficient programs.

(a) Be sure your program *works* before you worry about its efficiency.

(b) Make your program impervious to bad input data before you become concerned about efficiency.

(c) Don't sacrifice understandability for efficiency.

(d) Use simple techniques as much as possible. This will usually mean greater efficiency also.

(e) Time spent on selecting or developing a better algorithm is much more likely to increase program efficiency than is polishing the code.

(f) Before you make changes in your program that are intended to increase efficiency, analyze program measurements to make sure any proposed changes have the desired effect. An easy way to measure how time is being utilized in a program is to include statements that provide you with a count of the number of times each statement is executed.

3. Writing a program that is more debuggable.

(a) Develop the solution to your problem first in algorithm form. When that seems correct, translate it into program statements in whatever language you wish.

(b) Use control-flow instructions that proceed generally from beginning to end without hard-to-follow branching. Avoid GO TO's completely if you can.

(c) Write and test your program in small modules. Use subprograms freely.

(d) Don't patch up bad code. If it's bad, re-do it. (This usually means that you rewrite that portion of the algorithm that resulted in bad code.)

(e) Be extra careful of actions that may be executed more often or less often than is wanted. Such errors are usually the result of faulty loop initialization or incorrect placement of loop incrementing.

(f) Avoid comparisons of real (noninteger) numbers for strict equality. The method of generating one or both of the numbers being compared may introduce minute errors in the internal representation of the number, thus making for illegitimate inequality.

(g) Watch for subscripts that go beyond the bounds of acceptable values.

(h) Use variable names that mean something in the context of the program.

4. Tips for testing a program.

(a) Check your program carefully at those places where special actions could happen, as, for example, if a loop is not performed at all. If that is a legitimate possibility and your program is in For-Tran, use IF statements to branch around the loop.

(b) Test your program at allowable minimum and maximum values. Such boundary points are most apt to cause problems.

(c) Don't stop testing when you find one bug. Keep looking for more.

5. Writing well-structured programs.

(a) Use indentation to set off the beginning and ending of a DO loop.

(b) Use different depths of indentation to set off nested loops.

(c) Use indentation to identify IF–THEN–ELSE IF–ELSE structures.

(d) Use different depths of indentation to set off nested IF–THEN–ELSE IF–ELSE structures.

(e) Use indentation in comments to set off distinct topics.

(f) Avoid using GO TO statements entirely if possible.

SUMMARY

This appendix lists some helpful tips for developing good programs. Practice applying as many of them as you can as often as you can. Should you want a more detailed discussion of these topics, many of them are covered in either or both of the references mentioned at the beginning of this appendix.

A Comparison of Five Versions of ForTran

Listed here are some common features of ForTran whose handling in a given version is apt to be different from its treatment in another version. An X under a column heading means that the feature in that version is treated as described. Otherwise the feature is treated differently or is not available in that version. This list is not as complete as it might be. There are other less common features that vary from one version to another. Consult the appropriate reference manuals for further information.

Feature	1966 Standard ForTran	1977 Standard full set	1977 Standard subset	WATFIV	MNF
1. Free-form (list directed) input/output using asterisk. READ*,v1,v2,...,vn PRINT*,v1,v2,...,vn		X			X
2. Free-form (list directed) input/output without asterisk. READ,v1,v2,...,vn PRINT,v1,v2,...,vn				X	
3. Block IF		X	X	X	X
4. ELSE IF		X	X		X
5. Mixed-mode ELSE IF expressions		X	X	X	X
6. READ nn, list (nn is a reference number of a FORMAT statement).	X	X			

Feature	1966 Standard ForTran	1977 Standard full set	1977 Standard subset	WATFIV	MNF
7. Apostrophes to enclose character strings.		X	X	X	X
8. Character data by using H-format specification.	X	X	X	X	X
9. Asterisks to enclose character strings.					X
10. Arithmetic expression in output list.		X		X	X
11. Number of allowable subscripts.	3	7	3	3	3
12. Any acceptable integer allowed as arithmetic expression for subscript.		X	X		X
13. Any acceptable integer allowed as arithmetic expression for subscript except a subscripted variable or function.				X	
14. DO-loop index may be either integer or real.		X			
15. DO-loop parameters may be positive or negative; integer or real; constants, variables, or arithmetic expressions.		X			X
16. If DO-loop termination parameter value is greater than initial parameter value, loop will not be executed even once.		X	X		X
17. Comma in DO statement after statement number allowed but optional, e.g., DO 40,I=1,N		X	X		X
18. Multiple assignment statement involving arithmetic expression, e.g., V1=V2=V3=expression.				X	
19. Multiple assignment statement involving only constant, e.g., V1=V2=V3=constant.		X	X	X	X
20. Allowable to end a DO loop with a transfer statement.				X	
21. Comma optional before index in computed GO TO, e.g., GO TO(s1,s2),I		X	X		X
22. Any integer arithmetic expression allowed as index for computed GO TO.		X			X

Feature	1966 Standard ForTran	1977 Standard full set	1977 Standard subset	WATFIV	MNF
23. READ using an H format.	X				
24. OPEN, CLOSE, and INQUIRE statements.		X			
25. CHARACTER data type.		X	X	X	
26. Concatenation operator.		X			
27. Substrings.		X			
28. PROGRAM statement.		X			X
29. Alternate entry and return points for subroutines.		X			
30. Variable as the array size specifier in DIMENSION or type-declaration statements, e.g., DIMENSION X(N) INTEGER A(K)		X		X	X

APPENDIX D

ForTran Library Functions

Following is a list of library functions that are included in ForTran 77. Abbreviations used are these:

R   means real value (floating point).
I   means integer value (fixed point).
D   means double precision real value.
C   means complex value, $X+iY$.
CH means character value.
L   means logical value.

Symbolic name	Number of arguments	Type of arguments	Type of result	Description and limitations
ABS	1	R	R	Absolute value of $X = \begin{cases} X \text{ if } X \geqslant 0 \\ -X \text{ if } X < 0 \end{cases}$
IABS	1	I	I	Same as for ABS but in integer form.
DABS	1	D	D	Same as for ABS but in double precision.
CABS	1	C	R	$CABS(X) = \sqrt{real^2(X) + imag^2(X)}$
FLOAT	1	I	R	Converts integer to real.
IFIX	1	R	I	Converts real to integer.
AINT	1	R	R	Produces the largest signed integer less than or equal to the argument.
INT	1	R	I	Same as for AINT but result is integer.
IDINT	1	D	I	Same as for INT but argument must be double precision.
AMOD	2	R	R	$AMOD(X,Y) = X - INT(X/Y) *Y$.

Symbolic name	Number of arguments	Type of arguments	Type of result	Description and limitations
MOD	2	I	I	Same as for AMOD but with integers.
DMOD	2	D	D	Same as for AMOD but with double precision.
AMAX0	≥2	I	R	Finds the largest of the specified integer arguments and gives real results.
AMAX1	≥2	R	R	Same as AMAX0 but arguments are real.
MAX0	≥2	I	I	Same as AMAX0 but result is integer.
MAX1	≥2	R	I	Same as AMAX0 but arguments are real and result is integer.
DMAX1	≥2	D	D	Same as AMAX0 but arguments and result are double precision.
AMIN0	≥2	I	R	Finds the smallest of the specified integer arguments and gives real results.
AMIN1	≥2	R	R	Same as AMIN0 but arguments are real.
MIN0	≥2	I	I	Same as AMIN0 but result is integer.
MIN1	≥2	R	I	Same as AMIN0 but arguments are real and result is integer.
DMIN1	≥2	D	D	Same as AMIN0 but arguments and result are double precision.
SIGN	2	R	R	$SIGN(X,Y)$=sign of Y with $|X|$.
ISIGN	2	I	I	Same as SIGN but with integers.
DSIGN	2	D	D	Same as SIGN but with double precision values.
DIM	2	R	R	$DIM(X,Y) = X - AMIN1(X,Y)$
IDIM	2	I	I	Same as DIM but with integers.
DDIM	2	D	D	Same as DIM but with double precision values.
SNGL	1	D	R	Converts double precision real argument to the most significant single precision.
DBLE	1	R	D	Converts single precision real argument to double precision.
REAL	1	C	R	Produces the real part of a complex argument.
AIMAG	1	C	R	Produces in real form the imaginary part of a complex argument.
CMPLX	2	R	C	Converts two real arguments into complex form.
EXP	1	R	R	$EXP(X)=e^X$
DEXP	1	D	D	Same as EXP in double precision.
CEXP	1	C	C	$CEXP(Z)=e^X(COS(X) + i\text{*} SIN(Y))$ where $Z = X + i \text{*} Y, i = \sqrt{-1}$.
ALOG	1	R	R	Produces the natural log of a real, non-negative argument.

Symbolic name	Number of arguments	Type of arguments	Type of result	Description and limitations
DLOG	1	D	D	Same as ALOG in double precision.
CLOG	1	C	C	CLOG(Z)=.5*ALOG(X²+Y²) + i*ATAN (Y/X), where Z=X + i*Y.
ALOG10	1	R	R	Produces the common log of a real, non-negative argument.
DLOG10	1	D	D	Same as ALOG10 in double precision.
SQRT	1	R	R	Produces the square root of a real argument.
DSQRT	1	D	D	Same as SQRT in double precision.
CSQRT	1	C	C	$CSQRT(Z)=\sqrt{X^2+Y^2}(COS(T) + i*SIN(T))$ where T=(1/2)*ATAN(Y/N) and Z=X + iY.
SIN	1	R	R	Produces the trigonometric sine of a real argument where the argument is in radians.
DSIN	1	D	D	Same as SIN in double precision.
CSIN	1	C	C	CSIN(Z)=SIN(X) * cosh(Y) + i*COS(X)* sinh(Y) where Z=X + iY.
ASIN	1	R	R	Produces in radians the arcsine of a real argument.
DASIN	1	D	D	Same as ASIN in double precision.
COS	1	R	R	Produces the trigonometric cosine of a real argument where the argument is in radians.
DCOS	1	D	D	Same as COS in double precision.
CCOS	1	C	C	CCOS(Z)=COS(X)cosh Y− i*SIN(X)* sinh Y where Z=X + iY.
ACOS	1	R	R	Produces in radians the arccosine of a real argument.
DACOS	1	D	D	Same as ACOS in double precision.
ATAN	1	R	R	Produces in radians the arctangent of a real argument.
DATAN	1	D	D	Same as ATAN in double precision.
ATAN2	2	R	R	Produces in radians the arctangent of (X/Y) where Y is not zero.
DATAN2	2	D	D	Same as ATAN2 in double precision.
TANH	1	R	R	Produces the hyperbolic tangent of a real argument.
DTANH	1	D	D	Same as TANH in double precision.
ICHAR	1	CH	I	Converts character code of a character argument to its integer equivalent.
CHAR	1	I	CH	Converts an integer argument to its equivalent character.

Symbolic name	Number of arguments	Type of arguments	Type of result	Description and limitations
LEN	1	CH	I	Produces the length (number of characters) of a character-string argument, including blanks.
INDEX	2	CH	I	INDEX(CH1,CH2)=position in character string CH1 where substring CH2 begins.
LGE	2	CH	L	LGE(CH1,CH2)=true or false depending on whether or not the code for CH1 is greater than or equal to the code for CH2.
LGT	2	CH	L	Same as LGE but the comparison is "greater than."
LLE	2	CH	L	Same as LGE but the comparison is "less than or equal."
LLT	2	CH	L	Same as LGE but the comparison is "less than."

Although the preceding list of library functions includes all those in ForTran 77, many versions of ForTran include others. We mention five of the most common. Before you use these in a program, check that your system includes them.

Symbolic name	Number of arguments	Type of argument(s)	Type of result	Description and limitations
RAND	0 or 1	I	R	Argument (if required) determines the starting point of a sequence of pseudo-random numbers uniformly distributed between 0 and 1. A negative argument (if required) causes a different sequence every time.
AND	$\geqslant 2$	L	L	Produces a result in logical form when the logical AND is performed on two or more arguments.
OR	$\geqslant 2$	L	L	Produces a result in logical form when the logical OR is performed on two or more arguments.
COMPL	1	L	L	Produces a result in logical form when the logical COMPLEMENT is performed on the argument.
SHIFT	2	R,I	R	SHIFT(X,N) causes a circular shift to the left of N bits in the internal representation of X. SHIFT(X,−N) causes an arithmetic shift to the right of N bits in the internal representation of X.

Many versions of ForTran make available some library subroutine subprograms. To use these, a CALL statement must be used in your program. Here are four of the most common library subroutines. Before using these in a program, check local availability.

EXIT   Terminates the program.

DATE   Produces the current date from the computer's internal clock.

IDATE   Produces the year and day of the year as a single five-digit number, YYDDD, where YY refers to the two-digit year designation and DDD the three-digit day-of-year.

CLOCK   Produces the time of day as HHMMSS where HH is the two-digit hour designation, MM the two-digit minute designation, and SS the two-digit second designation.

APPENDIX E
Flowcharting

In this book we have emphasized problem-solving techniques and, as a step-by-step process of solving a problem, the writing of algorithms. Many computer scientists like to see algorithms displayed in the form of a diagram called a *flowchart* rather than as pseudocode used throughout this book. Thus, a flowchart is a diagram of the steps one expects to take in solving a given problem, showing the sequential relationships among those steps. Flowcharts use special symbols with which we must become familiar. Also, there are certain general procedures that make it easier to know how to construct a flowchart. Someone has called a flowchart "a blueprint of an algorithm for solving some problem." In this appendix we present some information that should help you learn how to develop such blueprints. We begin by considering the special symbols of flowcharting.

FLOWCHARTING SYMBOLS

There are quite a number of symbols (twenty-five or more) used by people who construct flowcharts. However, there is no common agreement on the meaning of many of these symbols, so in an introductory book such as this one it is appropriate to limit the number of symbols used. We shall consider six symbols, the ones on which there is virtually complete agreement as to meaning.

1. *Terminal symbol*: This symbol, an ellipse, is used to indicate the beginning or end of a solution path. To specify which of the two is being

designated, words such as START, FINISH, or STOP are written inside the symbol as follows:

2. *Input-output symbol*:

This symbol, a parallelogram, indicates that information not determined by steps within the plan itself (input) is brought into the solution plan. It is also used to indicate the displaying of results in some form useful to a human being (output). Here are some examples:

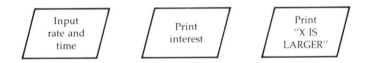

As these examples show, the precise operation being executed (input or output) is apparent from information written inside the parallelogram.

3. *Computation symbol*:

The rectangle is used to indicate an arithmetic process or the manipulation of information or objects previously available in the solution plan. As with the other symbols, specific operations are written inside the rectangle.

Here are some samples:

| Compute INTEREST by multiplying RATE by TIME | Let Z = R*(A+B) | Run cold water over eggs in kettle held over sink |

As you can see from the three examples, the size of the symbol has no effect on its meaning. All three rectangles, regardless of their size, refer to the execution of an arithmetic process or of manipulation of objects. For any flowcharting symbol, its size is determined by how much is to be written inside it. The shape of the symbol determines the kind of operation being represented.

4. *Decision symbol*:

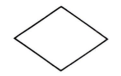

The diamond is used to show that a test for the existence of a specified condition is to be made and the results of that test determine the next path to take. The test is usually indicated by writing inside the diamond a question that can be answered yes or no, or by writing inside the diamond a statement that is either true or false. Here are some examples:

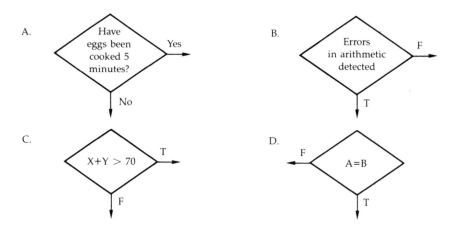

Note the following facts from the preceding examples:
a. The size of the diamond has no effect on its meaning.
b. Anything may be written inside the diamond as long as it indicates the testing for some condition.
c. Exits from the diamond are at any of the vertices although typically the top vertex is used for entering the diamond.

In example A, if the answer to the question written inside the diamond were "no," the path to follow in the flowchart would be down from the diamond. If the answer had been "yes," the path to follow would be to the right. In example C, if the sum of two quantities referred to as X and Y were greater than 70, the path to follow in the flowchart would be to the right. If not, the path to follow would be downward.

5. *Connector symbol:*

The circle is used to indicate continuation *to* some part of the flow-chart not in the immediate vicinity or *from* some part of the flowchart not in the immediate vicinity. For example, if the flowchart were so large that it had to be written on more than one page, then a connector symbol would be used at the end of one page and the beginning of the next. A terminal symbol would not be used in this case because such a situation does not constitute the end of the entire flowchart. As with other symbols, size has no effect on the meaning of the symbol. Typi-cally, a number or letter or combinations thereof are written inside the circle. To properly show the continuation to and from, two circles would have the same information inside them. Consider these exam-ples:

A.   B.

It should be apparent from these examples that there must always be at least two connector symbols with the same information inside. One symbol is the exit point from some path in the flowchart and the second symbol is the continuation of that logical path at some other location in the flowchart. This will be more clearly apparent as we consider entire flowcharts.

6. *Replacement symbol:*

This symbol, the left-pointing arrow, is used inside the computation symbol to indicate the replacement of information in the location named at the head of the arrow by information in the location named at the tail of the arrow. Here is an example:

The symbol means to replace the contents of a storage location called A by the result of adding the contents of a location named X to three times the contents of a location named Y. Note that only a single memory location is given at the head of the arrow, though more than one location with appropriate arithmetic operation symbols may appear at the tail side of the arrow. You will notice in examples of complete flowcharts that arrows are used to connect the various flow-charting symbols. These arrows are not replacement symbols. Their function is to indicate the direction of the path along which to proceed in order to arrive at the completion of the task. Only when a left-pointing arrow appears inside a computation symbol is it to be interpreted as a replacement symbol.

DEVELOPING FLOWCHARTS

Now that we know some fundamental symbols of flowcharting, we turn our attention to the task of writing a flowchart.

Problem E.1. Write an algorithm for cooking two five-minute eggs.

Algorithm E.1:

1. Place kettle on burner and put three inches of water in it.
2. Turn on burner and heat water to boiling.
3. Carefully place two eggs in the boiling water and note the time.
4. Keep water boiling on turned-down burner.
5. Have eggs been in boiling water five minutes? If not, go to step 4. If so, go to step 6.
6. Run cold water over eggs in the same kettle held under sink faucet.
7. Remove eggs from kettle.
8. Stop.

From this example it may be surmised that a good algorithm consists of (1) a finite number of steps performed a finite number of times, (2) steps that are not ambiguous, and (3) a procedure that actually accomplishes the desired result. It should also be clear from this example that an algorithm can be written for a nontechnical kind of task, one for which a computer would not likely be used. However, since this book is concerned with computers, we will consider a second example more apt to be solved using a computer.

Flowchart E.1

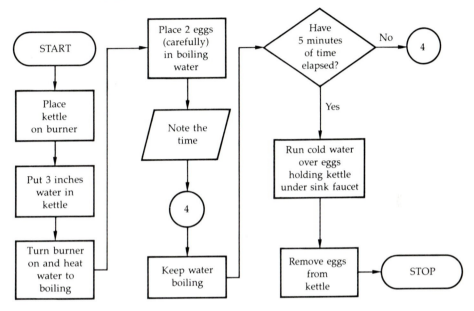

As you study Flowchart E.1, you will notice that this method of describing a solution to a problem shows clearly the relationships among the several steps. The different symbol shapes make it obvious when specific kinds of tasks must be done in relation to other tasks. You will notice that lines joining the various symbols are drawn either vertically or horizontally, never diagonally. This is a matter of custom but probably does have its foundation in the fact that the resulting diagram is more readable to more people than it otherwise would be. Note also that the symbols are carefully drawn. Of course, you would expect neat illustrations in a printed book, but even hand-drawn flowcharts should have this feature. This is possible because a template, a plastic sheet with cut-outs of the various shapes, is used. The person constructing the flowchart uses the template to trace out whatever symbol is desired at each point in the diagram.

Problem E.2. Write an algorithm for finding the sum of the products of an undesignated number of pairs of numbers such that each pair of numbers is read from a card and the product computed. For each subsequent pair of numbers do the same operations and keep a running accumulated sum of the products. Continue this process until a card with two zeroes is encountered, at which time write the accumulated sum on a piece of paper.

Algorithm E.2:

1. Write zero in a place reserved for writing the accumulated sum.
2. Read a pair of numbers from a card.
3. Are both numbers zero? If so go to step 7. If not, go to step 4.
4. Compute the product of the two numbers.
5. Add this product to the last number written showing the accumulated sum, and replace the last accumulated sum with this new sum.
6. Go to step 2.
7. Write the accumulated sum.
8. Stop.

Flowchart E.2

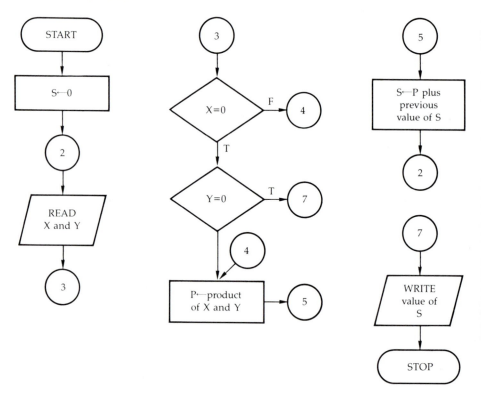

As you refer to Flowchart E.2, let us examine some of its characteristics. Notice the use of connector symbols. In order to fit the flowchart neatly in the available space the flowchart was divided into three vertical sections on the page. Connector symbols may be used in any way convenient to the flowchart developer. Their function is to make it possible to provide branching from one place to another and to enable reference to be made to a particular part of the flowchart. For example, refer to connector ④. (Notice how easily a connector symbol makes it to pinpoint a specific part of the flowchart.) Look at the computation symbol (rectangle) following connector ④. A replacement symbol is used in that computation symbol to indicate that a unit of information called P is computed by multiplying the value of X by the value of Y. As is the case in Algorithm E.2, there is one loop in Flowchart E.2. Connector ② and the five nonconnector symbols following it constitute the loop. In this example, transfer out of the loop occurs when both X and Y are zero—that is, at the second diamond following connector ③. Input data used in this way are referred to as trailer data. The purpose of trailer data is to cause an exit from a loop which includes an input step. Recall that a loop makes it possible to repeat steps already specified so that tedious rewriting of those steps can be avoided.

Problem E.3.   Read a set of data related to a classroom where each student is given an identification number and three test scores. For each student, determine whether the average of the three test scores is more than 65. If it is, print the student identification number and the word "pass." If the average is 65 or less, print the student identification number and the word "fail." Continue reading student data until the identification number 99999 is read, then stop.

Flowchart E.3

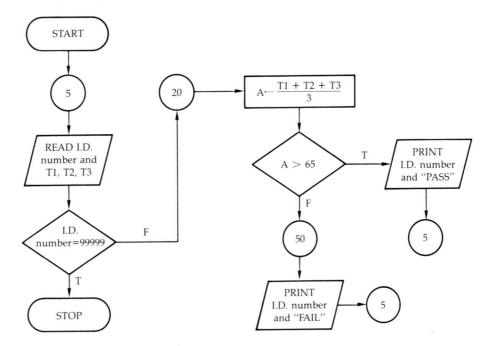

Flowchart E.3 demonstrates how an algorithm with a few steps can handle a large amount of data. In fact, this flowchart displays the same logical plan for solution whether there are data for 10 students or 100 students. The key unit of data that will bring the user of Flowchart E.3 to the STOP symbol is the encountering of a student identification number of 99999. Remember that input data used specifically to stop any further input are called trailer data, or sometimes "dummy" data. The assumption in this case is that no real student will have been assigned an identification number of 99999, so when that number is sensed no attempt is made to process those data and no more data are read.

Let's test Flowchart E.3 with some trial data. Trial data are typically small sets of data but contain enough variety to try out all branches in the flowchart. Suppose we use the following:

I.D. Number	T1	T2	T3
10005	75	60	60
10010	58	63	67
10020	80	85	88
10040	65	58	67
99999	0	0	0

Now we refer to Flowchart E.3 and when data are read from the preceding table, we assume reading proceeds from left to right beginning with the top row, then row two, and so on. The first symbol to specify an action is the input symbol following connector ⑤. There we are asked to read four numbers, I.D. Number, T1, T2, and T3. Draw four boxes having those names, like this:

I.D. Number	T1	T2	T3

Taking the data in the order previously described and placing them in appropriate boxes we get

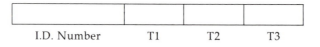

10005	75	60	60
I.D. Number	T1	T2	T3

The next action required in the flowchart is to compare the contents of the box "I.D. Number" with the number 99999. In this case they are not equal so we proceed to connector ⑳. Following that is a computation symbol calling for the establishment of a box named "A," which will contain the average of the numbers in T1, T2, and T3. When this computation is complete, we have five boxes of information, as follows:

10005	75	60	60	65
I.D. Number	T1	T2	T3	A

Continuing in the flowchart, the next step is a comparison of the contents of box "A" with the number 65. In the present situation the contents of "A" is 65, hence not greater than 65, so we proceed to connector (50) and on to the output symbol, which asks us to print the contents of box "I.D. Number" and the word "FAIL." The results would be a line of printed information (output) like this:

```
10005      FAIL
```

Following the output symbol, the flowchart calls for a branching to connector (5). That brings us back to the beginning and we repeat the input operation, this time using as data the numbers 10010, 58, 63, and 67, respectively. When that operation is complete, the boxes described previously would now have these contents, since we assume the former contents have been erased.

10010	58	63	67
I.D. Number	T1	T2	T3

Proceeding in the flowchart to the computation of the contents of box A yields the five boxes as follows:

10010	58	63	67	62.67
I.D. Number	T1	T2	T3	A

Note that whenever a box is assigned new contents, we assume previous contents have been erased first. This time, again, when the number in box A is compared with 65, the next step is to execute the output operation following connector (50). We have now produced two lines of output:

```
10005      FAIL
10010      FAIL
```

The flowchart next specifies a return to connector ⑤. As before, we read the next line of information from the table of data, after which the four boxes have the following contents:

10020	80	85	88
I.D. Number	T1	T2	T3

Box I.D. Number does not contain 99999, so box A is computed to give it a value of 84.33. Continuing in the flowchart calls for the comparison of the contents of box A with 65. This time A is greater, so the flowchart specifies the printing of the line:

```
10020    PASS
```

All of the printed output to this point consists of three lines, as follows:

```
10005    FAIL
10010    FAIL
10020    PASS
```

Following the steps in the flowchart returns us to connector ⑤ and to the input symbol where, this time, we read the fourth line of input data. The four boxes would look like this:

10040	65	58	67
I.D. Number	T1	T2	T3

Again, since box I.D. Number is not equal to 99999, we compute the number 63.33 to be placed in box A. Next, when the number in box A is compared with 65, the flowchart directs us to connector ㊿ and on to the printing of the line

```
10040    FAIL
```

Returning to connector ⑤ and on to the input symbol results in the four familiar boxes having these contents:

99999	0	0	0
I.D. Number	T1	T2	T3

Comparing the contents of box I.D. Number with 99999 indicates this time that equality exists, so the flowchart sends us to the STOP symbol, telling us the job is completed. All the output generated by this flowchart, given the table of data shown earlier, are these four lines:

```
10005        FAIL
10010        FAIL
10020        PASS
10040        FAIL
```

If, in this example, a teacher were providing test results for four students, it is clear that the output would consist of a list of students identified by unique numbers and their grades of pass or fail. It should also be apparent that the same flowchart could have been used to process data for 30 students, or 100 students, or any number of students. The example used data for only four students to keep it reasonably short. It should be pointed out, further, that the simple grading scheme of "pass" or "fail" could rather easily be changed to an A, B, C, D, F arrangement by inserting more decision symbols with appropriate comparisons, each followed by an appropriate output symbol.

SOME HELPS IN CONSTRUCTING FLOWCHARTS

Many people, when first introduced to flowcharts, find their development a frustrating experience. This reaction probably has to do with the fact that the process is new, and seems to have little, if any, relationship to previous experiences. To help you avoid that unpleasantness often experienced by first-time students of flowcharting, the following list of procedures is given:

1. Know the flowcharting symbols and their meanings.
2. Remember that a flowchart is simply a method of describing a plan that accomplishes a task.
3. Before starting the flowchart, make sure you have a precise formulation of the task to be done. You must know what job it is that you are attempting.
4. Before starting the flowchart identify the data available at the start of the task and the information to be produced by completing it.
5. Before starting the flowchart, list any facts, formulas, relationships, and so forth that seem to apply to the task. It is better to have too many of these than too few, since you can always discard those concepts that later turn out to have no significant bearing on the task to be done.

6. Begin the construction of the flowchart introducing known data as they are needed to generate further information. Proceed to the computation of new information from that which was previously introduced, in turn, using computed information to produce still more information. Throughout this process, refer to your list of facts, formulas, and so on.

7. Use enough steps to keep the transition from step to step smooth, but not so many as to make the final result incomprehensible because of too much detail.

8. Keep in mind the task to be accomplished, so that the flowchart terminates when the job is done.

9. Whenever possible, test your flowchart with sample data, selecting the sample so as to try out all paths in the flowchart.

10. When errors in logic appear at the time of testing your flowchart, make appropriate changes and try the sample data again until the flowchart properly handles it.

These procedures should prove helpful as you learn about flowcharting. However, there is no substitute for experience. You are, therefore, urged to construct your own flowcharts for all examples given in this chapter without referring to the corresponding flowchart in the book. Please remember that flowcharts constructed by two different people will probably not look exactly the same. We all use different logical processes in arriving at given conclusions, and this fact will usually demonstrate itself in different flowcharts. There is no reason why any two flowcharts solving a given problem should be the same. The important consideration is that both of them provide a plan for accomplishing the given task and communicate that plan to anyone willing to use them.

Selected references

Bohl, Marilyn, *Flowcharting Techniques,* Science Research Associates, Chicago, 1971.

Schriber, Thomas J., *Fundamentals of Flowcharting,* John Wiley and Sons, New York, 1969.

APPENDIX F

Answers to Selected Exercises

CHAPTER 2

2. c. punctuation mark   **e.** correct   **h.** correct   **j.** correct
l. too long   **o.** correct   **q.** begins with numeral
t. special character   **w.** punctuation mark   **y.** correct   **bb.** correct
ee. correct   **hh.** punctuation mark   **jj.** correct   **ll.** begins with $
oo. correct   **ss.** correct   **vv.** correct   **yy.** correct   **b1.** correct
e1. correct

5. c. NUM1=NUM2/(−NUM3)
f. READ*,A,B (A+B is an invalid variable name)   **i.** correct
l. correct   **o.** A200=A100*2 (Invalid variable names)   **r.** correct
u. PRINT*,'A+B','=',A+B (Equal sign in a PRINT is a string constant so must be enclosed in single quotes.)
x. A=B*AX*(AX+1) (Missing an arithmetic operator)
aa. READ*,X (Must be at least one variable in a READ)   **dd.** A=A**(−B)
gg. correct   **jj.** correct (Blanks ignored on either side of OR)
ll. CHARACTER A*2,B*3   **nn.** CHARACTER A*2,B,C*3   **qq.** correct
tt. correct   **ww.** correct (If A,B, and C are all declared type character)
zz. A=A(1:10) (If A is declared type character with length at least 10)

6. c. W=(−B+SQRT(B**2−4*A*C))/(2*A)
e. P=37*((A**2)**(1./3.)+(A**3)**.5)/((A*(B**.5))**.5+A)
g. Z=(A**(B**2)+B**(A**2))*(A+B+A*B/(A+B))

7. a. 125.0   **d.** 0 (The integer quotient 5/9 is zero so the entire product is zero.)
g. 0 (The integer quotient 10/200 has the value 0.)   **j.** 50.05   **m.** 1.0

8. a. WHAT IS YOUR STUDENT NUMBER?
YOUR STUDENT NUMBER IS 810001

9. b. LINE 1: PRINT*, 'HEADING LINE'
LINE 5: PRINT*,ONE,'+',TWO,'EQUALS',THREE (EQUALS is undefined and plus sign is a string constant and must be enclosed in quotes)

14. PROGRAM FOLD
     REAL SOFAR
     INTEGER COUNT
     COUNT = 0
     SOFAR = 0.015625
  20 IF(SOFAR.LT.36.0) THEN
         SOFAR = SOFAR * 2
         COUNT = COUNT + 1
         GOTO 20
     ENDIF
     PRINT*, 'FOLD PAPER',COUNT,'TIMES TO GET HEIGHT OF + 36 INCHES.'
     STOP
     END

CHAPTER 3

1. **h1.** Simple logical IF   **h4.** Compound logical IF
    k. Sentinel data and counters
2. **b.** IF(A.GT.B)GOTO 10 (The number 10 alone is not a valid statement.)
3. **a.** IF(A+B.GT.A)C=A+B (No arithmetic operation permitted left of =.)
    d. IF(A+B.LE.A)GOTO 10 (EL is invalid relational operator)
    g. correct (A must be a character variable of size 2.)
4. **c.** IF(A*B.GT.A+B.AND.A−B.GT.A+B)A=B (Needs relational expression to left of .AND.)   **f.** IF(A.OR.B.OR..NOT.C.OR.D)GOTO 10 (Needs relational operator between first relational expression and the .NOT.)   **i.** correct   **l.** correct (THEN A is a valid variable name.)   **o.** correct (GOTO is a valid variable name.)
5. **c.** correct   **f.** LINE 8: IF statement should be an ELSEIF statement
    i. correct
6. **c.** The output is a list of student ID numbers with their corresponding score and grade.   **g.** If N>=7, prints 'THERE ARE (N-6) SEVENS IN N' otherwise prints 'THERE ARE NO SEVENS IN N'. Note that from the program logic point of view lines 6 and 7 are redundant and reference number 15 could be placed on line 8.
8. correct
9. **b.** Logic error: see problem 10, line 7, ELSEIF(L.EQ.LLL)THEN
    e. Logic error: see problem 13
12. Insert after LINE 7:
     PRINT*,A,'*X*X+',B,'*X+',C'=0'
     (Delete lines 10 and 15.)
     Insert after LINE 16:
     ELSE
     PRINT*,'COMPLEX ROOTS TO QUADRATIC EQUATION'
     STOP
15. Replace lines 5-8 with:
         M = 10
         STR(1:1)=ST(M:M)
      10 M=M−1
         STR = STR(1:10−M)//ST(M:M)
         IF(M.NE.1)GOTO 10

CHAPTER 4

1. **a.** procedure   **d.** See page 194.   **g.** See page 195.
    j. See pages 198-200.   **m.** See page 200.

3. c. 1. Initialize A,B,&C to one and count to zero
        2. Find the square of A and call it X
        3. Find the value of (B**2 + C**2) and call it Y
        4. If X = Y then increment count by one and output A,B,C
        5. If count = 25, goto step 10
        6. If C <=B, increment C by one
        7. Elseif B<=A, increment B by one
        8. Else increment A by one
        9. Goto step 2
       10. Stop algorithm
    e. 1. Input the real number, NUM
        2. Initialize Exp = 0
        3. If(NUM< 0) goto step 8
        4. If (NUM < 10.0) goto step 12
        5. NUM = NUM/10.
        6. Increment EXP by one
        7. Goto step 4
        8. If (NUM > 1) goto step 12
        9. NUM = NUM * 10
       10. Decrement EXP by one
4. b. 5. 10.
8. ALGORITHM:
     1) Initialize program
     2) Declare variables:
            REAL a,b,c,d,x,xl,x2
     3) Read in values:
            READ*,a,b,c
     4) IF(a.EQ.0.0) THEN
            x=-c/b
     5) ELSE
            d=b**2-4*a*c
     6)   IF(d.GT.0)THEN
                x1=(-b+SQRT(d))/(2*a)
                x2=(-b-SQRT(d))/(2*a)
     7)   ELSEIF(d.EQ.0)THEN
                x1=x2=(-b)/(2*a)
     8)   ELSE
                PRINT*,'THERE ARE NO REAL ROOTS'
     9)   ENDIF
    10) ENDIF
    11) Stop algorithm
    PROGRAM:
    PROGRAM QUAD
    REAL A,B,C,D,X,X1,X2
    READ*,A,B,C
    IF(A.EQ.0.0)THEN
        X = -C/B

(continued on next page)

```
ELSE
    D = B**2−4*A*C
    IF(D.GT.0.0) THEN
        X1 = (−B+SQRT(D))/(2*A)
        X2 = (−B+SQRT(D))/(2*A)
        PRINT*,'THE REAL ROOTS ARE',X1,X2
    ELSEIF(D.EQ.0.0)THEN
        X1=(−B)/(2*A)
        X2=X1
        PRINT*,'THE REAL ROOTS ARE',X1,X2
    ELSE
        PRINT*,'THERE ARE NO REAL ROOTS'
    ENDIF
ENDIF
STOP
END
```

CHAPTER 5

1. **a.** An implied loop may be used in a read,write,or print statement.
d. If a DIMENSION statement is used, it must be before any other program statement in which the subscripted variable appears.   **g.** There may be any number of levels of nesting in a nested DO loop.   **k.** Transfer into a DO loop can only be done through its DO statement.   **n.** One may use any number of DO loops in a given program.   **q.** true   **s.** There may be a DO statement and a CONTINUE statement in order to have a DO loop.
v. Dimension statements specify the size of the subscripted variables.
x. true

2. **a.** DIMENSION A(10),B(20)   **d.** DIMENSION A(10,20),B(5,6,10)
g. correct   **k.** correct (Negative subscripts are legal.)
m. CHARACTER A(5),B,C(10)*10
p. correct (Assume A has an integer data type.)
s. correct (Assume I has an integer value.)   **v.** correct (Assume I has an integer data type.)   **aa.** PRINT*,A(10,20),B(5) (Contradicting number of subscripts for A)   **dd.** correct   **gg.** WRITE(1,*),A,(B(I,J),J=1,5) (Only one assignment per line in ForTran 77)   **jj.** correct (Initializes only the 5th element of array B)   **mm.** correct   **pp.** correct   **ss.** DO 9 J = 1,5

3. **a.** LINE 2: READ*,(A(I),I=1,10)   (A(6) through A(10) are undefined)
LINE 4: 10 SUM = SUM + A(I)   (Once A is designated an array, it must have a subscript.)
d. DO loop creates zero index, one way to correct it is:
DO 10 I - 1,9 10 A(I+1,1) = A(I,2)
g. The subscript range for B is 1 through 12. When I=0,B(0) is out of range. One way to correct this is:
LINE 9: B(I+1) = A(0,I)
LINE 10: B(I+7)= A(1,I)
j. When I=1,B(1,0) and B(0,1) in LINE 4 are out of range.
One way to correct this is: LINE 3: DO 10 I = 2,10

4. **a.**   1 2 3 4 5 6 7 8 9 10
2 4 6 8 10 12 14 16 18 20
.
.
.
10 20 30 40 50 60 70 80 90 100
prints multiplication table up to 10

c.

5. **b.** The program does not sum the prime factors together. One way to correct this
    is: LINE 6: IF(N/J.EQ.N/FLOAT(J)) SUM = SUM + J
    LINE 9: PRINT*,'THE SECOND PERFECT NUMBER IS',SUM
    d. The program given begins printing the least significant digit first.
    The following changes are suggested:
    DO loop should be:
    DO 10 I = 10,1,−1
       X = N/(10**I)
       IF(X.NE.O)PRINT*,X
    10 CONTINUE

CHAPTER 6

1. **a.** It is possible to write a complete ForTran program without using a FORMAT
    statement.   **d.** It is permissible to use both the H-format specification and
    quotation marks in the same FORMAT statement.   **g.** true   **j.** true
    m. true   **p.** true   **s.** 10 READ(1,20)A,B (Format number and statement
    number cannot be the same.)
    v. FORMAT statement 10 as referenced in the PRINT statement
       PRINT 10,A,B,'BE CAREFUL'
    needs to include format specifications for the variables A,B and character string.
    x. The statement READ*,A,B will cause the input of one or more data cards on a
    batch computer system.
2. **a.** correct (23 is a FORMAT reference number.)   **d.** correct (READ 20 is a legal
    variable name.)   **h.** 10 FORMAT(1X,F10.6,I6) (I must correspond to I-format
    specification in the FORMAT statement.)   **j.** 20 FORMAT(F4.2,I2,F6.2) (or else
    the contents of records THIS & IS will be reported as if they are character
    data.)   **m.** READ(1,10,END=20)A,B (END must not reference a
    FORMAT.)   **p.** correct (PRINTS is a legal variable name.)
    t. 20 FORMAT(2X,F8.2) (READ 30 is a legal variable name and has a real data type
    by default.)   **v.** PRINT 20,A,B   20 FORMAT('IS THIS A',F6.2) (or else
    the contents of records A & B will be reported as if they were character data.)
    y. PRINT*,PRINT5,READ5 (PRINT5 & READ5 are legal variable names.)
    bb. X=10.
        Y=10.
    dd. 10 FORMAT(F3.2,I6)
            PRINT 20,NOT,FORMAT
        20 FORMAT(I6,F6.2)
    (Format specifications must match variable types.)
    ff. B=3.0 (B must be initialized to real value if it is to be added to A.)
        PRINT*,A,A+B
    hh. READ(6,10,END=20)A,B (END must not reference a FORMAT and no
        arithmetic permitted in READ.)

kk. READ 10,A,B
PRINT 10,X,B
oo. CHARACTER *6 X (Need a variable for the A-format specification)
WRITE(6,20) X

3. **a.** PRINT 10,A
10 FORMAT(/61X,F6.2)
d. PRINT 1
PRINT 2
PRINT 3

.

.

.

PRINT 29
PRINT 30
g. PRINT 10,'NAME','STREET','CITY','STATE','TELEPHONE'
10 FORMAT(A4,2X,A6,2X,A4,2X,A5,2X,A9,2X)
c. At the top of a new page, prints two real numbers in the form:

———.——

———.——

e. I
—— 1.00—— 1.00——1.00
—— 2.00—— 4.00——8.00

.

.

.

—— 10.00—— 100.00——****** (Asterisks mean specified format does not allow enough space.)

.

.

g. WHAT_IS_YOUR_FIRST_NAME?
WHAT_IS_YOUR_AGE_____JANE?
_____JANE_YOU_MUST_BE_12_YEARS_OLD
i. 'X'+'Y'=?
WHAT IS YOUR ANSWER?
If answer is correct, if X<=50, prints 'GOOD', if X>50 prints 'SUPER'. If answer is wrong, prints 'INCORRECT! TRY AGAIN'.

5. **b.** correct   **e.** A is not defined before LINE 4. Suggested change is:
LINE 4: IF(NUM .EQ. 'O') STOP

CHAPTER 7

1. **a.** A COMMON statement must appear in a program after the type declaration statements and before the executable statements.   **e.** true
h. Any INTEGER statements in a program must precede any DATA statements which initializes variables in the INTEGER statement.   **k.** true
n. A function defined by a function statement must have at least one argument.
p. true   **s.** true   **v.** true   **x.** true   **z.** A function subprogram can return more than one value to the calling program through the subprogram name, argument list or COMMON variables.
cc. If an array is to appear in a COMMON statement,
it need not appear in a preceding DIMENSION statement
if in the COMMON statement it appears with constant subscript(s)
specifying the array size.

2. **c.** correct (DATA is a valid variable name.)   **e.** correct (I OR J is a legal variable name.)   **h.** correct (Assume all variables declared as character strings.)
    k. correct (Assume I has previously been assigned an integer value.)
    n. correct (DATA,OR,& DIM are legal variable names.)
    q. NAME(G,X) = (3. * G)/X (This is a function definition statement.)
    t. correct (Assume subroutine SUB1 has only 2 parameters.)   **v.** correct
3. **b.** LINE 4 & LINE 5: Y is defined as an array. It must have a subscript when used and cannot be used as a DO loop control variable.   **e.** LINE 4: ARY is an array which conflicts with its appearance in a function definition statement in this line.   **g.** LINE 1: Comma missing. LINE 2: Cannot use a variable in a GOTO unless it has appeared previously in an ASSIGN statement.
    i. LINE 1: Must declare the dimension of the array before the DATA statement.
    k. correct   **n.** correct (However, F(X) in the CALL statement is not the same as F(X) in line 3.)   **p.** correct (Only if A and B are declared type integer.)

CHAPTER 8

1. **b.** It is permissible to use both the DATA statement and COMMON statement in the same program.   **c.** true   **f.** true   **i.** true   **k.** true
    n. More than one PARAMETER statement may appear in a program.
    p. There is a means of providing for more than one entry point to a subprogram through the ENTRY statement.   **r.** Variables specified in COMPLEX statements are automatically defined as two REALS.
    u. If an arithmetic expression includes one complex variable then all variables will be treated as complex variables.   **w.** true   **y.** true
2. **a.** SUBROUTINE SUB(X,Y)   **d.** F(X)=X*X+1 (The word FUNCTION is not required in a function definition statement.)   **g.** correct   **j.** correct
    l. LINE 2: ENTRY SUB1 (ENTRY statement always needs an entry name.)
    n. LINE 3: 10 FORMAT(2F10.2,A6) (Complex variables use two real numbers for each variable. It is assumed A is a character variable.)
    p. LINE 1.5: CHARACTER A
    LINE 4:10 FORMAT(10(F10.2,F10.2,A1))
    (Complex variables use two real numbers for each variable.)
3. **a.** 2 7 4.5   **c.** 2 3 5 5 6 9 10 12   **f.** _F_T_T
4. **b.** LINE 8: Too many variables to correspond to LINE 2.
        LINE 10: EQUIVALENCE statement will cause memory problems.
    d. correct

Index